CELLULAR
IMMUNOLOGY
LAB**F**AX

The LABFAX series

Series Editors:

B.D. HAMES Department of Biochemistry and Molecular Biology, University of Leeds, Leeds LS2 9JT, UK

D. RICKWOOD Department of Biology, University of Essex, Wivenhoe Park, Colchester CO4 3SQ, UK

MOLECULAR BIOLOGY LABFAX

CELL BIOLOGY LABFAX

CELL CULTURE LABFAX

BIOCHEMISTRY LABFAX

VIROLOGY LABFAX

PLANT MOLECULAR BIOLOGY LABFAX

IMMUNOCHEMISTRY LABFAX

CELLULAR IMMUNOLOGY LABFAX

Forthcoming titles

ENZYMOLOGY LABFAX

BACTERIOLOGY LABFAX

CELLULAR IMMUNOLOGY

EDITED BY
P.J. DELVES

Department of Immunology,
University College London Medical School,
Arthur Stanley House, 40–50 Tottenham Street,
London W1P 9PG, UK

*β*IOS
SCIENTIFIC
PUBLISHERS

ACADEMIC PRESS

All rights reserved by the publisher. No part of this book may be reproduced or transmitted, in any form or by any means, without permission in writing from the publisher.

First published in the United Kingdom 1994 by
BIOS Scientific Publishers Limited
St Thomas House, Becket Street, Oxford OX1 1SJ, UK

ISBN 0 12 208885 9

A CIP catalogue entry for this book is available from the British Library.

This Edition published jointly in the United States of America by Academic Press, Inc. and BIOS Scientific Publishers Limited.

Distributed in the United States, its territories and dependencies, and Canada exclusively by Academic Press, Inc., 525 B Street, Suite 1900, San Diego, California 92101-4495 pursuant to agreement with BIOS Scientific Publishers Limited, St Thomas House, Becket Street, Oxford OX1 1SJ, UK.

Typeset by Marksbury Typesetting Ltd, Midsomer Norton, Bath, UK.
Printed by Information Press Ltd, Oxford, UK.

The information contained within this book was obtained by BIOS Scientific Publishers Limited from sources believed to be reliable. However, while every effort has been made to ensure its accuracy, no responsibility for loss or injury occasioned to any person acting or refraining from action as a result of the information contained herein can be accepted by the publishers, authors or editors.

PREFACE

This volume in the LABFAX series follows the general style set by its predecessors, namely a large body of information of use to laboratory scientists has been condensed into a relatively small amount of space. As in any subject, the boundaries around the term 'cellular immunology' are hard to fathom, but it is hoped that much valuable information required by researchers studying the cells of the immune system will be found within. Given the amount of information contained in many of the tables, rather than quote original references we have generally concentrated on directing the reader to review articles or other sources which enable an entry into the literature on a particular subject. Where data are lacking, blank spaces have been left in the tables to enable the reader to insert the relevant information as it becomes available.

Standardized genetic nomenclature dictates that genes are identified using italicized characters and that their protein products are referred to using plain text. However, with the exception of the chapter which covers MHC nomenclature (Chapter 10), we have stayed with the approach normally used in the majority of immunological publications, namely using plain text also for the genes whilst making it clear that it is the gene that is being referred to.

The major credit for this work rests with the contributors who have expended considerable energies going through source material in order to produce the tables and figures within their chapters. I would like to thank them all not only for their contributions but also for the fact that their timely submission of manuscripts allowed this book to be produced within the deadlines originally envisaged. Thanks are also due to David Briggs, Ray Hicks, John Kearney, Peter Koder, Dennis Osmond and David Webster for advice and comments, and to Jonathan Ray at BIOS for all his help during the preparation of this volume. Finally, to Jane, Joe, Tom and Jessica – thanks for everything.

Peter J. Delves

CONTENTS

3. LEUKOCYTE DEVELOPMENT   33

4. TISSUE AND CELL CULTURE   45

5. CELL LINES AND HYBRIDOMAS   65

6. ASSAYS OF CELLULAR ACTIVITY 95

7. CELL-SURFACE ANTIGENS 115

8. ANTIGEN RECEPTORS 155

9. ANTIGEN PROCESSING AND PRESENTATION   173

10. THE MAJOR HISTOCOMPATIBILITY COMPLEX   195

11. SOLUBLE IMMUNOREGULATORY MOLECULES   215

CONTRIBUTORS

M. BINKS

Department of Immunology, University College London Medical School, Windeyer Building, Cleveland Street, London W1P 6DB, UK

F.M. BRENNAN

Kennedy Institute of Rheumatology, Sunley Building, 1 Lurgan Avenue, London W6 8LW, UK

K.H. BROOKS

Department of Microbiology, Giltner Hall, Michigan State University, East Lansing, MI 48824-1101, USA

B.M. CHAIN

Department of Immunology, University College London Medical School, Windeyer Building, Cleveland Street, London W1P 6DB, UK

P.J. DELVES

Department of Immunology, University College London Medical School, Arthur Stanley House, 40–50 Tottenham Street, London W1P 9PG, UK

R.G. FERNANDEZ-BOTRAN

Department of Microbiology, University of Texas Southwestern Medical Center, 5323 Harry Hines Blvd, Dallas, TX 75235, USA

B.D. FLORENCE

Department of Pathology and Infectious Diseases, Royal Veterinary College, University of London, Royal College Street, London NW1 0TU, UK

D.L. GIBBONS

Kennedy Institute of Rheumatology, Sunley Building, 1 Lurgan Avenue, London W6 8LW, UK

M.R. GOLD

Biomedical Research Centre, The University of British Columbia, 2222 Health Sciences Mall, Vancouver, British Columbia V6T 1Z3, Canada

J. HAU

Department of Pathology and Infectious Diseases, Royal Veterinary College, University of London, Royal College Street, London NW1 0TU, UK

D.R. KATZ

Department of Immunology, University College London Medical School, Windeyer Building, Cleveland Street, London W1P 6DB, UK

P.M. LYDYARD

Department of Immunology, University College London Medical School, Arthur Stanley House, 40–50 Tottenham Street, London W1P 9PG, UK

S.J. MARTIN

Division of Cellular Immunology, La Jolla Institute for Allergy and Immunology, 11149 North Torrey Pines Road, La Jolla, CA 92037, USA

J. McCULLOCH

Samuel Lunenfeld Research Institute, Mount Sinai Hospital, University of Toronto, 600 University Avenue, Toronto, Ontario M5G 1X5, Canada

S. OZAKI

2nd Department of Internal Medicine, Kyoto University Faculty of Medicine, 54 Shogoin Kawaharacho, Sakyo-ku, Kyoto 606, Japan

D. RAYNER

Department of Laboratory Medicine and Pathology, University of Alberta, 5B4.02 WC Mackenzie Health Sciences Center, Edmonton T6G 2R7, Canada

Y. SATTA

Max-Planck Institut für Biologie, Abteilung Immunogenetik, Corrensstr. 42, D-7400, Tübingen, Germany

L. SEALY

Department of Immunology, University College London Medical School, Windeyer Building, Cleveland Street, London W1P 6DB, UK

N. TAKAHATA

National Institute of Genetics, The Graduate University for Advanced Studies, Mishima 411, Japan

A.R. VENKITARAMAN

MRC Laboratory of Molecular Biology, Hills Road, Cambridge CB2 2QH, UK

ABBREVIATIONS

Ab	antibody
ADA	adenosine deaminase
ADCC	antibody-dependent cell-mediated cytotoxicity
Ag	antigen
8-AG	8-azaguanine
AIDS	acquired immunodeficiency syndrome
AIM	activation inducer molecule
α_2-M	α_2-macroglobulin
AMP	adenosine monophosphate
anti-Ig	anti-immunoglobulin
APC	antigen-presenting cells
ATCC	American Type Culture Collection
ATP	adenosine triphosphate
BALT	bronchus-associated lymphoid tissue
BGP	biliary glycoprotein
BiP	binding protein
BLA	Burkitt's lymphoma-associated antigen
BL-CAM	B-lymphocyte cell-adhesion molecule
BrdU	5-bromo-2'-deoxyuridine
CALLA	common acute lymphoblastic leukemia antigen
cAMP	cyclic adenosine monophosphate
CBP	carbohydrate-binding protein
CD	cluster of differentiation
cDNA	complementary DNA
CEA	carcinoembryonic antigen
cGMP	cyclic guanine monophosphate
CIMR	Coriell Institute for Medical Research–NIGMS Human Genetic Research Cell Repository
CML	cell-mediated lymphocytotoxicity
ConA	concanavalin A
CTL	cytotoxic T lymphocyte
DAF	decay accelerating factor
DAG	diacylglycerol
DMEM	Dulbecco's modified Eagle's medium
DMSO	dimethyl sulfoxide
DPP4	dipeptidylpeptidase IV
EAE	experimental autoimmune encephalomyelitis
EBNA	EBV nuclear antigen
EBV	Epstein–Barr virus
ECACC	European Collection of Animal Cell Cultures
ECF	eosinophil chemotactic factor
ECMR	extracellular matrix receptor
EDTA	ethylenediaminetetraacetic acid
EGF	epidermal growth factor
ELAM	endothelial leukocyte adhesion molecule

ELISA	enzyme-linked immunosorbent assay
ELISPOT	enzyme-linked immunospot
E5N	ecto-5′-nucleotidase
ER	endoplasmic reticulum
ERK	extracellular-signal regulated kinase
FA	Freund's adjuvant
FACS	fluorescence-activated cell sorter
FAL	3-fucosyl-N-acetyl-lactosamine
FBS	fetal bovine serum
FCA	Freund's complete adjuvant
FcR	Fc receptor
FCS	fetal calf serum
FIA	Freund's incomplete adjuvant
FITC	fluorescein isothiocyanate
FMLP	formyl-Met-Leu-Phe
FNR	fibronectin receptor
GALT	gut-associated lymphoid tissue
G-CSF	granulocyte colony-stimulating factor
GDP	guanosine diphosphate
GM–CSF	granulocyte–macrophage colony-stimulating factor
GNEFs	guanine nucleotide exchange factors
GPI	glycosyl-phosphatidylinositol
GTP	guanosine triphosphate
GTPase	guanosine triphosphatase
GVHD	graft versus host disease
HAT	hypoxanthine/aminopterin/thymidine
HBSS	Hank's balanced salt solution
H&E	hematoxylin and eosin
HEPA	high-efficiency particulate air
Hepes	N-2-hydroxyethylpiperazine-N'-ethane-sulfonic acid
HGPRT	hypoxanthine guanine phosphoribosyl transferase
HLA	human leukocyte antigen
HPA	hemolytic plaque assay
HPLC	high-pressure liquid chromatography
HRF	homologous restriction factor
Hsp	heat-shock protein
HT	hypoxanthine and thymidine
HTLV-1	human T-cell leukemia virus-1
Ia	I-region-associated antigen
ICAM	intercellular adhesion molecule
ID	intradermal
IELs	intraepithelial lymphocytes
IFN-γ	interferon-γ
IgSF	immunoglobulin superfamily
IL-7	interleukin-7
IM	intramuscular(ly)
InsP	inositol phosphate
$Ins(1,4,5)P_3$	inositol 1,4,5-triphosphate
IP	intraperitoneal(ly)
IP_3	inositol 1,4,5,-triphosphate
IV	intravenous(ly)

K	killer
KL	*kit* ligand
LAK	lymphokine-activated killer
LAM	leukocyte adhesion molecule
LBP	lipopolysaccharide binding protein
LCA	leukocyte common antigen
LCL	lymphoblastoid cell line
LDA	limiting-dilution analysis
LDL	low-density lipoprotein
LFA	leukocyte function-associated antigen
LGL	large granular lymphocytes
LHR	lymph node homing receptor
LIF	leukemia inhibitory factor
LPS	lipopolysaccharide
LRP	lipoprotein receptor related protein
LTB_4	leukotriene-B_4
LU	lytic units
mAb	monoclonal antibodies
MAFA	mast-cell function-associated antigen
MALT	mucosa-associated lymphoid tissue
MAP	mitogen-activated protein
MCP	membrane cofactor protein
M-CSF	macrophage colony-stimulating factor
2-ME	2-mercaptoethanol
MEK	MAP kinase kinase
MEM	minimum essential medium
8-MG	8-mercaptoguanosine
MGF	mast-cell growth factor
MHC	major histocompatibility complex
MIP	macrophage inflammatory protein
MIRL	membrane inhibitor of reactive lysis
MLR	mixed lymphocyte reaction
MRP	motility-related protein
MTT	3-(4,5-dimethylthiazol-2-yl)-2,5-diphenyltetrazolium bromide
NBT	nitroblue tetrazolium
NCA	nonspecific cross-reacting antigen
NCAM	neural cell adhesion molecule
NEP	neutral endopeptidase
NFPR	*N*-formylated peptide receptor
NK	natural killer
PAF	platelet activating factor
PALS	periarteriolar lymphoid sheath
PBL	peripheral blood lymphocytes
PBMC	peripheral blood mononuclear cells
PBR	peptide-binding region
PBS	phosphate-buffered saline
PC	phosphatidylcholine
PCD	programmed cell death
PCR	polymerase chain reaction
PDGF	platelet-derived growth factor
PECAM	platelet–endothelial cell-adhesion molecule

PEG	polyethylene glycol
PHA	phytohemagglutinin
PI	phosphatidylinositol
PIP_2	phosphatidylinositol 4,5-bisphosphate
PKC	protein kinase C
PLC	phospholipase C
PLD	phospholipase D
PMA	phorbol 12-myristate 13-acetate
PMNs	polymorphonuclear leukocytes
PtGP	platelet glycoprotein
PTKs	protein tyrosine kinases
PWM	pokeweed mitogen
RFLP	restriction fragment length polymorphism
RIA	radioimmunoassay
RPMI	Roswell Park Memorial Institute
SAC	*Staphylococcus aureus* Cowan strain I
SC	subcutaneous(ly)
SCF	stem-cell factor
SCID	severe combined immunodeficiency
SDS-PAGE	sodium dodecyl sulfate polyacrylamide gel electrophoresis
SH2	*src* homology 2
SIV	simian immunodeficiency virus
SLE	systemic lupus erythematosus
SLF	steel factor
SOD	superoxide dismutase
SPF	specific pathogen free
SRBCs	sheep red blood cells
SSCP	single-strand conformation polymorphism
SSO	sequence-specific oligonucleotides
SV40	simian virus 40
TAPA	target of an anti-proliferative antibody
TBV	total blood volume
TCR	T-cell receptor
TdT	terminal deoxyribonucleotidyl transferase
6-TG	6-thioguanine
$TGF\alpha$	transforming growth factor α
THAM	thymocyte-activating molecule
TK	thymidine kinase
TLC	thin-layer chromatography
TM	transmembrane
TNF	tumor necrosis factor
TRAP	TNF-related activation protein
TSA	thymic shared antigen
TX	thromboxane
u-PAR	urokinase-type plasminogen activator receptor
UT	untranslated
UV	ultraviolet
VCA	viral capsid antigen
VCAM	vascular cell adhesion molecule
VLA	very late antigen
VNR	vitronectin receptor

CHAPTER 1
LABORATORY ANIMALS
B.D. Florence and J. Hau

Although the mouse has been the most popular animal in immunology research, many other species have also been studied and served as a model for man (1). Immunology is unique in comparison with other scientific disciplines in its use of animals for the production of biological materials. A vast number of animals, predominantly rabbits, but also goats, sheep, pigs, horses, chickens, etc., are used as a source of polyclonal antibodies. Guinea-pigs have, for many years, been used for production of complement factors, and mice are the most popular animals for monoclonal antibody production.

1. WELFARE OF LABORATORY ANIMALS

Whenever laboratory animals are used it is essential to make sure that they are dealt with in a way that ensures their maximal good welfare (2–4). Alternatives to using live animals should be investigated whenever possible, e.g. *in vitro* production of monoclonal antibodies instead of *in vivo* production. If the use of animals is unavoidable, the animals should be of the best possible health status, maintained in a well-controlled good environment, looked after by competent staff and used as quickly as possible. The importance of dedicated and well-educated technical staff cannot be overemphasized. Monitoring of animal welfare relies on a sound knowledge of animal husbandry and behavior. Only skilled personnel are able to recognize subtle changes in animal behavior which may indicate distress and pain.

It is important constantly to be aware of possible ways to use more refined techniques, even if this results in the use of an increased number of animals. If, for instance, a certain amount of antibody of a given kind is to be produced, it is often better to avoid potentially aggressive adjuvants, such as Freund's complete adjuvant (FCA), by using another adjuvant and an increased number of immunization animals. Another way to reduce stress and discomfort is to use sedatives and analgesics. When animals are euthanized this should also be done as humanely as possible. Animals used in immunological research are often exsanguinated during general anesthesia. For approved procedures for humane killing the reader should consult the relevant guidelines (e.g. *Recommendations for Euthanasia of Experimental Animals*, Commission of the European Communities Joint Research Centre, Environment Institute).

2. HEALTH MONITORING

When using animals for immunological research it is obviously important to use healthy animals and animals of similar health status for each experiment to obtain reliable results. Animals are often bought as 'SPF' (free of a number of specified pathogenic organisms). Particularly in immunological research there is a need to avoid any uncontrolled stimulation or suppression of the immune system, and to have knowledge about the extent of exposure of the animals to immunogens in the environment. Therefore it is often desirable to 'barrier' house the animals thereby minimizing the risk of infections from other animals and staff. In some experiments it may be necessary to use animals with a known defined microflora. These gnotobiotic animals can be bought, but they are expensive and they have to be housed in isolators to maintain their microbiological status. Regardless of the type of animals used, it is essential to introduce a health monitoring program to ensure that the animals remain free of pathogenic micro-organisms.

Health monitoring of laboratory animal colonies is based on three principles:

(i)   A few animals can be sampled for examination, and the results of the screening taken as representative for the entire colony.
(ii) If an animal is found to be infected with a particular micro-organism, the whole colony must be considered infected.
(iii) If no animals are found to be infected, the colony is considered free of this organism.

If these principles are to be effective, it is necessary to employ statistically valid screening systems. For details about sample size, statistical treatment of data and screening frequency the reader is referred to Hansen (5).

3. ANESTHESIA AND ANALGESIA

The increasing concern for animal welfare and the knowledge that animals experience pain has made the proper use of analgesics a standard practice in the use of animals for research. The important features of analgesia and anesthesia are that they are easy to administer, reliable, reversible and safe for the operator and animal. Combinations of the new drugs currently available now make it possible to have most, if not all, of these features. This is referred to as using 'balanced anesthesia' (see *Table 4*).

3.1. Pre-anesthesia
Conditioning (habituation) and acclimatization
This usually involves a sufficient period of time (usually at least 7 days) to allow the animal to become familiar with the accommodation, routines and handling. Frequent and empathic handling of small rodents has a taming effect which facilitates easier and less stressful induction of anesthesia.

Pre-anesthetic examination
A clinical examination to establish that the animal is free of disease and able to cope physiologically with the anesthesia is an equally important part of the total procedure. Health screening reports should also be referred to. These should include any changes in food and water consumption and changes in the body weight of the animal.

Pre-medication
Pre-anesthetic medication (*Table 1*) must be recommended whenever possible, because this:

(i)   decreases fear and thus stress of the animal prior to and during anesthesia;
(ii) provides chemical restraint, thus enabling easier and less stressful handling;
(iii) reduces the risk of anesthesia by reducing salivary and bronchiole secretions and blocking vaso-vagal reflexes, which would normally result in bradycardia or cardiac arrest;
(iv) reduces the amount of anesthetic required, and thus increases the margin of safety of anesthesia;
(v)   facilitates a smooth recovery from anesthesia.

Emla$^{\circledR}$ cream applied topically over the site of intravenous (IV) injection aids in a smooth, uneventful induction of anesthesia. This local anesthetic is useful prior to venesection of rabbit ear veins.

Table 1. Pre-medication/sedation: dosages of recommended drugs

Drug	Mice	Rat	Guinea-pig	Rabbit	Dog	Primate
Atropine	0.05 mg kg^{-1} IP	0.05 mg kg^{-1} IP	0.05 mg kg^{-1} IP	–	0.05 mg kg^{-1} SC	0.05 mg kg^{-1} SC
Hypnorm	0.1–0.3 ml kg^{-1} IP 1:10 diln	0.2–0.5 ml kg^{-1} IP	1.0 ml kg^{-1} IM	0.2–0.5 ml kg^{-1} IM	0.1–0.2 ml kg^{-1} IM	0.3 ml kg^{-1} IM
Acepromazine	–	–	–	–	0.2 mg kg^{-1} IM	–
Glycopyrrolate	–	–	–	0.1 mg kg^{-1} SC	–	–
Ketamine	–	–	–	–	–	5–25 mg kg^{-1}

For additional information see Flecknell (6).
Abbreviations: diln, dilution; IM, intramuscularly; IP, intraperitoneally; SC, subcutaneously.

3.2. Anesthesia

There are two methods for the delivery of an anesthetic, either by inhalation or injection. There are several routes for injections: these are subcutaneous (SC), intramuscular (IM), intraperitoneal (IP) or intravenous (IV). In small rodents, anesthesia by gaseous inhalation is easily achieved using a Perspex induction chamber and gas scavenging mask. However, it is important to provide supplementary heat or insulation during anesthesia as small animals can become hypothermic very quickly while under anesthesia. Their body temperature as well as their other vital signs need to be monitored continuously during this time (*Table 2*). Methods of access to injection sites for administration of substances and fluids, as well as for sampling body fluids, are described for rats in Waynforth and Flecknell (7).

Table 2. Vital signs

	Body temperature (°C)	Respiratory rate (×/min)	Heart rate (×/min)
Mouse	36–37	84–230	500–600
Rat	38–39	85–110	320–480
Guinea-pig	38.4–39.9	69–104	150–400
Rabbit	38.5–40	32–60	205–237

Anesthesia by inhalation

Induction with gas anesthesia is best achieved in an induction chamber connected to a conventional gas anesthetic machine set at the appropriate levels (*Table 3*). It enables good observation of the depth of anesthesia during induction. The excess gas should be scavenged into the exhaust ventilation. Following induction of anesthesia, the animal may be intubated with an appropriate endotracheal tube (6) or placed in a face mask of appropriate size. For small animals under 2 kg use an Ayres T-piece or the Jackson/Reese modification of this circuitry. The concentration of anesthetic vapor required for maintenance is less than for induction (*Table 3*). Free gas flow as delivered from the anesthetic machine should equal two times the minute volume of the animal's lungs. Minute volume is the tidal volume (the volume of one expiration) × respiration rate per minute (c. 150–200 ml kg^{-1} min^{-1}).

Halothane can only be used with a correctly calibrated flow meter and gas delivery circuitry. Methoxyflurane, because of its relatively low blood solubility, can be safely and satisfactorily used in the 'ether jar' as well as with a calibrated flow meter and gas delivery circuitry. The ether jar apparatus should be constructed in such a way that the animal does not come in direct contact with the drug (7). Methoxyflurane is the drug of choice for inexperienced operators, or if close monitoring of anesthesia is not possible, as when the surgeon and anesthetist are one and the same person.

Table 3. Anesthesia by inhalation: dosages of recommended drugs

Anesthetic	Induction (%)	Maintenance (%)
Halothane	3–4	0.5–2.0
Methoxyflurane	3–4	0.4–1

Anesthesia by injection
The new injectable anesthetic combinations now available (*Table 4*) can be reversed with specific antagonists (*Table 5*) enabling quick, smooth recoveries and correction of inadvertent overdosing. They are generally considerably less expensive to use than gaseous anesthetics as they have no requirement for the costly delivery systems, although an oxygen supply is recommended because of respiratory depression.

3.3. Analgesia
Analgesia is achieved by the reduction in neural signals in pain pathways. This can be accommodated by several types of agents (*Table 6*):

(i)   Local anesthetics: lignocaine injection; *Emla® cream topically.
(ii)   Opioids: fentanyl; *buprenophine.
(iii) Non-steroidal anti-inflammatory drugs: phenylbutazone; *flunixin.
(iv)   Corticosteroids: dexamethazone.
(v)   Alpha$_2$-agonist: xylazine (poor in rodents but good in ruminants); medetomidine.
*Recommended preference.

4. ANIMAL MODELS IN IMMUNOLOGY

Jenner's recognition of the similarity between cowpox, a nonfatal disease of cows, and smallpox of humans may have been the first use of an animal model for immunological purposes in dealing with a disease of humans. Since then many similarities in some of the diseases of animals and humans have been observed. Animal models are of two general types:

(i)   spontaneous, i.e. the animal develops the disease condition without external intervention; or
(ii)   induced experimentally, i.e. using infectious agents, irradiation, surgery, drug treatment, specific cell-depletion, genetic manipulation as with transgenics and knockouts, or some other form of interventive modulation of the animal's normal biology.

Inheritable genetic changes can be perpetuated in breeding colonies and may be commercially available (8). The study of immunology has been enhanced immensely in recent years by the development of spontaneous models such as the nude mouse and, more recently, the SCID mouse in inbred lines (9). These have given immunological researchers some very powerful tools.

The effective use of irradiation and reconstitution of these immunodeficient models has provided an excellent 'in vivo' test-tube for the study of the cellular immune response (10). Although animal models may not be identical images of the human disease, they often share some significant similarities, enabling study of the role of the immune system in specific disease processes.

4.1. Induced animal models
A brief list of induction methods, together with selected examples, follows.

Experimental infection:
(i)   Simian immunodeficiency virus (SIV) infection in macaques to study human immunodeficiency disease (acquired immunodeficiency syndrome, AIDS) (11).
(ii)   Infection of nude and SCID mice with *Mycobacterium leprae* (12).
(iii) Theiler's murine encephalitis virus in SJL/J mice as a model of multiple sclerosis (13).

Irradiation. Sub-lethal ionizing radiation of the animal, usually at about 2.5 Gy (250 rad) (1 G = 100 Rad); a dose of 3–10 Gy causes stem-cell death and is generally eventually lethal.

Table 4. Anesthesia by injection: dosages of recommended drugs

Drug	Mouse	Rat	Guinea-pig	Rabbit	Dog	Primate
Hypnorm (Fentanyl/Fluanisone)/Midazolam (diln 1:1:2 parts H_2O)	10 ml kg^{-1} IP	2.7 ml kg^{-1} IP	8 ml kg^{-1} IP	0.3 ml kg^{-1} IM + 2 ml kg^{-1} IP	Not used	Not used
Ketamine/Xylazine	150 mg kg^{-1} IP 10 mg kg^{-1} IP	90 mg kg^{-1} IP 10 mg kg^{-1} IP	40 mg kg^{-1} IP 5 mg kg^{-1} IM	25 mg kg^{-1} IM 5 mg kg^{-1} IM	5 mg kg^{-1} IV 1–2 mg kg^{-1} IV	10 mg kg^{-1} IM 0.5 mg kg^{-1} IM

These are the most current, most easily used and recommended drugs and methods of analgesia and anesthesia for laboratory animals. Previously used compounds, such as ether, chloroform and pentobarbitone are no longer considered to be safe or effective, and therefore cannot be recommended for use. Other, newer drugs are also available, but most must be given IV. This requires a certain level of expertise, and often sophisticated equipment for maintenance for any extended period of time. For additional information see Flecknell (6).

Abbreviations: diln, dilution; IM, intramuscularly; IP, intraperitoneally; IV, intravenously.

Table 5. Antagonists for injectable anesthetics: dosages of recommended drugs

Agent	Anesthetic	Dosage
Respiratory stimulant		
Doxapram	All	5–10 mg kg^{-1} IM, IP, IV
Specific antagonists		
Nalbuphine	Hypnorm	2–4 mg kg^{-1} IM
Buprenorphine	Hypnorm	0.01–0.05 mg kg^{-1} SC, IV
Naloxone	Hypnorm	0.01–0.1 mg kg^{-1} IV, IM, IP
Atipamerzole	Xylazine/medetomidine	1.0 mg kg^{-1} IM, IP, SC

Abbreviations: IM, intramuscularly; IP, intraperitoneally; IV, intravenously; SC, subcutaneously.

Surgical alteration. Lymphadectomy, thyroidectomy, thymectomy, splenectomy, hypophysectomy, allograft, xenograft, cannulation, etc., see ref. 7.

Drug or chemical induction. Cobra venom factor (structural and functional analog of complement component C3 – causes decomplementation by inducing continuous activation of the complement cascade and enables the study of the role of complement in immune responses). Levodopa induction of autoimmune hemolytic anemia in mice (14). Various immune modulatory drugs, such as steroids and cyclosporin A.

Specific cell manipulation. Selective removal of specific cell types using monoclonal antibodies; T-cell vaccination to induce resistance to experimental autoimmune encephalitis (EAE) (15).

Genetic manipulation:
(i)   Chimeras. Animals in which the cells of two different individuals (donor and recipient) coexist without any alloreactivity. They are thus useful in studying the phenomena of tolerance. The characteristic is not inheritable as the germ line belongs to only one, e.g. hu-PBL-SCID created by injecting human peripheral blood lymphocytes (PBL) into SCID mice (16).
(ii)   Transgenics. Animals which have had foreign DNA introduced into their germ line. The characteristics determined by this DNA are thus inheritable. This introduction can occur by microinjection or by the use of viral carriers. The change in the animal's genome results in the animal producing the protein product encoded by the transgene. Such animals, mostly mice (although there are transgenic cattle, sheep and hogs), are given a specific nomenclature, e.g. C57BL/6J-TgN(XX)Y, where Tg = transgenic, N = nonhomologous insertion, XX = insert designation, Y = lab code, on mouse strain C57BL/6J; or GenPharm® TSG p53, which is a commercially available transgenic mouse carrying the human tumor suppressive gene p53.
(iii) Knockout. The removal or inactivation of a portion of genetic material that governs a specific function of the DNA of an individual animal. The lack of this specific DNA enables the study of the various processes that might be affected by the lack of that gene's influence, e.g. inactivated interferon-γ gene.
(iv) Congenic inbreeding. Congenics are strains produced by the introduction of foreign genetic material into an inbred isogenic strain by repeated backcrossing (10 + times) of the offspring, carrying the donated gene, to the isogenic parent. The resulting animal is very similar to the original strain, except for the introduced gene plus associated genetic contaminants. This contamination may be reduced by additional backcrosses.

Table 6. Analgesia: dosages of recommended drugs

Drug	Mice	Rat	Guinea-pig	Rabbits	Dogs	Primates
Buprenorphine	0.05–0.1 mg kg^{-1} SC; 12 h	0.01–0.05 mg kg^{-1} SC; 12 h	0.05 mg kg^{-1} SC; 8–12 h	0.01–0.05 mg kg^{-1} SC, IM; 12 h	0.01 mg kg^{-1}; 8–12 h	0.01 mg kg^{-1} IM; 12 h
Flunixin	2–5 mg kg^{-1} SC, IM	2–5 mg kg^{-1} SC, IM	–	1.1 mg kg^{-1} SC, IM; 12 h	1 mg kg^{-1} *per os* (oral + IM); daily	2.5–10 mg kg^{-1} IM; daily

For additional information see Flecknell (6).
Abbreviations: IM, intramuscularly; SC, subcutaneously.

Reconstitution. Replacement of deficient cells or tissues. For example, the hu-PBL-SCID mouse is one reconstituted with human peripheral blood leukocytes; introduction of syngeneic donor bone marrow or fetal liver into unirradiated SCID mice results in reconstitution of primary and secondary lymphoid tissues normally deficient in these animals; neonatal thymus or congenic T cells injected into nude mice restores the T-cell-dependent functions previously reduced or lacking.

4.2. Spontaneous animal models

Most spontaneous models occur either by serendipity or by mutation of the genome of the animal. Once such a mutation has been recognized it can be perpetuated and modified by many of the previously mentioned genetic methods. Co-isogenic mutations occurring in an inbred strain are the most useful because they only differ from the original strain by the genetic material at one locus. It then becomes a substrain. An example of this is the AKR *nu-str* strain which differs only by the streaker (*str*) allele. The benefit is that there are no genetic contaminants to deal with as in congenic strains. Such mutations have given rise to many of the important mouse genotypes in common use today for immuno-logical studies and include the nudes (*nu*), SCIDs (SCID), X-ids (*xid*), beiges (*bg*), motheatens (*me*), and many others. By genetic manipulation, various combinations of these have been created, e.g. *bg/nu/xid* strains on BMX mice (18). The availability of a wide variety of inbred strains of such markedly different properties have made these mice a powerful instrument of current immunological research. It is beyond the scope of this chapter to list all of the strains, congenic strains, and allotypes available. See Festing (8) for a complete listing.

Most of the mouse models have been established in the common inbred strains. They nearly all originate from the following strains; A; AKR; BALB/c; CBA; C3H; C57BL; DBA; SJL; or their congenics (19). These were all derived from *Mus musculus* as a result of generations of various breeding programs in both the pet trade as well as in the laboratory. Similar mutations, such as the nude gene, have occurred in the rat and the guinea-pig also, but are not used nearly as widely as the mouse. All these models must be evaluated carefully as one can not assume that the conditions observed in these animals are identical to the conditions of the human disease. *Tables 7* and *8* show examples of some of the more common spontaneous models in current use.

5. TRANSPLANTATION IMMUNOLOGY

The genes that control graft rejection in mammals are known as histocompatibility genes. They determine the major histocompatibility (MHC) class I and class II, and the minor histocompatibility allotypes. An allotype represents the protein product of an allele which may be detected as an antigen by the immune system of another member of the same species. A haplotype is a set of linked genes located on a single chromosome. Allotypic recognition will result in an immune response leading to graft rejection. Differences in a single MHC class I or class II allotype, or even a very few of the minor histocompatibility genes, may result in graft rejection. Rejection in an unsensitized recipient is typically within 14 days. A second graft of the same genetic disparity will be rejected in 5–6 days or less, even at a different site than the original graft. This is due to the memory, specificity and systemic nature of the immune response. There are sites, such as the chamber of the eye, the cornea, the brain and the cheek pouches of hamsters, which, due to poor lymphatic drainage, are protected from certain immune effector functions, particularly delayed hypersensitivity and complement-fixing antibodies. Thus at these sites allotype grafts last much longer.

Table 7. Spontaneous models of immunodeficiency (20–23)

Animal/strain	Gene		Features	Human syndrome
Mouse				
Nude	*nu*	Chr. 11	Athymic	Di George
SCID	*scid*	Chr. 16	T and B cell def.	SCID
Motheaten	*me*	Chr. 6	T, B and NK cell def.	
Motheaten viable	*me*^{v}	Chr. 6	T, B and NK cell def.; hypergamma-globulinemia	
X-id	*xid*	Chr. X	CD5 and B cell def., impaired responses to polysaccharides	X-linked agamma-globulinemia variant
Beige	*bg*	Chr. 13	NK cell def.	Chediak–Higashi
Dominant spotting	*W*	Chr. 5	Mast cell def., anemia	Red cell aplasia
Steele	*Sl*	Chr. 10	Mast cell def., anemia	
Wasted	*wst*	Chr. 2	Thymus atrophy	
Dominant hemimelia (Lasat)	*Dh*		Asplenic, athymic	
Complement deficient	*Hc*	Chr. 2	C5 def.	
Osteopetrotic, micropthalmic	*op, mi*	Chr. 3, 6	Macrophage def.	Osteopetrosis
LPS	*lps*	Chr. 4	Low responses to lipopolysaccharides	
Rat				
Nude Rowlet	*nu*		Athymic	Di George
Nude New Zealand	*nuz*		Athymic	Di George
BB-DP	*dp*		T-cell lymphopenia, IDDM	
F344 (Fischer)	*op/ia*		Osteopetrosis	Osteopetrosis
Guinea-pig				
Complement def.	*Ss/Slp*		C4 def.	
Dog				
Various	*X-scid*		SCID	SCID
Gray collie			Cyclic neutropenia	Cyclic neutropenia
Horse				
Arabian	*scid*		SCID	SCID

Abbreviations: def., deficient; IDDM, insulin-dependent diabetes mellitus; NK, natural killer; SCID, severe combined immunodeficient.

Table 8. Spontaneous animal models of autoimmune disease (24–31)

Disease	Species	Strain
Systemic lupus erythematosus	Mouse	MRL/MP-lpr/lpr
	Mouse	NZB × NZW F1
	Mouse	BXSB
	Mouse	Motheaten
	Mouse	Palmerston-North
	Mouse	Swan
Rheumatoid arthritis	Mouse	MRL/MP-lpr/lpr
Sjögren's syndrome	Mouse	NOD
	Mouse	C3H-lpr/lpr
Progressive systemic sclerosis/scleroderma	Chicken	UCD line 200
	Mouse	TSK (tight skin)
Myasthenia gravis	Rat	F344 (Fischer)
	Dog	Jack Russell terrier
Ulcerative colitis	Primate	Cotton-top tamarin
Autoimmune hemolytic anemia	Mouse	NZB
Autoimmune orchitis	Mink	Russian
Vitiligo	Chicken	DAM (delayed amenalosis)
Autoimmune thyroiditis	Rat	Buffalo
	Chicken	Obese
	Dog	Argonne beagle
	Primate	Marmoset
Insulin-dependent diabetes mellitus	Mouse	NOD
	Rat	BB/W
	Dog	Non-obese keeshond
	Primate	Black Celebes macaque

5.1. Graft versus host reaction

This occurs when allospecific T cells, which are transferred with the graft, react against histo-compatibility differences of the host tissue in an immunologically compromised host, resulting in severe organ dysfunction, especially of the skin, liver and the gastrointestinal tract. Without immunosuppression, the incidence of acute graft versus host disease (GVHD) in man is 30–70% when the bone-marrow donor is a HLA-matched sibling, and it approaches 100% with an HLA-identical unrelated donor. The incidence of this reaction can be decreased by removal of T cells from the graft and/or the use of immunosuppressive drugs such as corticosteroids or cyclosporin A.

6. IMMUNIZATION OF ANIMALS FOR ANTIBODY PRODUCTION

6.1. Polyclonal antibody production

The rabbit is the most popular species for production of polyclonal antibodies. There is an element of conservatism in this, but rabbits are docile, therefore they are easy to handle and take blood samples from. The husbandry is simple, and female rabbits are preferred as they can be kept in systems allowing the housing of a couple of rabbits together with room enough to jump around. This will prevent problems such as osteoporosis due to lack of exercise.

Rabbits used for immunization are usually not inbred, and significant differences in antibody response (titer) are often observed, even between closely related individuals immunized with the same immunogen. If only small amounts of antibody are needed, it may thus be feasible to immunize inbred mice or rats.

Chickens are an attractive alternative to mammals for antibody production because antibodies can be harvested from the egg yolk. Several simple purification procedures have been published (32) and consequently much larger volumes of antibodies are obtainable from an egg-laying chicken than from a rabbit. An additional advantage to the reduction in number of animals used is the absence of restraint and blood sampling. Chicken IgG, sometimes referred to as IgY, is somewhat different from mammalian IgG. The isoelectric point is approximately one pH unit lower (i.e. pH 6.5). The molecular weight is c. 190 kDa, the heavy chain being 71 kDa and the light chain 24.4 kDa. Chicken IgG does not bind to mammalian complement, mammalian Fc-receptors, protein A or protein G. Cell lines for chicken monoclonal antibody production have been developed, but there may be problems with the longevity of the cells.

6.2. Monoclonal antibody production

The species most commonly used for monoclonal antibody production is the mouse. There are many murine myeloma cell lines available (Chapter 5, *Table 16*) and mice, for reasons unknown, generally seem to respond better than rats. The BALB/c lines of mice are preferred by many, again because many of the myeloma cell lines have a BALB/c background. When spleen donor and fusion partner have the same genetic background it is usually unproblematic to produce ascites with a good concentration of immunoglobulins. If other strains of mice are used, the ascites production should be performed in the F_1 hybrid strain with BALB/c, or in immunologically incompetent mice (irradiated, athymic or SCID).

Immunization with soluble antigens for monoclonal antibody production does not differ from immunization with a view to bleed the animals for polyclonal antibody production. Antigens of low molecular weight can be made more immunogenic by conjugation to a carrier such as bovine serum albumin, bovine gammaglobulin, ovalbumin or keyhole limpet hemocyanin.

Immunization with intact cells does not usually require an adjuvant, since they tend to be immunogenic in themselves. The cells to be used are washed extensively by centrifugation in phosphate-buffered saline (PBS) or serum-free medium. They are counted and resuspended at a concentration of 10^8 cell ml^{-1}. The immunization is carried out by injecting 100 µl IP; booster injections, 1–3, are administered every 3 weeks and the final boost prior to fusion is also given IP. The development of antibody titer should be monitored throughout the immunization period; if all mice on a given antigen are good responders, the splenic cells may be pooled.

6.3. Antigen purity and dose

When preparing antigens for immunization, it is usually desirable to obtain as high a degree of purity as possible. This is not crucial for producing monoclonal antibodies, but in polyclonal production it is essential if laborious absorption procedures of the antisera are to be avoided. If the antigen preparation is not pure, the animal is likely to produce antibodies with specificities against the impurities. This is increasingly pronounced in prolonged immunization schemes. Consequently this makes it attractive to use as short a duration of the immunization scheme as possible. However, if antibodies of high affinity are required, it may be necessary to compromise and use a longer immunization period.

The dose of antigen used is usually from c. 10 μg to 250 μg, depending on the antigen and animal species used. Very low or very high doses may result in the induction of tolerance in the animal. Although the dose of the antigen is not directly related to the resulting antibody titer, it is important that the concentration of the antigen in the solution to be injected is not very low, since this tends to give poor results. A concentration of 1 mg ml^{-1} usually results in a good response.

6.4. Adjuvants

Adjuvants (from the Latin *adjuvare*, to help) are usually employed to enhance the immune response. Although a plethora of adjuvants have been developed and tested, Freund's adjuvants (FA) still seem to be the most popular for experimental purposes. The main reason for this is probably that they are well established as potent adjuvants, widely available and inexpensive. FA is employed in two forms. Complete FA (FCA) consists of a mineral oil, an emulsifier and mycobacterial material. Incomplete FA (FIA) consists of mineral oil and emulsifier without bacterial material. FCA is far more aggressive to the animal than FIA, and causes extensive granuloma and often abscess formation. FIA is much more lenient, but at the same time often an efficient adjuvant. FCA must never be given intravenously or in repeated doses. Technical staff administering FCA should be warned about the hazards associated with accidental injection of FCA in humans (granuloma formation, very slow healing and often long-lasting pus secretion from the injection site). Medical advice should be obtained when this happens.

6.5. Routes of administration

Intradermal (ID) route

The ID route should not be used unless the purpose is to induce a cell-mediated response. It is very difficult to perform a proper ID injection in small animals because of their thin skin. The site must be closely shaven and a very small (tuberculin) needle used. In rabbits volumes deposited should be < 50 μl. In rodents smaller than the guinea-pig, ID injections should not be attempted.

Subcutaneous (SC) route

The SC route is the one recommended for immunization. Inoculations are usually performed dorsally in the neck region using multiple sites. In the rabbit it may be practical to deposit the immunogen solution more caudally, because rabbits are usually lifted by taking a scruff of the neck skin. The volume used should not exceed 0.1 ml at each of two sites in mice, 0.2 ml at each of two sites in rats and 0.25 ml at each of four sites in guinea-pigs, rabbits, sheep and goats. If FCA has been used, the region of deposit may produce abscesses. If these are considered to be painful the animal should be given proper medical treatment.

Intramuscular (IM) route

Although we recommend SC administration, IM immunizations are still used. In rabbits injections should be given into the middle of the biceps femoris (thigh) muscle. Care should be taken not to deposit the antigen mixture close to the periosteum. The volume should be <0.5 ml/site. Intramuscular injections should not be performed on animals smaller than rabbits.

Intravenous (IV) route
Not recommended for mixtures containing FCA.

Intraperitoneal (IP) route
Not recommended for mixtures containing FCA.

Footpad injections
Not recommended for mixtures containing FCA. If footpad injections are justified for scientific reasons, only one footpad must be used.

6.6. A standard procedure for the production of monoclonal antibodies
A mouse is given a primary immunization with the antigen solution emulsified in Freund's incomplete adjuvant (FIA) (1:1) either IP or preferably SC for soluble immunogens. If Freund's complete adjuvant is used, it must be used only once (the primary injection) and not IP. Immunization with whole cells does not usually require the use of adjuvants. The total dose volume with either soluble antigen or cells should not exceed 0.1 ml. On day 21 a booster injection is given, followed by a final booster on day 42, given IV for soluble antigen or IP for whole cells. Frequently, no adjuvants are used for the final boost.

6.7. A standard procedure for the production of polyclonal antibodies
The antigen solution is emulsified in FIA (1:1) and injected SC in areas of loose skin. The volumes should be 0.1 ml at each of two sites in mice, 0.2 ml at each of two sites in rats and 0.25 ml at each of four sites in guinea-pigs, rabbits, sheep, goats and chickens. If FCA is used, it should only be used once. Booster injections with or without FIA should be given every 14 days until a good response has been obtained. If a good response has not been obtained following three immunizations, it may be more reasonable to kill the animal and repeat the procedure in another individual.

6.8. Bleeding
The total blood volume (TBV) of laboratory animal species averages 65 ml kg^{-1}, and animals can be bled up to 15% TBV every month. Animals should be bled from a peripheral vein, and, if other sites are used, the animals should be anesthetized. For more details on immunization and production of antibodies, the reader is referred to Erb and Hau (33).

7. REFERENCES

1. Gershwin, M.E. and Merchant, B. (1981) *Immunological Defects in Laboratory Animals.* Plenum Press, New York.

2. Fox, J.G., Cohen, B.J. and Loew, F.M. (1984) *Laboratory Animal Medicine.* Academic Press, London.

3. Hallman, R.E.W. and Gorman, N.T. (1989) *Veterinary Clinical Immunology.* W.B. Saunders, London.

4. Svendsen, P. and Hau, J. (eds) (1994) *Handbook of Laboratory Animal Science.* CRC Press, Boca Raton, FL.

5. Hansen, A.K. (1993) *Scand. J. Lab. Anim. Sci.,* **20,** 11.

6. Flecknell, P.A. (1987) *Laboratory Animal Anaesthesia.* Academic Press, London.

7. Waynforth, H.B. and Flecknell, P.A. (1992) *Experimental and Surgical Technique in the Rat (2nd edn).* Academic Press, London.

8. Festing, M.F.W. (1993) *International Index of Laboratory Animals (6th edn).* Festing, Leicester, UK.

9. Elkon, K.B. and Ashany, D. (1993) *Brit. J. Rheumatol.,* **32,** 4.

10. Dick, J.E. (1991) *Mol. Gen. Med.*, **1**, 77.

11. McClure, H.M., Anderson, D.C., Ansari, A.A. and Klumpp, S.A. (1992) *Path. Biol. Paris*, **40**, 694.

12. Yogi, Y., Nakamura, K., Inoue, T., Kawatsu, K., Kashiwabara, Y., Sakamoto, Y., Izumi, S., Saito, M., Hioki, K. and Nomura, T. (1991) *Nippon Rai Gakk. Zass.*, **60**, 139..

13. Lindsay, M.D., Patrick, A.K., Prayoonwiwat, N. and Rodriguez, M. (1992) *Mayo Clin. Proc.*, **67**, 829.

14. Sharon, R. and Noar, D. (1992) *Int. J. Immunopharmacol.*, **14**, 1241.

15. Cohen, I.R. (1992) in *Encyclopedia of Immunology* (I.M. Roitt and P.J. Delves, eds). Academic Press, London, p. 1438.

16. Mosier, D.E., Picchio, G.R., Baird, S.M., Kobayashi, R. and Kipps, T.J. (1992) *Cancer Res.*, **52**, (19), 5552s.

17. Rygaard, J. (1994) in *Handbook of Laboratory Animal Science* (P. Svendsen and J. Hau, eds). CRC Press, Boca Raton, FL, Chapter 39.

18. Mule, J.S., Jicha, D.L. and Rosenberg, S.A. (1992) *J. Immunother.*, **12**, 196.

19. Lennon-Pierce, M., Lane, P.W., Davisson M.T. and Mobraaten, L.E. (1992) in *Encyclopedia of Immunology* (I.M. Roitt and P.J. Delves, eds). Academic Press, London, p. 1102.

20. Bancroft, G.J. (1992) in *Encyclopedia of Immunology* (I.M. Roitt and P.J. Delves, eds). Academic Press, London, p. 790.

21. Bonsma, M.J. and Caroll, A.M. (1991) *Ann. Rev. Immunol.*, **9**, 323.

22. Schneider, G.B., Relfson, M. and Ellis, T.M. (1992) *J. Bone Miner. Res.*, **7**, 941.

23. Felsburg, P.J., Somberg, R.L. and Perryman, L.E. (1992) *Immunodefic. Rev.*, **3**, 277.

24. Roitt, I.M. and Delves, P.J. (eds) (1992) *Encyclopedia of Immunology*. Academic Press, London.

25. Cohen, P.L. and Eisenberg, R.A. (1992) *Immunol. Today*, **13**, 427.

26. Merino, R., Fossati, L. and Izui, S. (1992) *Springer Semin. Immunopathol.*, **14**, 141.

27. Bach, J.F. (1992) *C. R. Acad. Sci.*, **111**, **314** (Suppl. 9), 45.

28. Crisa, L., Mordes, J.P. and Rossini, A.A. (1992) *Diabetes Metab. Rev.*, **8**, 4.

29. Muryoi, T., Andre-Schwartz, J., Saitoh, Y., Daian, C., Hall, B., Dimitriu-Bona, A., Schwartz, R.S., Bona, C.A. and Kasturi, K.N. (1992) *Cell. Immunol.*, **144**, 43.

30. Targan, S.R. *et al.* (1992) *Gastroenterology*, **102**, 1493.

31. Johnson, B.C., Morton, J.I. and Trune, D.R. (1992) *Otolaryngol. Head Neck Surg.*, **106**, 394.

32. Jensenius, J.C., Anderson, I., Hau, J., Crane, M. and Koch, C. (1981) *J. Immunol. Meth.*, **46**, 63.

33. Erb, K. and Hau, J. (1994) in *Handbook of Laboratory Animal Science* (P. Svendsen and J. Hau, eds). CRC Press, Boca Raton, FL, Vols I and II.

CHAPTER 2
TISSUES AND CELLS OF THE IMMUNE SYSTEM
D. Rayner

Lymphocytes are the specific parenchymal cells of the immune system. They are distributed in tissue compartments (the lymphoid tissues) and in circulating compartments (the blood and lymph). This organization:

(i)   provides environments for lymphocyte development (lymphopoiesis) and activation, and
(ii) allows lymphocytes to mobilize to sites of antigen challenge or inflammation.

1. PRIMARY LYMPHOID TISSUES

Primary or *central lymphoid tissues* (*Table 1*) generate mature virgin lymphocytes from immature progenitor cells. Many primary lymphoid tissues are associated with the gut mucosa or other endodermal derivatives such as the thymus and liver. Absence or destruction of primary lymphoid tissues can result in selective loss of cellular or humoral immunity. The thymus is the site of T-lymphocyte development in all higher vertebrates, whereas the bursa of Fabricius is a highly specialized organ where B-lymphocyte development occurs in avian species.

1.1. Thymus

A thymus (1, 8, 9) is present in all higher vertebrates, and in mammals is a soft, bilobed organ in the upper anterior mediastinum. During embryogenesis, the epithelial framework of the human thymus develops from branchial cleft ectoderm and pharyngeal pouch endoderm. Thymic tissue involutes with age (see *Table 2* and refs 14, 15): the human thymus reaches maximal size around puberty, and eventually the lymphoid tissue becomes largely replaced by fat.

Table 1. Major primary lymphoid tissues

Tissue	Cell lineage	Embryology	References and comments
Thymus	T	Foregut (3rd and 4th pharyngeal pouches)	1
Intestinal epithelium	T	Midgut/hindgut	Intraepithelial lymphocytes (2)
Bursa of Fabricius	B	Hindgut	Birds only (3)
Ileal Peyer's patch	B	Midgut	Ruminants (4)
Liver[a]	B, T[b]	Midgut	Fetal/neonatal B lymphopoiesis (5)
Bone marrow	B	Mesoderm	5, 6
Omentum and peritoneum	B	Mesoderm	B1 cells (7)

[a]Clonable B precursors are also present in the neonatal spleen (5).
[b]Some thymus-independent T-cell receptor-$\gamma\delta$+ T-cell development is believed to occur in the liver (2).

Table 2. Murine lymphoid tissues: cell content and size[a]

	Thymus	**Spleen**	**Lymph nodes**
Nucleated cells ($\times 10^7$)	10–30	5–20	1–3[b]
T:B ratio[c]	100:1	1:1	4:1
Weight at 6 weeks (mg)[d]	64	63	39[e]
Weight at 14 weeks (mg)	35	110	49
Weight at 62 weeks (mg)	15	83	33

[a]Refs 10–12.
[b]Cell yield per lymph node.
[c]MALT T:B ratios vary considerably from 0.2:1 (Peyer's patch) to 2.6:1 (BALT) (13).
[d]CBA mice, mean of both sexes. Female lymphoid organs are generally larger.
[e]Pooled weight: mesenteric, cervical, axillary and inguinal nodes.

The two main thymic lobes are divided into lobules. The thymic *cortex* is easily distinguishable from the *medulla* on sections, because of its close cellular spacing which causes darker staining (*Figure 1a*). At the boundary between them is a network of small blood vessels which give off branches to both the cortex and the medulla. The arterial supply of the thymus is from the internal thoracic and inferior thyroid arteries, with drainage to the corresponding veins and the left brachiocephalic vein. The innervation of the thymus is similarly complex (from the cervical sympathetic, vagus, phrenic and descending cervical nerves) and includes noradrenergic and cholinergic fibers.

Figure 1. Lymphoid tissues: (a) thymus (human newborn, hematoxylin–eosin, $\times 8$); (b) lymph node (human, $\times 8$); (c) spleen (mouse, $\times 32$); (d) Peyer's patch (mouse, $\times 32$). Dark-staining regions in these sections are areas of high lymphocyte density: the thymic cortex, lymph node cortex and paracortex, splenic white pulp and Peyer's patch lymphoid follicles.

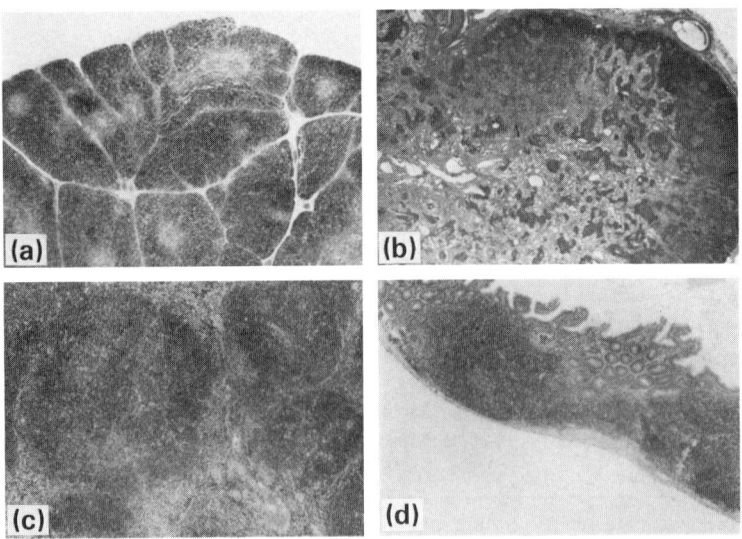

Developing T lymphocytes (thymocytes) are the most numerous cells in the thymus. The supporting epithelial cells, which express MHC class I and class II are less conspicuous on histologic sections, except where they accumulate as spheres of cells in the medulla, called Hassall's corpuscles. Thymic nurse cells (16) are extraordinary epithelial cells in the thymic cortex, which have the capacity to enclose 10–20 maturing thymocytes; recent evidence suggests the thymic nurse cells may have a neuroendocrine function. Other cell types in the thymus include macrophages, dendritic cells and a few B lymphocytes, which are mainly CD5$^+$ (see Section 4.1).

The anatomic layers of the thymus are zones of increasing T-cell maturation (see Chapter 3). Mature T cells leave the thymus through the postcapillary venules of the blood circulation, or through the efferent lymphatics. Considerable cell loss ($> 90\%$) occurs during thymocyte cell maturation, due to programmed cell death of:

(i) potential self-reactive clones (with affinity for self antigen $+$ MHC), and
(ii) unusable clones (unable to associate with self MHC).

The thymic epithelial cells elaborate several polypeptide hormones (thymosin-α1, thymopoietin, thymulin and thymic humoral factor), which are believed to promote thymocyte differentiation and also act on peripheral T cells (1). The thymic epithelial cells and other stromal cells (endothelial cells and macrophages) also produce other soluble factors, including cytokines and eicosanoids.

1.2. The bursa of Fabricius

This B-lymphopoietic tissue is restricted to birds (3). Although it has no mammalian homolog, it has provided a great deal of information about B lymphopoiesis.

The bursa is formed by an outpouching of the gut wall dorsal to the cloaca, and connected to it by an opening called the bursal duct. The bursa receives its blood supply through the pudendal and posterior mesenteric vessels, and is innervated by branches of the pelvic and intestinal nerves. Peptidergic nerve fibers extending near B cells and macrophages in the cortex of the bursa may participate in regulating B-cell development (17).

The specialized mucosal surface of the bursa is arranged as 12–15 infoldings (plicae), each of which encloses about 800 *bursal follicles*. Analogous to the thymus, the bursal follicles show a densely cellular cortex and a less cellular medulla. The bursal follicles are formed from a limited number of B-cell precursors which populate the bursa during embryogenesis. Following maturation, Ig gene rearrangement, and V region diversification through gene conversion, the B cells derived from the bursa colonize the secondary lymphoid tissues. The bursa involutes considerably around the time of sexual maturity, but continues to function as part of the gut-associated lymphoid tissue. (A nonfollicular T-cell-rich region, called the diffusely infiltrated area, is located adjacent to the bursal duct, and is thought to function as a secondary lymphoid tissue.)

The importance of the bursa to B-cell development was shown in 1956 by the observation that bursectomy *in ovo* caused severe hypogammaglobulinemia which persisted through the life of the adult bird. The technique of bursectomy remains useful in avian experimental models to achieve selective depletion of B cells. A natural equivalent, infectious bursal disease, is caused by viral destruction of the bursa.

1.3. Mammalian B lymphopoiesis

In mammals B-cell development is complex for a number of reasons:

(i) B-cell development from progenitor cells persists through adult life;

(ii) B cells develop in different locations at different stages of development and in different species (*Table 1*); and

(iii) there is evidence for more than one independent B-cell lineage.

In the fetus, the major site of B-cell generation is the liver. The bone marrow is the primary site of mammalian postnatal B-cell development (5, 6), although some B cells are bone marrow-independent (see Section 4.1). Marrow is present in the medullary cavity of all the bones, in close association with the cancellous or spongy bone. In humans, marrow becomes more fatty with increasing age, and cellular marrow becomes largely confined to the axial skeleton. Bone marrow is a complex tissue which serves as the source of all the blood cell lineages in addition to generating mammalian B cells. The stromal cells of the bone marrow are believed to promote hematopoiesis and lymphopoiesis through release of growth factors, including interleukin-7 (IL-7), and through membrane-bound growth factors.

2. SECONDARY (PERIPHERAL) LYMPHOID TISSUES

Secondary (peripheral) lymphoid tissues provide environments for clonal expansion and affinity maturation, and allow for lymphocyte homing and storage in appropriate sites (*Table 2*).

In general, secondary lymphoid tissues are compartmentalized into B-cell areas and T-cell areas. The basic unit of the B-cell areas is the *lymphoid follicle* (*Figure 2a and b*), an aggregate of B cells, together with accessory cells which support B-cell proliferation and maturation. Lymphoid follicles are found in lymph nodes, spleen, mucosa-associated lymphoid tissues, and can also be found in sites of chronic inflammation.

Following antigen stimulation, the primary follicle changes to a secondary follicle which contains a *germinal center* (18). This is a pale-staining region of large activated oligoclonal

Figure 2. Lymphoid follicles (human lymph node cortex, × 80). (a) hematoxylin–eosin; the arrow indicates a germinal center. (b) Anti-CD20 immunoperoxidase preparation of an adjacent section to demonstrate B cells.

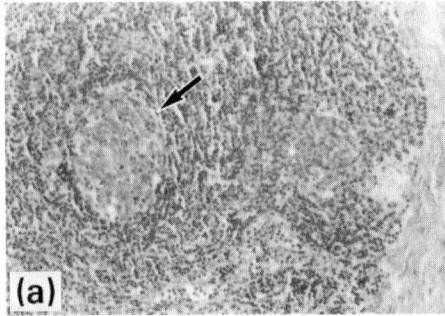
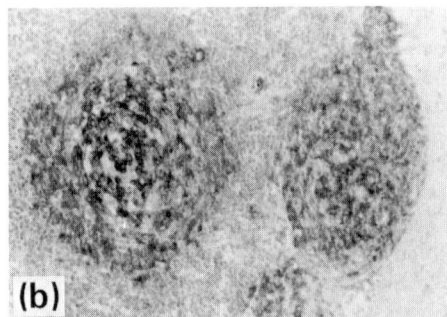

B cells (follicle center cells, sometimes called centroblasts and centrocytes), surrounded by a cellular mantle zone of small B lymphocytes. Follicle center cells proliferate near the base of the germinal center, and mature when they encounter antigen packaged in aggregates of immune complex ('iccosomes') bound to specialized Fc-receptor-bearing follicular dendritic cells (19). The germinal centers also contain T-helper cells, and distinctive macrophages (tingible body macrophages), which phagocytose nuclear debris from dead lymphocytes.

2.1. Lymph nodes

Lymph nodes are arranged in groups distributed along the lymphatic circulation. Since lymph vessels and blood vessels usually run together, the lymph nodes draining a region also tend to follow major blood vessels. Lymph nodes are located in the subcutaneous tissues, in tissue planes between muscles, and adjacent to the vessels of visceral organs (*Figure 3*).

Usually described as ovoid or bean-shaped, lymph nodes vary in size from about 0.1 to 2.5 cm in humans. In mice, normal peripheral lymph nodes range up to about 0.1 cm, although a

Figure 3. Lymph nodes of the mouse. Modified from ref. 20 with permission from Academic Press.

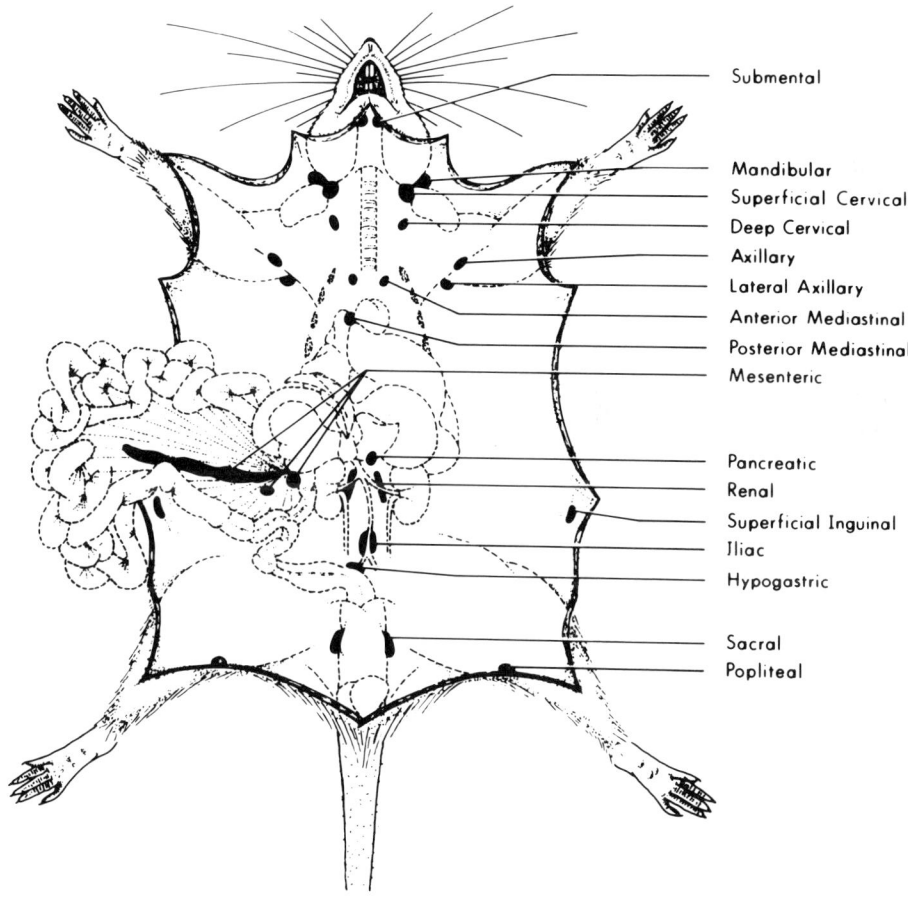

Submental

Mandibular
Superficial Cervical
Deep Cervical
Axillary
Lateral Axillary
Anterior Mediastinal
Posterior Mediastinal
Mesenteric

Pancreatic
Renal
Superficial Inguinal
Iliac
Hypogastric

Sacral
Popliteal

TISSUES AND CELLS OF THE IMMUNE SYSTEM

single elongated mesenteric lymph node is larger. For preparation of lymphocyte suspensions, the mesenteric, superficial cervical, inguinal, axillary and popliteal lymph nodes are easily located. In both humans and animals, lymph nodes draining sites of antigen challenge or inflammation may be considerably enlarged.

Lymph nodes have three regions (*Figure 1b*):

(i)   a superficial B-cell-rich *cortex*, containing lymphoid follicles;
(ii) a T-cell-rich zone between and deep to the follicles, called the *paracortex*; and
(iii) a still more deeply situated *medulla*, which is contiguous with the lymph node hilum, and contains lymphatic sinuses and blood vessels.

Lymph nodes have functional links with both the lymphatic and blood circulations. Afferent lymphatic vessels (Section 3) discharge lymph into a superficial space called the subcapsular or marginal sinus. The lymph passes through the sinuses of the lymph node, and exits through an efferent lymphatic at the lymph node hilum. The lining cells of the lymphatic sinuses within the lymph node (sinus-lining histiocytes) are phagocytic and help clear suspended material from the lymph.

High endothelial venules in the lymph node paracortex provide an interface with lymphocytes in the circulating blood. These specialized vessels have adhesion molecules which bind lymphocytes, and allow them to exit the blood circulation. *Interdigitating cells* in the paracortex are dendritic cells which express MHC class I and II determinants and are important antigen-presenting cells for T-cell responses (21, 22).

2.2. Spleen

The spleen (*Figure 1c*) is an unpaired organ found in all vertebrates, located in the upper left quadrant of the mammalian abdomen. Its blood supply is from the splenic artery, which branches before entering the splenic hilum, and then further ramifies along the trabecular fibrous tissue which forms the supporting skeleton of the spleen. Venous drainage is through the splenic vein, which is part of the hepatic portal system. (This means that the spleen is often secondarily enlarged in diseases of the liver.) The spleen is innervated by branches of the celiac plexus, and is penetrated by noradrenergic and peptidergic (substance P) nerve fibers, which may be involved in immunoregulation. Splenic tissue has two major compartments, red pulp and white, or lymphoid, pulp.

Red pulp
This is made up of:

(i)   blood sinuses lined by specialized phagocytic endothelium; and
(ii) the surrounding splenic cords, which are rich in macrophages.

Most (90%) of the human splenic blood flow is believed to be a slow-transit 'open' circulation, with the arterial supply feeding directly into the splenic cords; these in turn drain into the blood sinuses which have a specialized porous endothelium. The intimate arrangement of the blood with phagocytic cells allows the red pulp to clear senescent blood cells, circulating debris and micro-organisms.

White or lymphoid pulp
This is arranged in cuffs which encase small blood vessels, called *periarteriolar lymphoid sheaths* (*PALS*). The PALS include a T-cell-rich zone which directly surrounds the arteriole, and a more eccentrically located B-cell-containing lymphoid follicle. The boundary between the red pulp and the white pulp is called the marginal zone, and contains macrophages and extensions

of follicular dendritic cells that together present entrapped antigen to lymphocytes in the white pulp. The association of splenic lymphoid tissue with splenic blood vessels:

(i)   gives the immune system access to blood-borne antigens, and
(ii) is believed to provide a route for lymphocytes to re-enter the blood circulation.

The spleen is a major site of hematopoiesis in mice and other animals, and is believed to act as a storage organ for blood cells and platelets. In humans, hematopoiesis occurs in the spleen: (i) in fetal life, and (ii) in diseases characterized by abnormal expansion of the bone marrow.

2.3. Mucosa-associated lymphoid tissue (MALT)

Resident and recirculating populations of lymphocytes associated with the mucosal surfaces of the body form the mucosal immune system (8). MALT differs from other secondary lymphoid tissues in a number of ways:

(i)   Particularly in the gut, MALT contains a high proportion of B cells committed to IgA secretion and IL-4-secreting $CD4^+$ T cells which support the class switch to IgA.
(ii)   MALT lymphocytes tend to recirculate to their specific MALT, rather than entering the systemic lymphocyte pool.
(iii) MALT is adapted to sampling the antigens on the adjacent mucosal surface (unlike the spleen and lymph nodes, which receive antigens from the blood and afferent lymph respectively).

In general, the lymphoid aggregates that make up MALT are closely associated with specialized overlying mucosal surfaces. In the tonsils, a special fenestrated epithelium in the tonsillar crypts is believed to allow sampling of luminal antigens. In the gut, the epithelial lining overlying Peyer's patches contains specialized M (microfold) cells which are believed to serve a similar purpose.

In the human pharynx, mucosal lymphoid tissue is present as follicular aggregates called the palatine, pharyngeal and lingual *tonsils*. The tonsils together make up a ring of lymphoid tissue (Waldeyer's ring) which surrounds the upper end of the aerodigestive tract. Similar nasopharyngeal lymphoid tissue is present in rodents and other animals (13).

Gut-associated lymphoid tissue (GALT) forms the largest part of the mucosal lymphoid tissue and has several components. Peyer's patches are large (2–10 cm long in humans) aggregates of lymphoid follicles in the mucosa and submucosa of the small bowel (*Figure 1d*). They are most prominent in the ileum, and are predominantly made up of B cells. Similar aggregated lymphoid follicles form a sheath of lymphoid tissue around the appendix. Numerous single lymphoid follicles (at least 10^4 in the human colon) are present elsewhere in the submucosa and mucosa of the bowel.

GALT also has a diffuse component, made up of:

(i)   lymphocytes and other leukocytes distributed through the lamina propria of the gut mucosa; and
(ii) intraepithelial lymphocytes (IELs), situated between the epithelial cells themselves. About 60% of IELs are thymus-independent T cells, which have the phenotype $CD4^-$ $CD8_{\alpha\alpha}^+$, but do not express $CD8_\beta$ chains.

Bronchus-associated lymphoid tissue (BALT) is made up mainly of lymphoid aggregates situated between bronchi or bronchioles and pulmonary arterial branches. These aggregates have B-cell-rich and T-cell-rich areas which are inconstantly located in relation to surrounding structures. A diffuse population of lymphocytes is also present in the bronchial lamina propria.

3. LYMPHOCYTE RECIRCULATION AND HOMING

Lymphocytes are transported from place to place within the body in two parallel circulations: (i) the lymphatic circulation, and (ii) the blood.

Lymphatics are thin-walled low-pressure vessels lined by endothelium. Lymphatic capillaries communicate with the extracellular spaces of the peripheral tissues, and through their permeable walls take in interstitial fluid with suspended colloidal material. Lymphatics draining the small intestinal villi (lacteals) contain lymph with a high content of microscopic fat droplets (chyle). The splenic pulp, bone marrow, brain and spinal cord have no lymphatic drainage. Lymph draining proximally acquires lymphocytes (80% T cells), a few macrophages and dissolved proteins (including Ig) as it passes through lymph nodes. Following antigenic challenge, lymph out of a lymph node is transiently impeded, a process called lymph trapping.

Progressively larger lymphatic vessels join and anastomose to form two main lymphatic channels in the upper thorax: the thoracic duct and right lymphatic duct. These vessels drain lymph into the central blood circulation, through openings near the junctions of the subclavian and internal jugular veins. For experimental work, lymph can be collected by placement of an indwelling cannula in the thoracic duct (12).

Lymphocyte homing from the blood to lymphoid tissue occurs through specialized vessels which include the high endothelial venules of the lymph node paracortex, and the arterioles of the spleen. The distribution of blood-borne lymphocytes in the peripheral tissues is regulated by the expression of different adhesion molecules on lymphocytes which are capable of recognizing specific ligands (addressins) expressed on endothelial cells in different tissues. These 'homing molecules' are of several types, including lymphocyte lectins (selectins) and their endothelial oligosaccharide ligands, CD44, Ig superfamily members and integrins. Similar recognition mechanisms facilitate lymphocyte homing to sites of inflammation (23).

4. CELLS OF THE IMMUNE RESPONSE

4.1. Lymphocytes

Unstimulated lymphocytes show a characteristic round nucleus with coarse chromatin and little cytoplasm (*Figure 4a*). B and T lymphocytes (*Table 3*) are not reliably distinguishable by morphology, but can be separated easily by functional studies and surface antigen expression (24).

In different functional states, lymphocytes may change in appearance and expression of phenotypic markers. *Blast cells* are activated B or T lymphocytes which become enlarged, show an 'open' chromatin pattern, and are separable from resting lymphocytes by their lower buoyant density. (The term 'blast' can have different meanings depending on the context, and is also used for immature leukocytes such as leukemia cells.) *Plasma cells* (*Figure 4c*) are B cells which have followed a terminal differentiation pathway to become specialized for antibody secretion. In stained preparations, plasma cells have the well-known 'clock-face' eccentric nucleus, a paranuclear clear zone representing the Golgi apparatus, and basophilic cytoplasm reflecting their high content of ribosomes.

B lymphocytes exist in at least three separate lineages, which have been designated B1a, B1b and B2 (*Table 4*). B1 cells have been implicated in human and experimental autoimmune diseases (rheumatoid arthritis, Sjögren's disease) and are phenotypically similar to the CD5[+] neoplastic B cells of chronic lymphocytic leukemia.

Figure 4. Cells of the immune response: (a) small lymphocyte (left) and eosinophil (right); (b) monocyte; (c) plasma cells (arrowed) surrounded by lymphocytes. (a) and (b) are from human peripheral blood and (c) is from an imprint of an inflamed human lymph node. May–Grünwald–Giemsa, × 800.

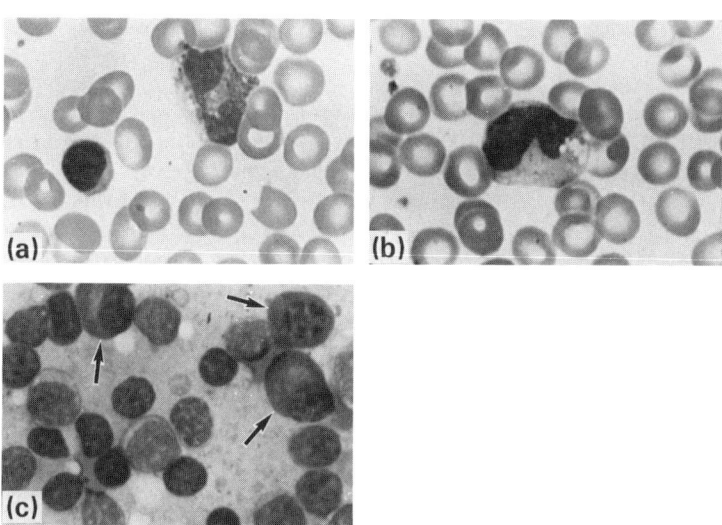

T lymphocytes can be subdivided in different ways. MHC restriction, CD4/8 expression and function tend to be highly correlated: T cells which recognize antigen bound to class I MHC molecules usually express CD8, and include most cytotoxic T cells. Conversely, MHC class II-restricted T cells usually express CD4 and function as helper cells for inflammatory or antibody responses.

In murine and human T-cell clones, $CD4^+$ T cells can be further subclassified based on their patterns of cytokine secretion (*Table 5*) (25, 26). The T_H1 and T_H2 subsets have different preferences for antigen-presenting cells and tend to down-regulate each other reciprocally through the actions of cytokines (IL-4, IL-10, IFN-γ).

T cells can also be grouped according to their class of antigen receptor. Most peripheral T cells express the $\alpha\beta$ heterodimeric (TCR-2) form of the T-cell receptor. A smaller population expresses the $\gamma\delta$ (TCR-1) form of the receptor, and is less well understood. These cells form resident intraepithelial populations in the gut, skin and reproductive tract which show absent or limited receptor diversity. They differ from $\alpha\beta$ T cells in that: (i) most $\gamma\delta$ cells are $CD4^-8^-$, and (ii) many $\gamma\delta$ clones are restricted to MHC products with low polymorphism (such as TL) rather than classical polymorphic class I and II MHC molecules (27).

Natural killer (NK) cells are lymphocytes which, by definition, do not participate in the specific immune response. They were initially identified by their spontaneous killing of tumor cells (or virus-infected cells), in the absence of prior sensitization. NK cells typically show the morphology of large granular lymphocytes and express CD16 and CD56 (humans) or NK-1.1/2.1 (mice), but do not express CD3 and do not rearrange or express Ig or TCR genes. Although they are not thymus-dependent, they show some phenotypic and functional similarities to T cells, and it appears that T and NK cells arise from a common immediate

TISSUES AND CELLS OF THE IMMUNE SYSTEM　　　　　　　　　　**25**

Table 3. B and T cells[a]

	B	T
Receptor	Ig (H_2L_2)	TCR ($\alpha\beta$ or $\gamma\delta$)
Antigen recognition	Native	Processed + MHC
Functions	Antibody secretion; antigen presentation	Regulation; cytotoxicity
Development	Liver (fetal); bone marrow (B2); fetal omentum (B1)	Thymus (most); extrathymic (gut, liver?)
Respond to cytokines[b]	IL-4, IL-5, IL-6 (growth and differentiation); IFN-γ	IL-1, IL-6 (activation); IL-2, IL-4 (proliferation and activation); IL-8 (chemotaxis); IL-12 (T_C differentiation)
Secrete cytokines[b]	IL-1, IL-6	IL-2, IL-3, IL-4, IL-5, IL-6, IFN-γ, TNFα, TNFβ, GM-CSF
Phenotypic markers[c]	CD5$^-$23$^+$ (B2), CD5$^+$23$^-$ (B1a) or CD5$^-$23$^-$ (B1b); B220; CD19,20,21,22	CD2,3,5,7[d], w90[e]; CD4$^+$ or CD8$^+$ (TCR$_{\alpha\beta}$ cells); CD4$^-$8$^-$ (most TCR$_{\gamma\delta}$ cells); CD4$^-$8$_{\alpha\alpha}^+$ (gut IELs)
MHC expression	Class I$^+$ and II$^+$	Class I$^+$; class II inducible in human (not mouse) T cells
Adherence (nylon wool)	Yes	No
Mitogens	LPS[f]; PWM[d]; dextran (weak)	Anti-CD2, anti-CD3; ConA; PHA; PWM[d]

[a]See Section 4.1 for refs and discussion of T and B cell heterogeneity. Abbreviations: ConA, concanavalin A; GM-CSF, granulocyte–macrophage colony-stimulating factor; IFN-γ, interferon-γ; LPS, bacterial lipopolysaccharide; PHA, phytohemagglutinin; PWM, pokeweed mitogen; TNF, tumor necrosis factor.
[b]Partial listing.
[c]Major surface markers, excluding activation antigens and antigens transiently expressed in development. See Chapter 7 for a detailed discussion of surface antigen expression.
[d]Human.
[e]Mature mouse and rat T cells. Expressed early in human T-cell development.
[f]Mouse.

progenitor (28). *Lymphokine-activated killer (LAK) cells* are produced by culturing lymphocytes with IL-2 *in vitro*. LAK cells show MHC-unrestricted cytotoxicity similar to that of NK cells, but are able to kill a broader range of targets. *Killer (K) cells* are lymphocytes (cytotoxic T or NK cells) which are able to mediate antibody-dependent cellular cytotoxicity (ADCC), by binding antibody-coated target cells through Fc receptors on their cell membranes. In fact, the cell populations mediating these three kinds of killing overlap considerably, and NK cells can act as K cells or LAK cells when given the right conditions.

4.2. Nonlymphoid cells

Nonlymphoid cells (*Table 6*) participate in host defences in three main ways:

(i)   They mediate the acute inflammatory reaction.
(ii) They perform accessory functions in the inductive phase of the immune response, through antigen presentation and cytokine secretion.

Table 4. Murine mature B-cell subsets

Cell type	Markers	Localization	Self-replenishing
B1a (Ly-1/CD5 B cells)	$CD5^+$, IgM^{high}, IgD^{low}, $CD23^-$, $B220^{low}$, $Mac-1^{low}$	40–50% of the B cells in the peritoneal and pleural cavities, but <5% of splenic B cells	Yes – derived from fetal omentum or peritoneum
B1b ('Ly-1 sister' cells)	$CD5^-$, IgM^{high}, IgD^{low}, $CD23^-$, $B220^{low}$, $Mac-1^{low}$	10–20% of the B cells in the peritoneal and pleural cavities, but <5% of splenic B cells	Yes – derived from fetal omentum or peritoneum
B2 ('conventional' B cells)	$CD5^-$, IgM^{low}, IgD^{high}, $CD23^+$, $B220^{high}$, $Mac-1^-$	>95% of spleen and lymph node B cells	No – continuously replenished from bone marrow

(iii) They mediate effector functions initiated by the specific immune response (e.g. in delayed-type hypersensitivity).

In addition to lymphocytes, circulating leukocytes (*Table 7*) include several kinds of cell which are derived from bone marrow progenitors, and which can be distinguished by morphology and function. These include monocytes (*Figure 4b*), which are the circulating members of the mononuclear phagocyte system, and the granulocytes (or myeloid cells) which are further subdivided into neutrophils, eosinophils and basophils. All the cells of the blood in fact are thought to derive from the same undifferentiated stem cell as lymphocytes. In addition, the growth and differentiation of these cells is regulated by interleukins and colony-stimulating factors which may be synthesized by lymphocytes.

Mononuclear phagocytes include the blood monocytes and the tissue macrophages (also called histiocytes). These cells are:

(i) phagocytic;
(ii) able to kill micro-organisms through oxidative and other mechanisms;
(iii) able to present antigen to MHC class II-restricted T cells; and
(iv) capable of secreting cytokines and other inflammatory mediators (*Table 6*).

Table 5. T_H1 and T_H2 populations

Type	Major cytokines	Response
T_H1	IFN-γ, IL-2, IL-3, TNF-α, TNF-β, GM-CSF	DTH; macrophage activation; help for IgG2a
T_H2	IL-3, IL-4, IL-5, IL-6, IL-10, IL-13, TNF-α, GM-CSF	Antibody (particularly IgM, IgG1, IgA, IgE); eosinophil and mast cell products
T_H0	IL-2, IL-3, IL-4, IL-5, IL-10, IL-13, IFN-γ, GM-CSF	Broad

Abbreviations: as in *Table 3*, plus DTH, delayed-type hypersensitivity.

TISSUES AND CELLS OF THE IMMUNE SYSTEM

Table 6. Nonlymphoid cells in host defense

	Bone marrow origin	Circulating cell	Cytokines and effector molecules[a]	Isolation method[b] (ref.)
Monocyte	Yes	Yes	Cytokines (IL-1, IL-6, IL-8, TNFα); activated oxygen; enzymes (proteases, lipases, hydrolases); PGs and LTs	31
Neutrophil	Yes	Yes	Activated oxygen; proteases and other enzymes; defensins	31
Eosinophil	Yes	Yes[c]	Activated oxygen; enzymes; PGs (E series); cationic proteins, TGFα	32
Basophil	Yes	Yes	Proteases; histamine; PGD2; LTs (C4, D4, E4); PAF; ECF	reviewed in 33
Platelet	Yes (mega-karyocyte)	Yes	Vasoactive substances (serotonin, histamine, TXA2), PDGF	34
Macrophage	Yes (circulating monocyte)	No	See monocytes	35
Dendritic cell	Yes	Yes[d]	–[e]	reviewed in 36
Mast cell	Yes	No	Cytokines[f] (IL-1, IL-3, IL-4, IL-5, IL-6, TNFα); other factors similar to basophils	reviewed in 33
Endothelial cell	No	No	Cytokines (IL-1, IL-6, IL-8, IFN-α, IFN-β); vasoactive factors (PGI$_2$, NO, endothelin)	37

[a]Partial listing. See also refs 8, 29. Abbreviations as in *Table 3* and: ECF, eosinophil chemotactic factor; LT, leukotriene; NO, nitric oxide; PAF, platelet activating factor; PDGF, platelet-derived growth factor; PG, prostaglandin; TGFα, transforming growth factor α; TXA2, thromboxane A2.
[b]Ref. 30 is a general source for leukocyte cell separation methods.
[c]Circulate briefly – mainly resident in tissues.
[d]Blood and lymph circulations.
[e]Stimulation of T cells by dendritic cells is thought to involve membrane-bound co-stimulators. Dendritic cells do not appear to synthesize IL-1 or IL-6.
[f]Shown *in vitro*.

Table 7. Circulating blood cell counts[a]

	Human	Mouse	Rat
Leukocytes (total $\times 10^9$/l)	4.5–11	5–11	6–18
lymphocytes (%)	25–33[b]	63–80[c]	65–83
monocytes (%)	3–7	1–14	1–4
neutrophils (%)	54–62	9–37[d]	14–27[d]
eosinophils (%)	1–3	0.3–4	0.1–4
basophils (%)	0–0.75	–[e]	–[e]
Platelets ($\times 10^9$/l)	150–350	250–1500	500–1000
Erythrocytes ($\times 10^{12}$/l)[d]	4.2–6.2	8.8–10.5	6.5–9.0

[a]Mouse and rat values are ranges from both sexes of several strains (38, 39). Human values are a normal reference range (40).
[b]For details of subpopulations see Chapter 3, *Table 6*.
[c]Approximately 70% T cells; 24% B cells.
[d]Higher in males.
[e]Circulating basophils are extremely rare in rats and mice.

Some tissues have resident populations of macrophages, such as the pulmonary alveolar macrophage or the Kupffer cells present in the liver sinusoids. Under some conditions (such as the presence of foreign material, or certain delayed hypersensitivity diseases), macrophages form aggregates (granulomas) which may include multinucleate cells (giant cells) formed by fusion of individual macrophages.

Dendritic cells (21, 22, 36) make up a family of bone marrow-derived leukocytes found at low frequency ($\sim 0.1\%$) in: (i) primary and secondary lymphoid tissues, (ii) the lymphatic and blood circulations, and (iii) the interstitium of nonlymphoid tissues. Cells in this group include the Langerhans cells of the epidermis, the veiled cells of the lymph and interdigitating dendritic populations in lymphoid organs.

Dendritic cells are not fixed, and they are able to acquire antigen in one site and move it to lymph nodes via the lymph. This relocation is accompanied by a maturational change towards a form which is highly adapted to stimulating primary T-cell responses in the lymph node. (A similar change occurs when cells such as Langerhans cells are put in tissue culture.) Most authors believe that the dendritic cell lineage is separate from the mononuclear phagocyte system, but this is not completely clear: for example, Langerhans cells show features consistent with monocyte origin, but 'committed' dendritic cells in lymphoid tissues do not. Dendritic cells express class I and II MHC antigens, are rich in adhesion molecules, and in their mature form do not express Fc receptors.

Granulocytes (*Tables 6, 7* and refs 29, 40) include three types of cell:

(i) neutrophils, which are motile, short-lived, phagocytic cells specialized for providing a 'first response' at sites of injury and inflammation;
(ii) eosinophils, which are recruited in anti-helminth responses, immediate hypersensitivity reactions, and nonspecific inflammatory responses; and
(iii) basophils, present in very low numbers in the circulation, which participate in immediate and some delayed-type hypersensitivity reactions.

Platelets are small (2–3 μm) cytoplasmic fragments derived from bone marrow precursor cells (megakaryocytes). Their main function is in hemostasis, but they can be activated by immune complexes through their Fc receptors (8). Platelets also play a role in inflammation and repair, through the secretion of vasoactive substances, and the fibrogenic cytokine platelet-derived growth factor (PDGF) (29). Vascular *endothelial cells* secrete:

(i)   platelet activating factor (PAF);
(ii) a range of cytokines similar to macrophages; and
(iii) vasoactive mediators (*Table 6*).

They are inducible for MHC class II expression by culture with IFN-γ.

Mast cells are tissue-resident cells which are considered similar to basophils, since they both have Fc receptors for IgE (Fc$_\varepsilon$RI) and release histamine on activation. However, mast cells:

(i)   arise from different precursor cells;
(ii) do not circulate;
(iii) express fewer surface antigens; and
(iv) differ from basophils in morphology and granule chemistry (41).

The major importance of mast cells is their role in immediate hypersensitivity, due to their capacity to bind: (i) IgE to Fc$_\varepsilon$RI receptors, and (ii) anaphylatoxin (C3a, C4a and C5a) to complement receptors on their cell membranes. Mast cells are heterogeneous (42): in rodents, two main populations (mucosal mast cells and connective tissue mast cells) can be distinguished by staining characteristics and granule morphology. In humans, different mast cell populations (T and TC mast cells) can be grouped by their content of different proteolytic enzymes, and different functional responses.

5. REFERENCES

1. Kendall, M.D. (1991) *J. Anat.*, **177**, 1.

2. Rocha, B., Vassalli, P.I. and Guy-Grand, D. (1992) *Immunol. Today*, **13**, 449.

3. Zapata, A.G. and Cooper, E.L. (1990) *The Immune System: Comparative Histophysiology*. Wiley, Chichester.

4. Reynolds, J.D. (1987) *Curr. Top. Microbiol. Immunol.*, **135**, 43.

5. Rolink, A., Haasner, D., Nishikawa, S.I. and Melchers, F. (1993) *Blood*, **81**, 2290.

6. Rolink, A. and Melchers, F. (1993) *Adv. Immunol.*, **53**, 123.

7. Kantor, A.B. and Herzenberg, L.A. (1993) *Ann. Rev. Immunol.*, **11**, 501.

8. Roitt, I.M., Brostoff, J. and Male, D.K. (1993) *Immunology (3rd edn)*. Mosby, London.

9. Boyd, R.L., Tucek, C.L., Godfrey, D.I., Izon, D.J., Wilson, T.J., Davidson, N.J., Bean, A.G.D., Ladyman, H.M., Ritter, M.A. and Hugo, P. (1993) *Immunol. Today*, **14**, 445.

10. Santisteban, G.A. (1960) *Anat. Rec.*, **136**, 117.

11. Crispens, C.G. (1975) *Handbook on the Laboratory Mouse*. Charles C. Thomas, Springfield.

12. Hudson, L. and Hay, F.C. (1989) *Practical Immunology (3rd edn)*. Blackwell Scientific Publications, Oxford.

13. Kuper, C.F., Koornstra, P.J., Hameleers, D.M., Biewenga, J., Spit, B.J., Duijvestijn, A.M., van Breda-Vriesman, P.J. and Sminia, T. (1992) *Immunol. Today*, **13**, 219.

14. Kendall, M.D., Johnson, H.R.M. and Singh, J. (1980) *J. Anat.*, **131**, 483.

15. Stocker, J.S. and Dehner, L.P. (1992) *Pediatric Pathology*. Lippincott, Philadelphia.

16. Wekerle, H., Ketelsen, U.P. and Ernst, M. (1980) *J. Exp. Med.*, **151**, 925.

17. McCormack, W.T., Tjoelker, L.W. and Thompson, C.B. (1991) *Ann. Rev. Immunol.*, **9**, 219.

18. Berek, C. and Ziegner, M. (1993) *Immunol. Today*, **14**, 400.

19. Tew, J.G., Kosco, M.H., Burton, G.F. and Szakal, A.K. (1990) *Immunol. Rev.*, **117**, 185.

20. Cook, M.J. (1983) in *The Mouse in Biomedical Research* (H.L. Foster, J.D. Small and J.G. Fox eds). Academic Press, New York, Vol. 3.

21. King, P.D. and Katz, D.R. (1990) *Immunol. Today*, **11**, 206.

22. Klinkert, W.E.F. (1990) *Immunol. Rev.,* **117,** 103.

23. Picker L.J. and Butcher E.G. (1992) *Ann. Rev. Immunol.,* **10,** 561.

24. Reynolds, G.J. (1982) *Lymphoid Tissue: A Histological Approach.* Wright, Bristol.

25. Mosmann, T.R. and Coffman, R.L. (1989) *Ann. Rev. Immunol.,* **7,** 145.

26. Fitch, F.W., McKisic, M.D., Lancki, D.W. and Gajewski, T.F. (1993) *Ann. Rev. Immunol.,* **11,** 29.

27. Haas, W., Pereira, P. and Tonegawa, S. (1993) *Ann. Rev. Immunol.,* **11,** 637.

28. Lanier, L.L., Spits, H. and Phillips, J.H. (1992) *Immunol. Today,* **13,** 392.

29. Rubin, E. and Farber, J.L. (1993) *Pathology (2nd edn).* Lippincott, Philadelphia.

30. Di Sabato, G., Langone, J.J. and Van Vunakis, H. (1984) *Methods Enzymol.,* **108.**

31. Boyum, A. (1984) *Methods Enzymol.,* **108,** 88.

32. Shult, P.A., Graziano, F.M., Wallow, I.H. and Busse, W.W. (1985) *J. Lab. Clin. Med.,* **106,** 638.

33. Valent, P. and Bettelheim, P. (1992) *Adv. Immunol.,* **52,** 333.

34. Kinlough-Rathbone, R.L., Packham, M.A. and Mustard, J.F. (1983) in *Measurements of Platelet Function* (L.A. Harper and T.S. Zimmerman, eds). Churchill Livingstone, Edinburgh.

35. McCarron, R.M., Goroff, D.K., Luhr, J.E., Murphy, M.A. and Herscowitz, H.B. (1984) *Methods Enzymol.,* **108,** 274.

36. Steinman, R.M. (1991) *Ann. Rev. Immunol.,* **9,** 271.

37. Warren, J.B. (1990) *The Endothelium. An Introduction to Current Research.* Wiley-Liss, New York.

38. Bannerman, R.M. (1983) in *The Mouse in Biomedical Research* (H.L. Foster, J.D. Small and J.G. Fox, eds). Academic Press, New York, Vol. 3.

39. Ringler, D.H. and Dabich, L. (1979) in *The Laboratory Rat* (H.J. Baker, J.R. Lindsey and S.H. Wiesbroth, eds). Academic Press, New York, Vol. 1.

40. Bick, R.L. (1993) *Hematology. Clinical and Laboratory Practice.* Mosby, St Louis.

41. Paul, W.E., Seder, R.A. and Plaut, M. (1993) *Adv. Immunol.,* **53,** 1.

42. Kitamura, Y. (1989) *Ann. Rev. Immunol.,* **7,** 59.

Tissues and Cells

CHAPTER 3
LEUKOCYTE DEVELOPMENT
P.J. Delves and P.M. Lydyard

1. HEMATOPOIESIS

Hematopoiesis is initiated in the yolk sac early in fetal development and is then taken over by the liver and spleen, before residing in the bone marrow, where it is retained for the lifetime of the animal (1, 2) (*Table 1*).

Table 1. Fetal hematopoiesis

Site	Man (gestation 280 days)	Mouse (gestation 20 days)
Yolk sac	Up to day 60	Up to day 16
Liver and spleen	Days 50–150	Days 12–20
Bone marrow	From day 79	From day 19

In the embryo, the first blood cells to be produced are large megaloblastoid nucleated red cells. Immature lymphocytes can be detected in fetal blood by day 64 in man and phenotypically mature lymphocytes by day 90. However, although progenitor cells for the phagocytic lineages are found during fetal life, mature neutrophils and monocytes are not produced until birth.

Multipotential hematopoietic stem cells constitute < 0.1% of the bone marrow cells but are self-renewing and give rise to all the myeloid and lymphoid lineages (3, 4). As stem-cell differentiation occurs, there is a progressive commitment to a given lineage. Although stochastic in nature, this process is controlled by external factors such as cytokines and interactions with stromal cells. Techniques were developed as long ago as the 1960s for the growth of murine bone marrow cells *in vitro* (5, 6). Dexter and colleagues (7) were later able to grow myeloid cells from bone marrow cultures. This depended on the growth of a confluent layer of adherent stromal cells (fibroblasts, fat cells, macrophages and endothelial cells) (see Chapter 4, *Table 3* and Chapter 5, *Table 7*) which secretes a number of hematopoietic growth factors, including granulocyte–macrophage colony-stimulating factor (GM-CSF), granulocyte colony-stimulating factor (G-CSF), macrophage colony-stimulating factor (M-CSF) and interleukins 6 (IL-6) and 7 (IL-7). The stromal cell layer also has to contact the progenitor cells physically in order for differentiation to occur. This is due in part to the fact that stem-cell factor and some of the other hematopoietic growth factors remain largely cell-surface-associated.

Whitlock and Witte introduced modifications to the Dexter system which allowed the generation of B lymphocytes from bone marrow cultures (8; see also Chapter 4, *Table 3* and Chapter 5, *Table 8*). Because these cultures contain self-renewing hematopoietic stem cells, they can be grown continuously for many months, thus permitting the analysis of cytokine production, adhesion molecules, and other facets of hematopoiesis. Clonogenic assays in soft agar have been used to measure the ability of individual precursor cells to produce colonies which are capable of differentiation into mature functional cells (3, 9–11). The major types of colony-forming cells, and the growth factors that support their development, are shown in *Figure 1*.

2. LYMPHOCYTE DEVELOPMENT

2.1. B cells

In mammals, B lymphocytes arise directly from hematopoietic stem cells in the bone marrow, whereas in avian species a specialized organ, the bursa of Fabricius, is the site of B lymphopoiesis (see Chapter 2). A scheme for murine B-cell development which includes representative markers that permit the different stages to be distinguished is shown in *Figure 2*. The earliest B lineage cells (pro-B) differentiate into pre-B cells upon successful

Figure 1. A general scheme of hematopoiesis. Baso, basophil; BFU, burst-forming unit (produces multiple clusters of colonies); CFU, colony-forming unit; E, erythrocyte; Eo, eosinophil; EPO, erythropoietin; G, granulocyte; G-CSF, granulocyte colony-stimulating factor; GEMM, granulocyte, erythroid, monocyte, megakaryocyte; GM-CSF, granulocye–macrophage colony-stimulating factor; IGF-1, insulin-like growth factor-1; LIF, leukemia inhibitory factor; M, monocyte; Mφ, macrophage; M-CSF, macrophage colony-stimulating factor; mega, megakaryocyte; N, neutrophil; SC, stem cell; SCF, stem-cell factor.

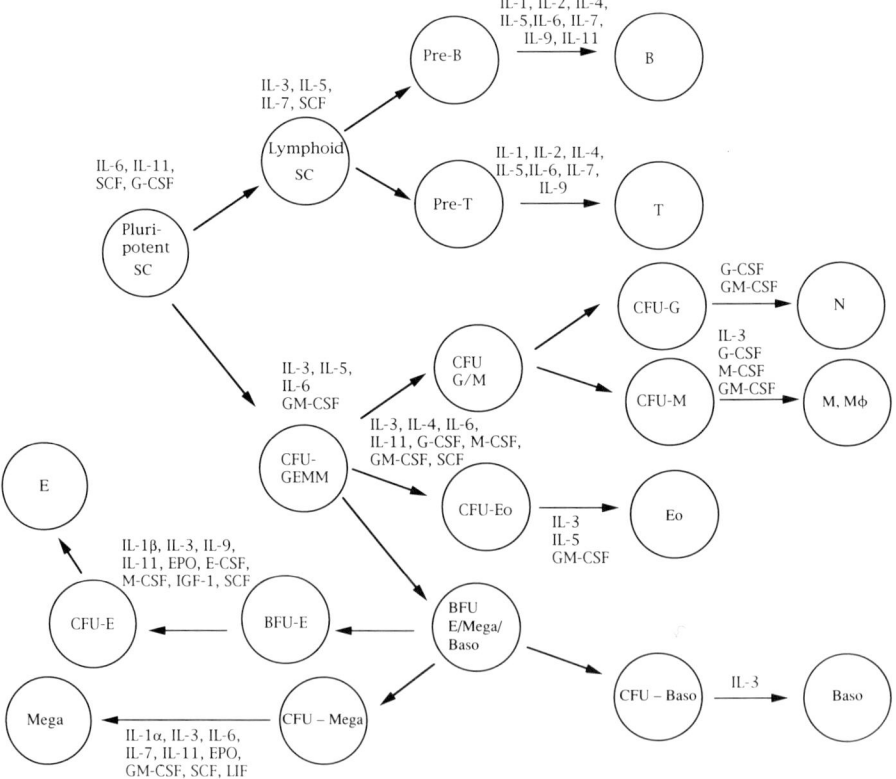

Figure 2. B-lymphocyte differentiation in the mouse. Human B-lymphocyte differentiation is thought to follow a broadly similar sequence of events. The total number of cells at each stage, and the approximate number of such cells produced per day, are shown. The Ig gene segments which have undergone rearrangement at each stage are indicated (V, variable; D, diversity; J, joining). Horizontal lines represent cell-surface expression of antigens which permit identification of each stage. Abbreviation: TdT, terminal deoxynucleotidyl transferase. For further information see ref. 12.

	Pro-B	Pre B-I	Pre B-II	Immat. B	Mature B
Total cells ($\times 10^7$)	0.1–0.2	7		2	10–50
Cells/day ($\times 10^7$)		0.5–1.5 (most die)		2–3 (most die)	0.3
IgH genes	Germ line	DJ	VDJ	VDJ	VDJ
IgL genes	Germ line	Germ line	Germ line	VJ	VJ
Ig expression		Cytoplasmic μ	Membrane $\mu/V_{pre-B} \lambda 5$	mIgM	mIgM + mIgD

TdT —————————————

CD45 ————————————————————————

RAG 1/2 ——————————————————————

CDw127 ————————————————————————

CD43 —————————————

CD23 ————————————

rearrangement of immunoglobulin heavy-chain genes. These cells average six cell divisions to produce a clone of 64 cells in bone marrow over 3–4 days, and express the surrogate light chain $V_{pre-B} \lambda 5$ (13). However, programmed cell death probably occurs in a majority of B-cell precursors due to either: (i) nonproductive immunoglobulin gene rearrangement, or (ii) interaction with antigen (e.g. self-antigen) of the newly expressed membrane immunoglobulin on these immature B cells (14). More mature B cells which have survived these deletional processes can be divided into three major populations (B1a, B1b and B2, see Chapter 2, *Table 4*) depending upon the presence or absence of the CD5 cell-surface antigen and upon the anatomical site to which they localize (15, 16).

Contact of mature B cells with specific antigen results in a rapid proliferation, with cell division occurring every 7–8 hours in lymphoid follicles, leading to germinal center development (17, and *Table 2*). The germinal centers provide an environment for somatic hypermutation of immunoglobulin genes and the development of memory B cells. Although

Table 2. Splenic germinal center development after primary stimulation by T-cell-dependent antigen (mouse)

Day	Appearance
1–4	Proliferating B cells in T-cell-rich periarteriolar lymphoid sheath (PALS)
3–8	Proliferating B cells in primary follicles
7–10	Germinal centers develop; proliferating B cells in basal dark zone, memory cell development thought to occur in apical light zone
7–14	Decline in PALS foci
14–28	Disappearance of germinal centers

plasma-cell precursors are also generated, these migrate out of the germinal centers to become antibody-secreting cells at locations such as the lymph node medullary cords, the splenic marginal zone, and the gut lamina propria. In the absence of antigen, memory B cells have a half-life of only 2–3 weeks. Antigen-bearing immune complexes persisting on follicular dendritic cells for long periods of time prolong clonal survival by maintaining cell division, and perhaps also the lifespan of individual memory B cells.

2.2. T cells

Cells that will mature into peripheral T lymphocytes need the environment of the thymus (18) in which to mature and be 'educated'. This process involves positive and negative selection procedures, such that the emergent T cells are able to recognize foreign antigenic peptides presented by self-MHC molecules. Two major waves of bone marrow-derived precursor cells (pro-T) seed the thymic rudiment on days 10 and 13 in the mouse (19), leading to a gradual development of TCR-expressing cells in the thymus during fetal life (*Table 3*).

Pro-T cells express low levels of CD4 upon arrival in the thymus, but expression is soon lost to create the CD4⁻CD8⁻ 'double-negative' population of pre-T cells (20). As these mature they progress from the subcapsular region of the thymus to the cortex, and then on to the medulla (*Figure 3*). Both $\alpha\beta$ and $\gamma\delta$ T cells are thought to arise from the double-negative population in the thymus, although the details of their divergent differentiation remain to be fully established (21). Extrathymic pathways of T-cell development have also been described for both $\alpha\beta$ and $\gamma\delta$ T cells, and may be a major differentiation pathway for intraepithelial T cells in the gut (22). Natural killer (NK) cell development is poorly understood, but it is known to be completely thymus-independent (23).

Table 3. Percentage of fetal thymocytes expressing TCR (mouse)

Days	TCR$_{\gamma\delta}$ (%)	TCR$_{\alpha\beta}$ (%)
14	7	0
15	27	0
16	40	2
17	28	12
18	20	26
19	10	48
20	5	70

Figure 3. T-lymphocyte differentiation in the mouse. Human T-lymphocyte differentiation is thought to follow a broadly similar sequence of events. Details relate to the αβ T-cell population. The anatomical location in the thymus at which the various stages are found is indicated at the top of the figure. The TCR α and β gene segments which have undergone rearrangement at each stage are indicated (V, variable; D, diversity; J, joining). Horizontal lines represent cell-surface expression of antigens which permits identification of each stage. Broken lines signify a low level of expression or, in the case of TCR, indicate β-chain found at the cell surface in the absence of α-chain. Abbreviation: HSA, heat-stable antigen.

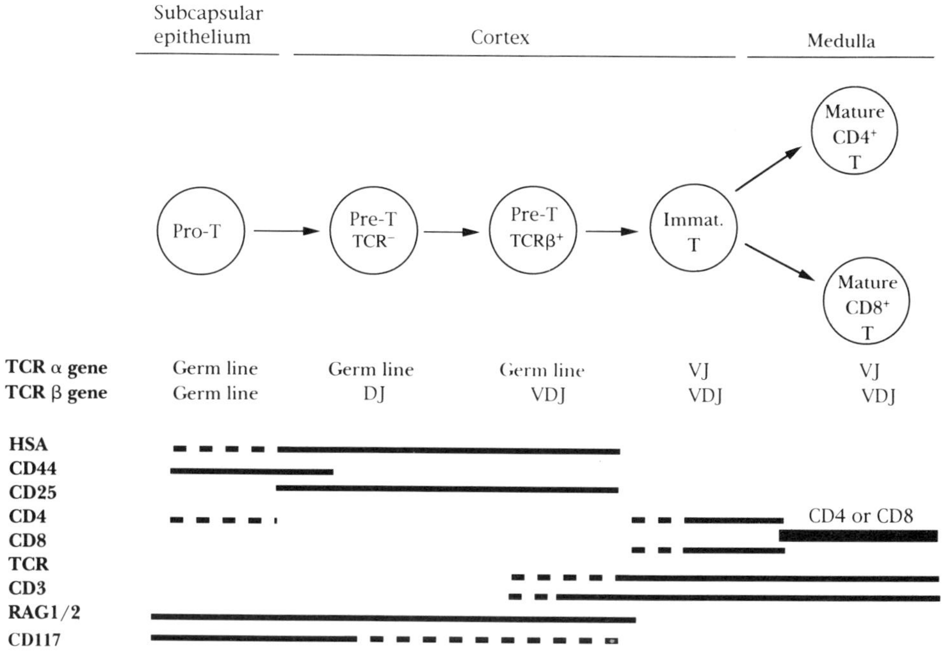

The thymic stromal cells (see Chapter 5, *Table* 9) produce several cytokines (e.g. IL-1, IL-6, IL-7, M-CSF, GM-CSF, TGFβ) and 'thymic hormones' (e.g. thymopoietin, thymulin). These mediators are thought to play a role in directing T-cell differentiation in the thymus. Thymic selection occurs at the 'double-positive' (CD4$^+$CD8$^+$) stage, and involves both positive selection of cells bearing TCR able to recognize 'self'-MHC molecules with a relatively low affinity, and negative selection of cells which recognize self-MHC (usually containing self-peptides) with high affinity (24, 25).

3. MARKERS OF CELLULAR ACTIVATION

As with many other cell types, activation of lymphocytes and myeloid cells is associated with an altered expression of cell-surface antigens, which presumably represent molecules required for further passage through the cell cycle and/or are associated with cellular function. It is thus possible to explore the activation status of these cells on the basis of cell-surface expression of 'activation antigens' (*Table* 4).

Table 4. Markers of cellular activation

Cells	Status	Useful markers
T	Naive	CD44^{low}, CD62L^{high}, CD45RA$^+$
	Memory	CD44^{high}, CD62L^{low}, CD45RO$^+$ (NB some RO$^+$ cells revert to RA$^+$)
	Activated[a]	CD9$^+$, CD11c$^+$, CD25$^+$, CD30$^+$, CD38$^+$, CD39$^+$, CD49a$^+$, CD69$^+$, CD70$^+$, CD80$^+$, CD95$^+$, CD96$^+$, CD98$^+$, CTLA-4$^+$, LAG-3$^+$, CD40L/gp39$^+$, MHC class II$^+$ (man)
B	Naive	HSA$^+$
	Memory	HSA$^-$
	Activated	CD25$^+$, CD28$^+$, CD30$^+$, CD38$^+$, CD39$^+$, CD43$^+$, CD69$^+$, CD70$^+$, CD80$^+$, CD95$^+$, CD98$^+$, BB18$^+$
NK	Activated	CD39$^+$, CD69$^+$, CD96$^+$, CD98$^+$, LAG-3$^+$
Monocytes	Activated	CD23$^+$ (man), CD25$^+$, CD105$^+$
Neutrophils	Activated	CD64$^+$

[a]Resting cells generally do not express, or only express low levels of, the markers mentioned for each cell type.

4. LEUKOCYTE LIFESPANS

Several approaches have been used in attempts to estimate the lifespan of immune cells, and there is still a degree of controversy regarding the various estimates (26, 27). A rough guide to some of the lifespans that have been estimated is given in *Table* 5.

5. AGE DEPENDENCE OF LYMPHOCYTE SUBPOPULATIONS

Recent studies in man have shown that the distribution of lymphocyte subpopulations varies with age. The results of one such representative study (28) are summarized in *Table* 6.

6. METHODS OF STUDYING LYMPHOCYTE DEVELOPMENT

A wide range of methodologies is available to study lymphocyte development both *in vivo* and *in vitro*. Some of the more commonly employed techniques are presented in *Tables* 7 and 8.

Table 5. Leukocyte lifespans (mouse)

Cell	Lifespan (intermitotic for dividing cells, survival for 'end' cells)
Nonrecirculating B cell (90–95%)	3–4 days
Recirculating B cell (5–10%)	4–6 weeks
Plasma cell – IgM	3 days
Plasma cell – IgG/IgA	up to 3 weeks
T cell	4–6 weeks
Interdigitating dendritic cell	5 days
Follicular dendritic cell	>1 year
Monocyte	1–2 days
Macrophage	50–70 days
Neutrophil	3–5 days

Table 6. Human peripheral blood lymphocyte populations as a function of age

	Cord blood		1 day–11 months		1–6 years		7–17 years		18–70 years	
	%	Number[a]	%	Number	%	Number	%	Number	%	Number
WBC	–	12.0	–	9.0	–	7.8	–	6.0	–	5.9
Lymphocytes	41	5.4	47	4.1	46	3.6	40	2.4	32	2.1
T cells (CD3$^+$)	55	3.1	64	2.5	64	2.5	70	1.8	72	1.4
CD4$^+$CD8$^-$	35	1.9	41	2.2	37	1.6	37	0.8	42	0.8
CD4$^-$CD8$^+$	29	1.5	21	0.9	29	0.9	30	0.8	35	0.7
CD4:CD8 ratio	1.2	–	1.9	–	1.3	–	1.3	–	1.2	–
CD3: % DR$^+$	2.0	–	7.5	–	9.0	–	12	–	10	–
% CD25$^+$	8.0	–	9.0	–	11	–	13	–	18	–
% CD57$^+$	0	–	1.5	–	3.0	–	5.5	–	10	–
CD4: % CD45RA$^+$	91	–	81	–	71	–	61	–	40	–
% CD62L$^+$	91	–	91	–	91	–	87	–	78	–
CD8: % CD57$^+$	0	–	7	–	10	–	17	–	29	–
B cells	20	1.0	23	0.9	24	0.9	16	0.4	13	0.3
% CD5$^+$	72	0.5	68	0.5	64	0.5	56	0.2	27	0.1
% CD23$^+$	35	–	50	–	61	–	63	–	64	–
% CD62L$^+$	57	–	66	–	79	–	90	–	87	–
% CDw78$^+$	49	–	22	–	30	–	32	–	32	–
NK cells	20	0.9	11	0.5	11	0.4	12	0.3	14	0.3

[a]Cell number $\times 10^3$ mm^{-3}. All values are given as the median of ≥ 16 individuals. Modified from ref. 28 with permission from Elsevier Science Publishers BV.

Table 7. *In vitro* methods of studying lymphocyte development

Method	Comments and references
Culture of cells or tissues containing hematopoietic cells under conditions which permit their differentiation	Includes various types of colony-forming cell assays (10)
Isolate lymphocytes from different tissues or organs at different times during development	Measure the emergence, and changes in distribution, of different populations. Data can most readily be obtained using flow cytometric analysis of cell-surface antigen expression (28)
Immunohistology of tissue sections	Permits localization of cell populations within lymphoid tissues at different times during development (29)
Organ culture of lymphoid tissues	For example, deplete hematopoietic cells from thymic lobes using 2′-deoxyguanosine and then reconstitute with the precursor cell population to be studied (30, and Chapter 4)
Reaggregate organ culture	Selected thymocyte and thymic stromal cell populations can be reconstituted *in vitro* (31)
Measure development of functional capacity	For example, proliferation, cytokine production, cytotoxic effector function (see Chapter 6)
Hydroxyurea deletion of dividing cells	Prevents entry of new cells into the population being studied (32)
Assess changes in levels of specific mRNA transcripts. e.g. RAG-1 and RAG-2 gene expression	For example, using sensitive Northern blotting techniques (33)
Measure changes in levels of enzymes, e.g. nonspecific esterase, TdT	Can be combined with other markers of cell differentiation (34)
Limiting dilution analysis	Used to determine precursor frequences at various stages of development (see Chapter 6)
Immortalize cells at different developmental stages using hybridoma technology, or establish T-cell clones driven by specific antigen or anti-CD3	Has limitations because some stages of cell differentiation are more amenable to immortalization than others (see Chapter 5)

Table 8. *In vivo* methods of studying lymphocyte development

Method	Comments and references
Label cells *in vivo* and follow their fate	The proportion of cells undergoing DNA replication during the S phase of the cell cycle can be assessed over a period of continuous labeling with tritiated thymidine or 5-bromo-2'-deoxyuridine (BrdU) (35)
Adoptive transfer using allotype-disparate cells or *in vitro* labeled cells, e.g. labeled with fluorescein isothiocyanate (FITC), integrated retroviral DNA sequences, etc.	Bone marrow chimeras can be created between F_1 mice and either parental strain, and the origin of emerging cells detected using H-2 haplotype-specific antibody (36)
Transgenic and gene knockout animals	For example, TCR, MHC, RAG-1/RAG-2, etc. Has proved very useful for analysis of positive and negative selection and repertoire generation (37)
Generation of chimeras between closely related species	Chick and quail cells can be distinguished by nuclear morphology. Chimeras can be created to investigate waves of colonization of thymus and bursa by bone marrow progenitors (38)
Adoptive transfer across species into immunodeficient animals	Human hematopoietic cells are transferred into severe combined immunodeficient (SCID) or *bg/nu/xid* (bnx) mice and their development monitored. Can combine with organ transplantation, e.g. fetal thymus (39)

7. REFERENCES

1. Linch, D.C. (1992) in *Encyclopedia of Immunology* (I.M. Roitt and P.J. Delves, eds). Academic Press, London, p. 248.

2. Dainiak, E.P. *et al.* (eds) (1990) *The Biology of Hematopoiesis.* Wiley-Liss, New York.

3. Ogawa, M. (1993) *Blood*, **81**, 2844.

4. Uchida, N. *et al.* (1993) *Curr. Opin. Immunol.*, **5**, 177.

5. Bradley, T.R. and Metcalf, D. (1966) *Aust. J. Exp. Biol. Med. Soc.*, **44**, 287.

6. Pluznik, D.H. and Sachs, L. (1965) *J. Cell Physiol.*, **66**, 319.

7. Dexter, T.M., Allen, T.D. and Lajtha, L.G. (1977) *J. Cell. Physiol.*, **91**, 335.

8. Whitlock, C.A. and Witte, O.N. (1982) *Proc. Natl. Acad. Sci. USA*, **79**, 3608.

9. Clark, S.C. (1993) in *Clinical Aspects of Immunology* (P.J. Lachmann, D.K. Peters, F.S. Rosen and M.J. Walport, eds). Blackwell Scientific Publications, Oxford, p. 327.

10. Visser, J.W.M., de Vries, P., Hogeweg-Platenburg, M.G.C., Bayer, J., Schoeters, G., van den Heuvel, R. and Mulder, D.H. (1991) *Sem. Hematol.*, **28**, 117.

11. Gordon, M.Y. (1993) *Blood Rev.*, 7, 190.

12. Rolink, A. and Melchers, F. (1993) *Curr. Opin. Immunol.*, **5**, 207.

13. Melchers, F., Haasner, D., Grawunder, U., Kalberer, C., Karasuyama, H., Winkler, T. and

Rolink, A.G. (1994) *Ann. Rev. Immunol.*, **12,** in press.

14. Matzinger, P. (1994) *Ann. Rev. Immunol.*, **12,** in press.

15. Kearney, J.F.(1993) *Curr. Opin. Immunol.*, **5,** 223.

16. Herzenberg, L.A. and Kantor, A.B. (1993) *Immunol. Today*, **14,** 79.

17. MacLennan, I.C.M. (1994) *Ann. Rev. Immunol.*, **12,** in press.

18. Boyd, R.L., Tucek, C.L., Godfrey, D.I., Izon, D.J., Wilson, T.J., Davidson, N.J., Bean, A.G.D., Ladyman, H.M., Ritter, M.A. and Hugo, P. (1993) *Immunol. Today*, **14,** 445.

19. Jotereau, F., Heuze, F., Salomon-Vie, V. and Gascan, H. (1987) *J. Immunol.*, **138,** 1026.

20. Dunon, D. and Imhof, B.A. (1993) *Blood*, **81,** 1.

21. Godfrey, D.I. and Zlotnik, A. (1993) *Immunol. Today*, **14,** 547.

22. Rocha, B., Vassalli, P. and Guy-Grand, D. (1992) *Immunol. Today*, **13,** 449.

23. Trinchieri, G. (1992) in *Encyclopedia of Immunology* (I.M. Roitt and P.J. Delves, eds). Academic Press, London, p. 1136.

24. Hugo, P., Kappler, J.W. and Marrack, P.C. (1993) *Immunol. Rev.*, **135,** 133.

25. Robey, E. and Fowlkes, B.J. (1994) *Ann. Rev. Immunol.*, **12,** in press.

26. Freitas, A.A. and Rocha, B.B. (1993) *Immunol. Today*, **14,** 25.

27. Osmond, D.G. (1993) *Immunol. Today*, **14,** 34.

28. Hannet, I., Erkeller-Yuksel, F., Lydyard, P., Deneys, V. and DeBruyère, M. (1992) *Immunol. Today*, **13,** 215.

29. Mackenzie, C.D. (1992) in *Encyclopedia of Immunology* (I.M. Roitt and P.J. Delves, eds). Academic Press, London, p. 783.

30. Coligan, J.E., Kruisbeek, A.M., Margulies, D.H., Shevach, E.M. and Strober, W. (eds) (1991) *Current Protocols in Immunology*. Wiley, New York, p. 3.18.1.

31. Anderson G., Jenkinson, E.J., Moore, N.C. and Owen, J.J.T. (1993) *Nature*, **362,** 70.

32. Ajchenbaum, F., Ando, K., De Caprio, J.A. and Griffin, J.D. (1993) *J. Biol. Chem.*, **268,** 4113.

33. Campbell, J.J. and Hashimoto, Y. (1993) *J. Immunol.*, **150,** 1307.

34. Zadeh, H.H. and Goldschneider, I. (1993) *J. Exp. Med.*, **178,** 285.

35. Butcher, E.C. and Ford, W.L. (1986) in *Handbook of Experimental Immunology* (D.M. Weir, L.A. Herzenberg, C. Blackwell and L.A. Herzenberg, eds). Blackwell Scientific Publications, Oxford, p. 57.1.

36. Philpott, K.L., Viney, J.L., Kay, G., Rastan, S., Gardiner, E.M., Chae, S., Hayday, A.C. and Owen, M.J. (1992) *Science*, **256,** 1448.

37. Arase, H., Arase, N., Ogasawara, K., Good, R.A. and Onoe, K. (1992) *Int. Immunol.*, **4,** 75.

38. Le Douarin, N.M. and Jotereau, F.V. (1977) *J. Exp. Med.*, **142,** 17.

39. Reisner, Y. (1992) in *Encyclopedia of Immunology* (I.M. Roitt and P.J. Delves, eds). Academic Press, London, p. 332.

CHAPTER 4
TISSUE AND CELL CULTURE
K.H. Brooks and R. Fernandez-Botran

1. GENERAL INFORMATION

The first two considerations when initiating research involving tissue culture are safety and sterility. A major source of contamination of research materials as well as laboratory-acquired infections is an aerosol of the contaminant. Two types of work area have been designed to deal with these problems. Laminar-flow benches which utilize a horizontal flow of sterile, filtered air across the work surface maintain sterility within the work area and prevent entry of contaminants from the room air. These benches are completely open at the front and thus allow easy access and a comfortable working area. The primary disadvantage of laminar-flow hoods is the total lack of protection of the researcher from contaminants present in the hood; in fact, the laminar-flow design forces potentially contaminated air directly toward the researcher. Thus, laminar-flow hoods cannot be used for any tissue culture which might contain a human pathogen. In contrast to the laminar-flow bench, biosafety cabinets provide sterility along with excellent protection for the investigator.

1.1. Biosafety cabinets

Design
Biosafety cabinets (1) use HEPA (High Efficiency Particulate Air) filters to maintain sterility. Although highly efficient, these filters should not be considered 100% effective. They are 99.97% efficient in removing particles of diameter 0.3 μm. HEPA filters are composed of a continuous sheet of glass fiber filter pleated over corrugated aluminum separators. This design leaves them very sensitive to damage by rough handling, punctures or Bunsen burners. Therefore, a newly installed biosafety cabinet or filter should not be used until it is certified for safety. The cabinet should be recertified at least annually, or after being moved, or after replacement of the filter. The lifespan of the filter is variable, but averages about 7–10 years.

Types of biosafety cabinets
Class I biosafety cabinets provide operator protection but little or no product protection. Air is drawn through the front opening across the work area into an exhaust plenum, where it passes through a HEPA filter and is exhausted either into the room or into a building exhaust duct. Class I biosafety cabinets are not suitable for tissue culture since sterility is not maintained within the cabinet.

Class II biosafety cabinets protect both the investigator and the cultures. There are two subtypes. Type A exhausts into the room, whereas type B cabinets are connected to a building exhaust system. The type A cabinets recirculate 70% of the air and 30% is exhausted into the room. Thus, explosive, flammable or toxic substances, including carcinogenic chemicals, should not be used in type A cabinets. Either type of class II biosafety cabinet is suitable for tissue culture, but because of the air barrier employed on the cabinet front, it should be installed away from high traffic areas, air supply, exhaust registers and doors. Also, do not position two cabinets opposite one another on both sides of an aisle. In addition, items should not be placed on the front grille, blocking the air flow, or stacked on top of the cabinet.

Class III biosafety cabinets provide total containment and are usually hooked up in series with other containment or processing units. These cabinets are not routinely used in immunological research.

Proper use of class II biosafety cabinets

(i)   Turn on lights and fan 10–15 min before use.
(ii)   Wash hands and arms with germicidal soap.
(iii) Put on a labcoat, preferably closed in front.
(iv)   Disinfect work surface with 70% ethanol.
(v)   Put all materials in the cabinet, at least 10 cm from the cabinet front.
(vi)   Do not use a Bunsen burner.
(vii) After everything is in the cabinet, wait 3–5 min to purge the cabinet of any residual airborne contaminants.
(viii) Work as far into the cabinet as possible, at least 10 cm from the front sash.

1.2. Other required equipment

In addition to a sterile work area, immunological tissue culture requires a centrifuge, incubator, pipettes, water bath, refrigerator, and usually some storage space in a $-70°C$ freezer. The centrifuge should have a working range from 200 to 5000 g and for most applications should be refrigerated. The incubator is normally used at $37°C$, but it is useful to be able to regulate the CO_2 percentage.

1.3. Methods of sterilization

All solutions, containers, instruments and pipettes need to be sterilized before use. There are several methods for sterilization (2) and none of these methods can be used for all the items needing to be sterilized. The advantages, disadvantages and primary applications for each of these methods are listed below.

Autoclave
Advantages. Kills everything.

Disadvantages. Excessive heat chemically alters solutions, media, etc.

Primary uses. Simple liquids, such as water and phosphate-buffered saline (PBS), or separation reagents, such as Sephadex G-10 in PBS; glass pipettes, microscope slides, pipette tips, valves for use on cell separation columns.

Standard conditions. 121°C under pressure for 30–45 min.

Filter sterilization
Advantages. Removes almost everything except mycoplasma, 0.2 μm filter removes about 50% of mycoplasma.

Disadvantages. Moderately expensive; microbial products such as endotoxin are not removed.

Primary uses. Media, fetal calf serum, cytokines, antibodies; any solution in a small volume which is to be added to cultures.

Ultraviolet (UV) irradiation
Advantages. Requires minimal equipment and is inexpensive.

Disadvantages. Requires long exposures and is relatively inefficient.

Primary uses. Supplemental sterilization of work surfaces or equipment which can not be autoclaved.

70% Ethanol
Advantages. Inexpensive, readily available, moderately effective.

Disadvantages. Short exposure to 70% ethanol is not 100% effective.

Primary uses. Sterilization of work surfaces, animal fur. Surgical instruments can be dipped in 70–95% ethanol and passed through a flame. Instruments should then be rinsed in sterile PBS or Hank's balanced salt solution (HBSS) to cool them and ensure that no ethanol comes in contact with the cells or tissue of interest. The instruments can be resterilized at several points during the procedure to prevent contamination of internal organs with organisms found on the skin.

1.4. Biosafety levels

Guidelines for four biosafety levels (3, 4) have been issued by the US Department of Health and Human Services. Biosafety level 1 guidelines are appropriate for work with defined and characterized strains of viable micro-organisms not known to cause disease in healthy adults. Biosafety level 2 guidelines are applied to work with a broad spectrum of indigenous, moderate-risk agents present in the community and associated with human disease of varying severity. These agents can be used safely on the open bench provided the potential for producing aerosols is low. Biosafety level 3 guidelines are applied to indigenous or exotic agents with a high potential for infection by aerosols, and where the disease may be serious or lethal. Biosafety level 4 guidelines apply to agents which pose a high individual risk of life-threatening disease. The assignment of micro-organisms to particular biosafety levels varies somewhat between different countries. The reader should therefore consult the relevant national guidelines for appropriate containment levels.

According to these guidelines, most cell and tissue culture employed in immunological studies should utilize biosafety level 2 or biosafety level 3 precautions. Primary and permanent cell lines from mouse, hamster, human, and rat should be treated as if they carry low-risk viruses. Human isolates from malignant tissues or those from tissues susceptive to or likely to harbor mammalian oncogenic viruses should be considered a moderate risk. Cells from herpes- and Epstein–Barr virus-transformed cultures would be handled as moderate risk viruses. All established human lymphocyte lines should be assumed to harbor Epstein–Barr virus. In addition, an individual should not handle lymphoid cells of a line derived from himself or a first-degree relative.

Biosafety level 2 guidelines applicable to cell and tissue culture

(i) Eating, drinking, smoking and applying cosmetics are not permitted in the work area.
(ii) Hands are washed before and after the cultures are handled.
(iii) A class II biosafety cabinet is used.
(iv) Mechanical pipetting devices are used; mouth pipetting is prohibited.
(v) Laboratory coats, gowns, smocks or uniforms are worn while in the laboratory.

In addition, needles and syringes should be discarded in a puncture-proof container. With human cultures, tissue culture flasks, centrifuge tubes, pipettes, etc., should be placed in a biohazard bag and autoclaved before disposal.

2. TISSUE CULTURE MEDIA

There are many media formulations in the literature and available commercially. In the authors' experience, the media listed below will be suitable for at least 90% of all cell lines and primary lymphocyte cultures. Occasionally, the investigator may chose to provide additional supplements to the medium.

2.1. Media formulations

The media formulations given in *Table 1* are provided by Whitaker M.A. Bioproducts, Walkersville, Maryland, USA.

Hank's balanced salt solution (HBSS) is used primarily to wash cells during cell preparation and separation. Since HBSS lacks essential amino acids and vitamins, it cannot support cell growth, but does maintain viability for at least 12 h. HBSS is much more cost-effective than using complete medium during cell preparation. Dulbecco's modified Eagle's medium (DMEM) is most often used to maintain hybridomas and long-term cell lines, whereas RPMI 1640 is required for primary lymphocyte cultures.

2.2. Supplements and media modifications

The two most common variables in media formulations are the buffering system and whether or not the media contains L-glutamine. This amino acid is required for lymphocyte growth but is relatively unstable at 4°C. Thus, it is common practice to store a stock of L-glutamine at -20°C and add it to the medium on the day of use. There are two common buffering systems: sodium bicarbonate and Hepes. Sodium bicarbonate acts as a physiological buffer, and cultures in sodium bicarbonate-buffered medium must be cultured in the presence of 5–10% CO_2. In contrast, Hepes maintains a physiological pH under atmospheric conditions. Hepes-buffered medium does contain some sodium bicarbonate since it is a physiological requirement for many cell types. A significant difference between RPMI 1640 and DMEM is the glucose content. Thus, a high glucose DMEM is available and may be useful for some cell lines.

Two critical supplements must be added to media used for lymphocyte cultures. The first is fetal bovine/calf serum (FBS/FCS). Primary lymphocyte cultures require 10% FCS whereas most cell lines can be maintained in 5% FCS. At FCS concentrations of 1–3%, some cell lines can be maintained for a few cell cycles. In the case of some hybridomas, special serum-free media or modified normal bovine serum can be used. In the authors' experience none of the serum-free media support optimal responses in primary lymphocyte cultures. The second supplement required by most lymphoid cells is 2-mercaptoethanol (2-ME). All murine primary monocyte and lymphocyte cultures require 50 μm 2-ME. There are a few cell lines, in particular T-cell lines, for which 2-ME is toxic. In addition, there are stock concentrates which can be added to media. These include essential amino acids, nonessential amino acids, vitamin mixtures and sodium pyruvate. For example, addition of nonessential amino acids and sodium pyruvate to RPMI 1640 enhances support of B-lymphocyte responses and some monocyte/macrophage lines.

2.3. Antibiotics

Four antibiotics are commonly used in cell and tissue culture: penicillin, streptomycin, amphotericin and gentamicin sulfate. These antibiotics can be used alone or in combination, to give the broadest protection against contamination. *Table 2* indicates the microbial specificity of each antibiotic and its effective dose.

Table 1. Components of common media (mg l^{-1})

Component	Hank's BSS	DMEM	RPMI 1640
L-Arginine		84	200
L-Asparagine			50
L-Aspartic acid			20
L-Cystine		48	50
L-Glutamic acid			20
Glycine		30	10
L-Histidine			15
L-Histidine HCl.H_2O		42	
Hydroxy-L-proline			20
L-Isoleucine		104.80	50
L-Leucine		104.80	50
L-Lysine.HCl		146.20	40
L-Methionine		30	15
L-Phenylalanine		66	15
L-Proline			20
L-Serine		42	30
L-Threonine		95.20	20
L-Tryptophan		16.00	5
L-Tyrosine		72.00	20
L-Valine		93.60	20
p-Aminobenzoic acid			1.00
D-Biotin			0.20
D-Calcium pantothenate		4.00	0.25
Choline chloride		4.00	3.00
Folic acid		4.00	1.00
i-Inositol		7.00	35
Nicotinamide		4.00	1.00
Pyridoxal.HCl		4.00	
Pyridoxine-HCl			1.00
Riboflavin		0.40	0.20
Thiamine-HCl		4.00	1.00
Vitamin B_{12}			0.005
$CaCl_2$ (anhydrous)		200	
$CaCl_2.2H_2O$	186		
$Ca(NO_3)_2.4H_2O$			100
$Fe(NO_3)_3.9H_2O$		0.10	
KCl	400	400	400
KH_2PO_4	60		
$Mg_2SO_4.7H_2O$	200	200	100
NaCl	8000	5400	6000
$NaHCO_3$	350	3700	2000
$NaH_2PO_4.H_2O$		125	
$Na_2HPO_4.7H_2O$	90		1512
Glucose	1000	1000	2000
Glutathione			1.00
Phenol red	20	15	5.00

Table 1. Continued

Component	Hank's BSS	DMEM	RPMI 1640
Sodium pyruvate		110	
Hepes		5957.50	

Abbreviations: BSS, balanced salt solution; DMEM, Dulbecco's modified Eagle's medium; RPMI, Roswell Park Memorial Institute media.
Formulae reproduced with permission from Whitaker M.A. Bioproducts.

Table 2. Specificity and dose of some common antibiotics

Effective against	Penicillin	Streptomycin	Gentamicin sulfate	Amphotericin
Gram-positive bacteria	+	+	+	
Gram-negative bacteria	+/−	+	+	
Mycoplasma		+	+	
Pseudomonas			+	
Yeast/mold				+
Effective dose	50 U ml^{-1}	50 µg ml^{-1}	50 µg ml^{-1}	125 ng ml^{-1}

3. ORGAN CULTURE

3.1. Long-term bone marrow cultures

Bone marrow contains multiple cell types that can influence lymphoid and myeloid cell development. Long-term bone marrow cultures are dependent upon the establishment of a heterogeneous feeder layer derived from the marrow cell populations (5, 6). The nonadherent cells (lymphoid and myeloid precursors) in these cultures die within 3 days of removal from the feeder layer (6). A feeder layer can be maintained for 14–18 weeks before the nonadherent cells must be transferred to a fresh feeder layer (7). Both macrophages and stromal cells seem to be required for establishment of pre-B-cell lines. Once established, the B-cell lines can be maintained on a stromal cell line rather than a bone marrow feeder layer (6). As indicated by *Table 3*, culture conditions determine whether the multipotential stem cells differentiate along the myeloid lineage (Dexter cultures, ref. 8) or the B-cell lineage (7, 9). In the Whitlock–Witte cultures, which select for B-cell differentiation, granulopoiesis ceases after 3–5 weeks (5).

Like the bone marrow population from which they are derived, bone marrow stromal cell lines (Chapter 5, *Table 7*) can be quite heterogeneous in their morphology and properties (10). A wide range of cytokines are produced by these lines, including M-CSF, GM-CSF, IL-6, IL-7, TGFβ and neuroleukin (6). Interestingly, none of the lines have been shown to produce IL-3. There are a few common features of bone marrow stromal cell lines (10):

Common features of bone marrow stromal cell lines

(i)   nonphagocytic;
(ii) express MHC class I antigens but no markers of hematopoietic cells;

Table 3. A comparison of long-term bone marrow culture systems

Culture conditions	Dexter	Whitlock–Witte
Media	DMEM	RPMI 1640
Serum	High % horse	Low % FCS
2-Mercaptoethanol	0	50 μM
Additive(s)	Corticosteriods	None
Temperature	33°C	37°C
Nonadherent cells present	Multipotential stem cells, differentiated myeloid cells	4–10 weeks 70–80% pre-B cells; > 10 weeks 80–100% mIgM$^+$ cells
Adherent cell components	Endothelial-like cells, dendritic-reticular cells, phagocytic mononuclear cells, and lipid-filled adipocytes	Fibroblasts, reticular-dendritic cells, epithelial-like cells, mononuclear cells

(iii) secrete extracellular matrix components and soluble mediators;
(iv) are not lineage restricted in their hematopoietic support capacity.

Table 4 outlines the basic properties of some of these stromal cell lines.

3.2. Thymic organ cultures
A more short-term culture system has been developed to study T-cell development (11–13). In this system, day 14 embryonic thymic lobes are cultured *in vitro* with deoxyguanosine,

Table 4. Properties of some bone marrow stromal cell lines (6)

Name	Description	Surface markers	Cytokines produced
MBA-2.4	Endothelial-like		M-CSF
B.AD	Fibroblastoid, phagocytic, pre-adipocyte		CSF
TC-1	Fibroblast/endothelial		M-CSF
S 17		MHC class II$^-$ CD90$^-$, CD11b$^-$, BP$^-$	IL-4
AC 4	Endothelial-like, pre-adipocyte	6C3$^+$, CD90$^-$	M-CSF
ALC	Fibroblastoid, pre-adipocyte		M-CSF G-CSF
30R		CDw124$^+$, 1-2% 6C3$^+$	
1xN/A6	SV40 transformed stromal cell clone		IL-7
BMS-2	Pre-adipocyte	CD90$^+$	M-CSF

which is toxic to immature thymocytes. These alymphoid thymic lobes are then repopulated with cells from fetal liver, bone marrow or embryonic thymus. A hanging-drop culture is used for the first 2 days until repopulation occurs, then T-cell development can be followed for 1-3 weeks in organ culture. It should be noted that deoxyguanosine also depletes interdigitating reticular cells and as a result tolerance induction is not observed in these cultures (14).

The advantage of the thymic organ culture is an intact microenvironment which includes the required stromal cells. Some thymic stromal cell lines (Chapter 5, *Table 9*) have been developed which can be used to study the thymocyte/stromal cell interaction. A few examples are described in *Table 5*.

Table 5. Properties of some thymic stromal cell lines

Cell line	Strain	Surface markers	Secreted factors	Comments	Ref.
IT 79 MTN C3	BALB/c	MHC class II$^+$		Forms thymic nurse cells in hanging drop cultures	15
TEC-L1	BALB/c		Thymosin-α1, thymosin-β3	Epithelial; forms thymic nurse cells	16
ET	C57Bl6x BALB/c F$_1$			IFN-γ induces MHC class II$^+$	17
MRL 104.8a	MRL lpr/lpr		Unknown T-cell growth factor	IFN-γ induces MHC class II$^+$	18

3.3. Splenic fragment cultures

Splenic fragment cultures (19, 20) are used to evaluate clonal antigen-driven humoral responses in an intact splenic environment. Cells are transferred into a mouse which, 24 hour previously, received 650 rad total body irradiation. One day after cell transfer the spleens are removed and dissected into cubes (about 30 per spleen). The splenic fragments are cultured in DMEM with 10% agammaglobulin horse serum and 5% chicken embryo extract. Soluble antigen is added for at least the first 3 days. Antibody levels in the culture supernatant can be monitored for several weeks.

4. CELL ISOLATION

4.1. Methods of tissue dissociation

After removing lymphoid tissue from an animal, the tissue can be dissociated to generate a cell suspension containing macrophages and lymphocytes. Methods suitable for each lymphoid tissue are indicated in *Table 6*.

Examples of how each method is performed
Needles. Splenocytes can be released by placing the spleen in about 10 ml of medium in a small Petri dish. One tip of the spleen can be snipped off and a bent dissecting needle used to

Table 6. Methods of tissue dissociation

Tissue	Method
Spleen	Forceps, needles
Thymus	Glass slides, mesh screen
Bone marrow	Syringe
Lymph nodes	Glass slides, mesh screen

squeeze the cells out, using a second bent needle to hold the spleen. When most of the cells are removed, the membranous sac can be discarded and a pipette used to dissociate clumps.

Forceps. One pair of forceps is used to stabilize the spleen while a pair of large-toothed forceps is used to puncture it. Repeated punctures and squeezing will result in release of nearly all the splenocytes into the medium in the Petri dish.

Stainless steel mesh screen. The lymphoid tissue is placed on a stainless steel mesh screen. A pestle, such as a syringe plunger, is used to force the tissue through the screen into medium in a Petri dish.

Glass slides. The lymphoid tissue is placed in a small Petri dish containing approximately 10 ml of medium. The tissue is positioned between the frosted ends of two microscope slides and the cells are squeezed from the tissue.

Syringe. An intact femur is removed and placed in a Petri dish containing medium, where it is rinsed and transferred to a fresh dish containing a few milliliters of the medium. The ends of the femur are snipped off and a syringe used to flush the marrow from the bone. Large aggregates can be dissociated with a Pasteur pipette.

4.2. Enumeration of cell number

Before cells can be placed in culture, an accurate cell count must be made. The first step in this process is to eliminate cell clumps. This can be accomplished by placing the cell suspension in a 15 ml conical centrifuge tube on ice for 10 min. Large clumps and debris will settle to the bottom and the cell suspension can be transferred to a new centrifuge tube. Additional media (20–30 ml) can be added and the cells pelleted by centrifugation. The cell pellet is resuspended in a known volume of medium, usually 5–20 ml. A dilution of the cells from 1:1 to 1:10 is made, depending on the relative cell density, in medium containing 0.2% trypan blue. The trypan blue is excluded by viable cells and thus the number of nonviable, blue cells can be subtracted from the total cell count to yield the viable cell count. Although this method is suitable for routine cell cultures, it has some failings. First, at least 100 cells must be counted on the hemocytometer. Secondly, the trypan blue requires about 5 min for optimal uptake, and some cells which exclude trypan blue may not be totally functional or maintainable in culture. A more accurate count may be obtained using a Coulter counter.

4.3. Washing and removal of dead cells

Prior to being placed in culture, lymphoid cells need to be washed at least three times. This helps to remove potentially harmful enzymes which may have been released during tissue dissociation. For similar reasons, it is often useful to remove dead cells. This is most effectively accomplished using a Ficoll®/Isopaque® solution (21, 22). The cell suspension is layered on to approximately five times its volume of Ficoll®/Isopaque® and centrifuged for

20 min at 500 g. The dead cells pellet in the bottom of the tube and the viable cells form a discrete band above. Other methods have been used to remove dead cells and debris. A particularly useful method with cell lines is to spin the cells through FCS. This is moderately effective in removing dead cells, and has the advantage of coating the cells in FCS, which can provide real help to a sick culture. The disadvantage of most methods of dead-cell removal is the significant loss of viable cells.

5. CELL SEPARATION TECHNIQUES

The experimental design will often require the removal of macrophages, or purification of one or more lymphocyte subsets, prior to initiation of the cultures. A variety of methods have been developed for this purpose. They can be grouped into two categories: positive and negative selection. As the name implies, with positive selection cells with a particular characteristic or surface marker are separated from all other cells lacking this characteristic. In negative selection, an unwanted cell population is removed from the cell population. Rarely is a single separation procedure useful in generating a sufficiently pure positively selected population and a negatively selected population. It is more common to use two different purification protocols on separate aliquots of a lymphoid population to generate the desired subsets. *Table 7* lists the most common separation techniques and the cell types to which they can be applied effectively.

Yield
None of the above cell separation techniques results in a 100% yield of the desired cell population. It is not uncommon to 'lose' 50% of the starting cell number. Most of these cells are lost during separation and subsequent cell washing, due to nonspecific adherence to the centrifuge tube. Techniques such as Percoll® gradient separation have a particularly low yield, as little as 10%, which can be minimized by careful monitoring of the gradient pH. The highest yield is obtained with antibody plus C'-mediated lysis used to remove an unwanted population. The yield from the fluorescence-activated cell sorter (FACS) is quite variable and dependent on such factors as percentage of positive cells in the starting population and the flow rate used. It may be possible to sort a population with a high percentage of positive cells at a slow rate, thus increasing the relative yield of sorted cells. However, more often, a low-frequency subset is desired. Thus, to obtain sufficient cells before viability is compromised, a

Table 7. Methods of cell separation and purification

Separation criteria	Method	Cell type	Refs
Size/density	Percoll	All	23
Size/density	1 g velocity sedimentation	All	24
Surface marker	Panning	B,T	25, 26
Surface marker	Antibody + C' depletion	T	27
Surface marker	Rosetting	B,T	27
Surface marker	Magnetic beads	B,T	27
Surface marker	FACS	All	28
Phagocytosis	Iron particles/magnet	Macrophage	29
Adherence	Sephadex G-10	Macrophage	30
Adherence	Nylon wool	T	26, 31
DNA/RNA content	FACS	All	28

FACS, fluorescence-activated cell sorter.

moderately rapid flow rate must be used. In this case, it is advisable to have a starting cell number several times greater than the theoretically required number.

Controls and purity assessment
FACS analysis and sorting requires that careful attention be paid to the appropriate controls. The critical control is the use of an irrelevant antibody of the same species and isotype as the antibody of interest. Parallel stained aliquots using the experimental and control antibodies will allow accurate detection of positive cells. When two- or three-color staining is utilized, additional controls are required, essentially every possible combination of each antibody and its control. This gives a total of eight aliquots of cells in the case of a simple two-color protocol. Without all of the possible combinations it is impossible to set the electronic gates accurately for positive cells with each color and compensate for emission bleedover into the other color.

Isotype-matched control antibodies may also be useful in a few of the other techniques, particularly panning. However, with most of the separation techniques purity assessment is used for quality control. In the vast majority of cases, including FACS, this is done by FACS analysis of the purified population. Occasionally, functional assays will be used. Two examples are ConA stimulation of a T-cell-depleted population to determine the percentage decrease in [^{3}H]thymidine uptake, and serine esterase staining for the presence of macrophages.

5.1. Dyes used in FACS analysis and cell sorting

The most common dye used for FACS analysis is fluorescein isothiocyanate. When two-color analysis is desired, phycoerythrin or Texas red is frequently used. All of these dyes are excitable at 488 nm. For three-color analysis, investigators often choose to use a dye excitable in the UV range since it is difficult to separate the emission spectra of three dyes which are all excited at 488 nm. An alternative to a UV-excitable dye is the use of the new conjugate dyes which 'relay' the fluorescent signal from a dye excited at 488 nm to a dye excited at a longer wavelength. The choice for the second and third dyes is largely dependent upon configuration of the particular FACS instrument and the density of the surface markers under study.

There are also various dyes that can be used for cell-cycle analysis as indicated in *Table 8*, but they differ in terms of the information obtained. For example, most of the dyes will provide

Table 8. Dyes for cell-cycle analysis and measurement of apoptosis (see also Chapter 6, *Table 11*)

Dye	Peak excitation	Peak emission	Uses	Ref.
Propidium iodide	488 nm	620–700 nm	Viability, DNA content, apoptosis	28,32
Acridine orange	488 nm	550 nm	DNA and RNA content	28,33,34
Mithramycin	425 nm	575 nm	DNA content	28,33 34
Hoechst 33342	350 nm	475 nm	DNA content, sorting of viable G_0/G_1 cells versus G_2/M cells	28,33,34

excellent information on the DNA content of the cells, but to determine the number of cells in G_0 versus G_1, acridine orange should be used; it emits green fluorescence when intercalated in DNA and orange fluorescence when intercalated in RNA.

6. CELL CULTURE

6.1. Cell density
The growth of most cell lines in culture slows down or stops upon reaching a particular cell density (usually $1\text{--}4 \times 10^6$ cells ml^{-1}). However, this effect is more pronounced with non-transformed cells (35). Allowing cells to attain such high densities (overcrowding) may often have adverse effects on the cultures and may select for unwanted variants, in the case of hybridomas or myeloma fusion partners, for example (36). Therefore such conditions should generally be avoided. However, culture of freshly obtained splenocytes or thymocytes is normally performed at relatively high densities ($10^6\text{--}10^7$ cells ml^{-1}). Thus, initial cell densities vary with different cell types and with the different applications of the cultures. *Table 9* gives examples of initial cell densities and applications.

Table 9. Initial cell densities for different cell types and applications in cellular immunology

Cell type/application	Initial cell density
Culture of indicator cell lines for bioassays (e.g. HT-2, CTLL, L-929)	5×10^4 cells ml^{-1}
General maintenance of cell lines, hybridomas	$1\text{--}3 \times 10^5$ cells ml^{-1}
Spleen cells, peripheral blood mononuclear cells (PBMC) for proliferation, cytokine secretion, antibody production assays	$2\text{--}5 \times 10^5$ cells ml^{-1}
Spleen cells, thymocytes for use as feeder cells or source of supernatants	$10^6\text{--}10^7$ cells ml^{-1}

6.2. Use of feeder cells
A number of culture systems in cellular immunology require the use of feeder cells. Depending on the particular application, both normal cells, obtained from different organs/tissues (thymus, spleen, bone marrow), or cell lines (3T3 fibroblasts, Epstein–Barr virus (EBV)-transformed B-cell lines) are used. One of the main reasons for the use of feeder cells is to raise the cell density in limiting-dilution cultures for the analysis of precursor-cell frequencies (LDA) or for cloning (37, 38). Additional functions of feeder cells are to provide accessory cell function, including both cell contact and secretion of growth factors or cytokines; to provide a source of antigen-presenting cells (APC) (39); and, in some systems, to provide a source of stimulation (antigen) for T cells (i.e. allogeneic feeder cells) (40). Depending on the particular system, the growth of feeder cells may have to be prevented by exposure to γ-irradiation or mitomycin C (37). *Table 10* gives examples of the use of feeder cells.

6.3. Inactivation of cells/cell substitutes
Certain applications involving the use of feeder cells, antigen-presenting cells or stimulator cells (e.g. T-cell proliferation assays, mixed lymphocyte cultures) require that those cells be inactivated (unable to proliferate) without affecting their accessory cell capabilities. The two most widely used methods for preventing proliferation of accessory/stimulatory cells

Table 10. The use of feeder cells and applications

Feeder cell	Cell type	Applications	Comments
Thymocytes	Mostly T cells	LDA, expansion and cloning of hybridomas	Source of IL-6. Thymocyte-conditioned medium or alternative sources of IL-6 can sometimes be used as substitutes (37)
Peritoneal cells	Macrophages/ lymphocytes	LDA, cloning of hybridomas, APC	Source of IL-1, IL-6 (37); use as APC requires inactivation
Spleen cells	Lymphocytes/ macrophages	LDA, cloning of hybridomas, APC	As above
Bone marrow	Bone marrow stromal cells	Growth of pre-B cells	Source of IL-7 (41)
NIH 3T3	Fibroblasts	LDA, growth of pre-B cells	
EBV-transformed lymphoblastoid cell lines	Transformed B cells	LDA, cloning, culture of T cells and NK cells	Potential source of IL-12 (42)

Abbreviations: APC, antigen-presenting cells; EBV, Epstein–Barr virus; LDA, limiting-dilution analysis.

are γ-irradiation or mitomycin C (38, 39). It should be pointed out, however, that the antigen-processing capabilities of some cell types are affected by irradiation, with B lymphocytes being more sensitive to irradiation than macrophages and cell lines (43). *Table 11* lists the most commonly used methods for the inactivation of cell proliferation.

Studies on the role of cell-to-cell contact in cellular activation, and identification of the surface molecules responsible for such interactions, are usually complicated by the secretion of cytokines and subsequent cytokine-mediated cellular activation. Inactivation of cells by a variety of methods, or the substitution of whole cells by plasma membrane preparations or soluble proteins, have been used to avoid interference by secreted factors. Whereas inactivation of cells by γ-irradiation or mitomycin C blocks the proliferative capacity of cells, cytokine secretion is preserved. *Table 12* lists some of the methods used to inactivate cells or to provide substitutes to whole, viable cells.

6.4. Use of culture supernatants

Culture supernatants from lymphocytic or monocytic cells and cell lines are often used as sources of growth factors/cytokines. It should be pointed out that in most cases these supernatants do not exclusively contain a single cytokine, but rather contain a complex mixture of several cytokines, and, consequently, the effects of a particular supernatant may not be identical to those of a particular purified or recombinant cytokine. Polyclonal activators or mitogens used to generate supernatants as sources of cytokines, usually have to be removed from supernatants in order to avoid potential mitogenic stimulation of recipient cells. Culture supernatants obtained from hybridomas are used as primary sources of monoclonal antibodies (mAb). However, the antibody concentration in these supernatants is not high (~ 10–$40\ \mu g\ ml^{-1}$) and additional purification of the mAb is desirable. Ascitic fluid generated by the inoculation of hybridomas in the peritoneal cavity of mice is an alternative, highly concentrated (5–10 mg ml^{-1}), source of mAb (36). *Table 13* gives a list of commonly used supernatant sources.

Table 11. Inactivation of cell proliferation

Treatment	Dose	Comments
γ-Irradiation (^{60}Co or ^{137}Cs source) (10)	500–1000 rad; cell density at $0.5–1 \times 10^7$ cells ml^{-1}	APC function of B cells is preserved
γ-Irradiation	1100–2000 rad	APC function of B cells is substantially decreased
γ-Irradiation	2000–3000 rad	APC function of B cells is abolished; APC function of macrophages and dendritic cells is preserved
γ-Irradiation	5000–10 000 rad	Doses required for transformed cell lines
Mitomycin C (38,44)	25–50 μg ml^{-1} for 30 min at 37°C; cell density at $0.5–1 \times 10^7$ cells ml^{-1}	Preserves APC function of B cells; removal of mitomycin C is critical
5-Bromo-2-deoxyuridine (BrdU) + UV light (300–400 nm) (38)	Optimal dose must be determined experimentally ~ 3 μg ml^{-1} for 24 h; UV light for 30 min	Selective killing of dividing cells; might have secondary effects on cells
Hot thymidine pulse, [^{3}H]methylthymidine (10–20 Ci mmol^{-1}) (38)	5–10 μCi ml^{-1}	Selective killing of proliferating cells; requires removal of labeled thymidine by washing or dilution with cold thymidine

Table 12. Inactivation of cells/viable cell substitutes

Method	Cytokine secretion	Cell division	Thymidine uptake	Surface marker mobility
γ-Irradiation	Yes	No	Minimal	Yes
Mitomycin C	Yes	No	Minimal	Yes
Fixation	Usually not (potential slow leakage under some conditions) (45)	No	No	No
Plasma membranes	No	No	No	No
Soluble proteins (hybrid proteins with Fc portion)	No	No	No	No, but can use secondary antibody to induce clustering

Table 13. Commonly used supernatant sources (46)

Cell/cell line	Stimulant	Cell density	Source of
EL-4 cells	PMA (10 ng ml^{-1})	10^6 cells ml^{-1}	IL-2, some IL-4 and IL-5
Spleen, lymph node cells, PBMC	ConA or PHA (2–5 μg ml^{-1})	10^6 cells ml^{-1}	IL-2, IL-4 (low amount)
Spleen cells (primary MLR)	Allogeneic spleen cells	$3–5 \times 10^5$ cells ml^{-1}	IL-2, IFN-γ (low amount)
Spleen cells (secondary MLR)	Allogeneic spleen cells	$3–5 \times 10^5$ cells ml^{-1}	IL-2, IFN-γ (high amount)
T$_H$1 cell lines (47)	Antigen + APC	$0.5–1 \times 10^6$ cells ml^{-1}	IL-2, IFN-γ, TNFα,β
T$_H$2 cell lines (47)	Antigen + APC	$0.5–1 \times 10^6$ cells ml^{-1}	IL-4, IL-5, IL-6, IL-10
Hybridomas	None	$0.5–1 \times 10^6$ cells ml^{-1}	mAb

Abbreviations: APC, antigen-presenting cells; ConA, concanavalin A; mAb, monoclonal antibodies; MLR, mixed lymphocyte reaction; PBMC, peripheral blood mononuclear cells; PHA, phytohemagglutinin; PMA, phorbol 12-myristate 13-acetate.

Proper storage of culture supernatants is critical in order to ensure biologic activity of cytokines or antibodies. *Table 14* gives recommended temperatures.

6.5. Polyclonal activators

Polyclonal activators are often used in cellular immunology to induce proliferation and/or cytokine secretion by T cells, or to induce proliferation and/or antibody production by B cells. *Tables 15* and *16* list the most commonly used polyclonal activators, their targets and optimal dose range. Unless specified, activators can be used in both murine and human systems.

Table 14. Storage time and temperature for culture supernatants

Temperature	Maximum recommended storage
4°C	Weeks to a month
−20°C	Months to a year
−70°C	Indefinitely

7. CELL AND TISSUE PRESERVATION

7.1. Fixation

Fixation of cells is sometimes required for inactivation purposes (described above) or for prolonged cell storage for use in immunoassays or immunofluorescence. Tissues also require fixation before immunofluorescence or immunohistochemical staining. Although several techniques are used for the fixation of cells and tissues, some of them might alter the conformation of membrane proteins, destroying antigenicity. Whether a particular antigen is susceptible to destruction by a given fixation method will have to be determined. *Table 17* lists some of the most commonly used methods for fixation of cells and tissues.

Table 15. Polyclonal T-cell activators (reviewed in ref. 44)

Activator	Dose	Effect	Comments
Anti-CD3 (soluble or plate-bound)	$0.1–5$ μg ml^{-1}	Proliferation, cytokine secretion	Soluble anti-CD3 requires accessory cells; plate-bound does not
Anti-TCRαβ (soluble or plate-bound)	$0.1–10$ μg ml^{-1}	Proliferation, cytokine secretion	Soluble anti-TCR requires accessory cells; plate-bound does not
Anti-TCRγδ (plate-bound)	$0.1–100$ μg ml^{-1}	Proliferation, cytokine secretion	Does not require accessory cells
Anti-CDw90	$1–50$ μg ml^{-1}	Proliferation, cytokine secretion	Soluble
Concanavalin A (ConA)	$1–10$ μg ml^{-1}	Proliferation, cytokine secretion	Requires accessory cells
Phytohemagglutinin (PHA)	$1–5$ μg ml^{-1}	Proliferation, cytokine secretion	Requires accessory cells
Staphylococcal enterotoxins A, B, E Superantigens	$1–100$ μg ml^{-1}	Proliferation, cytokine secretion	Requires accessory cells; specific for T cells bearing certain TCR V_β families
Phorbol esters (phorbol 12-myristate 13-acetate, PMA)	$1–10$ ng ml^{-1}	Proliferation, cytokine secretion	Protein kinase C (PKC) activator; does not require accessory cells
Ionomycin, A23187	$100–500$ ng ml^{-1}	Proliferation, cytokine secretion	Calcium ionophores, normally used with PMA

7.2. Cryopreservation

The long-term culture of cells or cell lines may result in the accumulation of phenotypic and genotypic changes in the cell population and unforeseen contamination with bacteria, mycoplasma, yeasts, etc. In order to avoid these problems, a laboratory must maintain a stable source of cells. Cells can be maintained almost indefinitely when frozen in the presence of a cryoprotective agent and stored at very low temperatures (liquid N_2).

Methods
Cells should preferably be frozen at their mid-log phase of growth. The cells are normally harvested, pelleted and resuspended in a freezing medium containing a cryoprotective agent. Dimethyl sulfoxide (DMSO) and glycerol (10%) are the two most commonly used cryoprotective agents. The composition of the freezing media and its content of serum is variable, with some laboratories using 10% DMSO in 90% heat-inactivated FCS or horse

Table 16. Polyclonal B-cell activators (reviewed in refs 48, 49)

Activator	Dose	Effect	Comments
Lipopolysaccharide (LPS) (mouse)	2–25 µg ml^{-1}	Proliferation, Ig secretion	Mostly IgM and IgG3
Pokeweed mitogen (PWM) (human)	Determine experimentally	Proliferation, Ig secretion	T-cell-dependent B-cell activation
Anti-Ig (IgM, IgD) (soluble or insoluble)	1–50 µg ml^{-1}	Proliferation, Ig secretion (with T-cell-derived cytokines)	Sepharose-bound anti-Ig is usually more effective
Staphylococcus aureus, Cowan strain I (SAC, protein A) (human)	Determine experimentally	Proliferation, Ig secretion (with T-cell-derived cytokines)	
Dextran sulfate (DXS)	10–100 µg ml^{-1}	Proliferation (with IL-5)	Weak mitogen, often used with IL-5 or with other mitogens, such as LPS
Epstein–Barr virus (EBV) (human)	Determine experimentally	Proliferation, Ig secretion, transformation	Bypasses normal B-cell regulation; T-cell-independent
Phorbol esters (PMA) + ionomycin	1–10 ng ml^{-1}	Proliferation	PKC activation and Ca^{2+} mobilization
8-Mercaptoguanosine (8-MG)	0.5 mM	Proliferation	Acts on T and B cells
Indolactam	0.01–1 µM	Proliferation	PKC activator

Table 17. Fixation of cells and tissues (38, 50)

Agent	Comment
Glutaraldehyde (0.25%)	Cross-links proteins; fixed cells stable at 4°C
Paraformaldehyde, pH 7.4 (1–2%)	Fixative of choice for immunofluorescence; fixed cells stable for ∼1 week at 4°C
Formaldehyde (10%)	Fixative for tissue; may destroy antigenicity of some membrane proteins
Ethanol (95–100%)	Fixative for tissues
Methanol, acetone, carbon tetrachloride	Other fixatives used for tissues
Low temperature (liquid N_2, acetone or 2-methylbutane and dry ice)	Snap-freezing of tissues for immunofluorescence

Table 18. Storage temperature and optimal storage time for cells

Temperature	Storage time
$-70°C$ (short term)	Optimal <6 months
$-135°C$	5–10 years
Liquid N_2 ($-175°C$)	Indefinitely

serum (38, 51); and other laboratories preferring 10% DMSO in a medium containing 10–20% heat-inactivated FCS. A serum-free formula containing methyl cellulose and DMSO has also been described (52). Cells are normally frozen at $-70°C$ and then transferred to an ultra-low-temperature freezer ($-135°C$) or to liquid N_2. The storage temperature determines how long the viability of the cells is preserved, see *Table 18*.

Thawing tips
Vials of cells must be thawed in a water bath at $37°C$, swirling gently. The contents of the vial are then transferred to a 15 ml centrifuge tube and diluted 1:10 with medium, added drop-wise over a period of 5 min. Following centrifugation (5 min at 500 g), the supernatant is removed and the cells cultured with fresh medium. After thawing, the appropriate growth factors should be added to factor-dependent cells, and T-cell lines or clones should be restimulated. It is recommended that drug-selected cell lines be initially cultured in the presence of the appropriate drug, although most hybridomas do not require reselection in hypoxanthine/aminopterin/thymidine (HAT)-containing media.

8. REFERENCES

1. Rayburn, S.R. (1990) *The Foundations of Laboratory Safety. A Guide for the Biomedical Laboratory.* Springer-Verlag, New York, Chapter 9.

2. Rayburn, S.R. (1990) *The Foundations of Laboratory Safety. A Guide for the Biomedical Laboratory.* Springer-Verlag, New York, Chapter 5.

3. US Department of Health and Human Services, Public Health Service, Centers for Disease Control and National Institutes of Health (1988) *Biosafety in Microbiological and Biomedical Laboratories.* US Government Printing Office.

4. American Industrial Hygiene Association (1985) *Biohazards Reference Manual*, Appendix III.

5. Whitlock, C., Denis, K., Robertson, D. and Witte, O. (1985) *Ann. Rev. Immunol.*, **3**, 213.

6. Kincade, P.W., Lee, G., Pietrangeli, C.E., Hayashi, S.-I. and Gimble, J. M. (1989) *Ann. Rev. Immunol.*, 7, 111.

7. Whitlock, C. A., Robertson, D. and Witte, O. N. (1984) *J. Immunol. Meth.*, **67**, 353.

8. Dexter, T.M. Allen, T.D. and Laitha, L.G. (1977) *J. Cell Physiol.*, **91**, 335.

9. Whitlock, C.A. and Witte, O.N. (1982) *Proc. Natl. Acad. Sci. USA*, **79**, 3608.

10. Dorshkind, K. (1990) *Ann. Rev. Immunol.*, **8**, 111.

11. Jenkinson, E.J., Franchi, L.L., Kingston, R. and Owen, J.J.T. (1982) *Eur. J. Immunol.*, **12**, 583.

12. Kingston, R., Jenkinson, E.J. and Owen, J.J.T. (1985) *Nature*, **317**, 811.

13. Jenkinson, E.J. and Owen, J.J.T. (1990) *Sem. Immunol.*, **2**, 51.

14. van Ewijk, W. (1991) *Ann. Rev. Immunol.*, **9**, 591.

15. Itoh, T., Doi, H., Chin, S., Nishimura, T. and Kasahara, S. (1988) *Eur. J. Immunol.*, **18**, 821.

16. Hiramine, C., Hojo, K., Koseto, M., Nakagawa, T. and Mukasa, A. (1990) *Lab. Invest.*, **62**, 41.

17. Palacios, R., Studer, S., Samarides, J. and Pelkonen, J. (1989) *EMBO J.*, **8**, 4053.

18. Tatsumi, Y., Kumanogoh, A., Saitoh, M., Mizushima, Y., Kimura, K., Suzuki, S., Yagi, H., Horiuchi, A., Ogata, M., Hamaoka, T. and Fujiwara, H. (1990) *Proc. Natl. Acad. Sci. USA*, **87**, 2750.

19. Klinman, N.R. and Aschinazi, G. (1971) *J. Immunol.*, **106**, 1338.

20. Klinman, N.R. (1972) *J. Exp. Med.*, **136**, 241.

21. Bøyum, A. (1968) *Scand. J. Clin. Lab. Invest.*, **21** (Suppl. 97), 1.

22. Kruisbeek, A.M. (1991) in *Current Protocols in Immunology* (J.E. Coligan, A.M. Kruisbeek, D.H. Margulies, E.M. Shevach and W. Strober, eds). Greene Publishing and Wiley-Interscience, New York, p. 3.1.4.

23. Mond, J.J. and Brunswick, M. (1991) in *Current Protocols in Immunology* (J.E. Coligan, A.M. Kruisbeek, D.H. Margulies, E.M. Shevach and W. Strober, eds). Greene Publishing and Wiley-Interscience, New York, p. 3.8.11.

24. Miller, R.G. and Phillips, R.A. (1969) *J. Cell Physiol.*, **73**, 191.

25. Mage, M.G., McHugh, L.L. and Rothstein, T.L. (1977) *J. Immunol. Meth.*, **15**, 47.

26. Wysocki, L.J. and Sato, V.L. (1978) *Proc. Natl. Acad. Sci. USA*, **75**, 2844.

27. Hunt, S.V. (1986) in *Handbook of Experimental Immunology* (D.M. Weir, L.A. Herzenberg, C. Blackwell and L.A. Herzenberg, eds). Blackwell Scientific Publications, Oxford, Vol. 2, Chapter 55.

28. Shapiro, H.M. (1988) *Practical Flow Cytometry*. A.R. Liss, New York.

29. Mishell, B.B. and Shiigi, S.M. (1980) in *Selected Methods in Cellular Immunology*. W. H. Freeman, San Francisco, p. 179.

30. Hathcock, K.S. (1991) in *Current Protocols in Immunology* (J.E. Coligan, A.M. Kruisbeek, D.H. Margulies, E.M. Shevach and W. Strober, eds). Greene Publishing and Wiley-Interscience, New York. p. 3.6.1.

31. Julius, M.H., Simpson, E. and Herzenberg, L.A. (1973) *Eur. J. Immunol.*, **3**, 645.

32. Telford, W.G., King, L.E. and Fraker, P.J. (1992) *Cytometry*, **13**, 137.

33. Darzynkiewicz, Z. and Crissman, H.A. (1990) *Methods in Cell Biology*. Academic Press, San Diego, Vol. 33.

34. Melamed, M.R., Lindmo, T. and Mendelsohn, M.L. (1990) *Flow Cytometry and Sorting*. Wiley-Liss, New York.

35. Ham, R.G. and McKeehan, W.L. (1979) *Methods Enzymol.*, **58**, 44.

36. Yokoyama, W.M. (1991) in *Current Protocols in Immunology* (J.E. Coligan, A.M. Kruisbeek, D.H. Margulies, E.M. Shevach and W. Strober, eds). Greene Publishing and Wiley-Interscience, New York, p. 2.5.1.

37. Miller, R.A. (1991) in *Current Protocols in Immunology* (J.E. Coligan, A.M. Kruisbeek, D.H. Margulies, E.M. Shevach and W. Strober, eds).

Greene Publishing and Wiley-Interscience, New York, p. 3.15.1.

38. Mishell, B.B. and Shiigi, S.M. (1980) *Selected Methods in Cellular Immunology*. W.H. Freeman, San Francisco.

39. Rosenwasser, L.J. and Rosenthal, A.S. (1978) *J. Immunol.*, **120**, 1991.

40. Strong, D.M., Ahmed, A.A., Thurman, G.B. and Sell, K.W. (1973) *J. Immunol. Meth.*, **2**, 279.

41. Namen, A.E., Schmierer, A.E., March, C.J., Overell, R.W., Park, L.S., Urdal, D.L. and Mochizuki, D.Y. (1988) *J. Exp. Med.*, **167**, 988.

42. Valiante, N.M., Rengaraju, M. and Trinchieri, G. (1992) *Cell. Immunol.*, **145**, 187.

43. Ashwell, J.D., DeFranco, A.L., Paul, W.E. and Schwartz, R.H. (1984) *J. Exp. Med.*, **159**, 861.

44. Kruisbeek, A.M. (1991) in *Current Protocols in Immunology* (J.E. Coligan, A.M. Kruisbeek, D.H. Margulies, E.M. Shevach and W. Strober, eds). Greene Publishing and Wiley-Interscience, New York, p. 3.12.1.

45. Streck, H., Gunther, C., Beuscher, H.U. and Rollinghoff, M. (1988) *Eur. J. Immunol.*, **18**, 1609.

46. Fitch, F.W. and Gajewski, T.F. (1991) in *Current Protocols in Immunology* (J.E. Coligan, A.M. Kruisbeek, D.H. Margulies, E.M. Shevach and W. Strober, eds). Greene Publishing and Wiley-Interscience, New York, p. 3.13.1.

47. Mosmann, T.R., Cherwinski, H., Bond, M., Giedlin, M.A. and Coffman, R.L. (1986) *J. Immunol.*, **136**, 2348.

48. Mond, J.J. and Brunswick, M. (1991) in *Current Protocols in Immunology* (J.E. Coligan, A.M. Kruisbeek, D.H. Margulies, E.M. Shevach and W. Strober, eds). Greene Publishing and Wiley-Interscience, New York, p. 3.10.1.

49. James, S.P. (1991) in *Current Protocols in Immunology* (J.E. Coligan, A.M. Kruisbeek, D.H. Margulies, E.M. Shevach and W. Strober, eds). Greene Publishing and Wiley-Interscience, New York, p. 7.11.1.

50. Kawamura, A., Jr. (ed.) (1977) *Fluorescent Antibody Techniques and Their Applications (2nd edn)*. University of Tokyo Press, Tokyo.

51. Yokoyama, W.M. (1991) in *Current Protocols in Immunology* (J.E. Coligan, A.M. Kruisbeek, D.H. Margulies, E.M. Shevach and W. Strober, eds). Greene Publishing and Wiley-Interscience, New York, p. A.3.15.

52. Waymouth, C. (1976) *TCA Manual*, **2**, 1.

CHAPTER 5
CELL LINES AND HYBRIDOMAS
S. Ozaki

1. CELL LINES

Various methods have been used to establish long-term cell lines (1–3). These are listed in *Table 1*. Both immortal and mortal cells can be maintained *in vitro*, the latter requiring periodical stimulation to grow. Characteristics of those cell lines used in immunological research are shown in *Tables 2–4*.

2. VIRAL- AND GENE-MEDIATED TRANSFORMATION

Papovaviruses such as polyoma virus or simian virus 40 (SV40) have transforming and immortalizing potential and have been used as model viruses to study virus-induced transformations (44, 45). SV40 has three virus-encoded proteins (VP1, VP2 and VP3) and two additional proteins coded for by the early region of SV40; small t and large T antigens. Among them, the large T antigen is a key molecule in transformation and immortalization. In practice, SV40-mediated transformation is carried out by transfection of the large T antigen gene by either the calcium-phosphate method (46, 47), electroporation (48) or micro-injection (49, 50). Plasmids with SV40 DNA, whose origin of DNA replication is defective, have been devised (51, 52). The SV40 large T antigen has been used for transforming a variety of human cells of nonlymphoid lineage, such as fibroblasts (53), keratinocytes (54), bronchial epithelial cells (55) and synovial cells (56).

Some viruses have been used to transform and immortalize human lymphoid cells. These include Epstein–Barr virus (EBV) (35, 57–59) and human T-cell leukemia virus-1 (HTLV-1)

▶ p. 70

Table 1. Methods for establishing cell lines

---

1. Primary culture of *in vivo*-transformed cells
2. Cytokine/growth factor-dependent cell lines
3. Viral-mediated transformation
   EBV
   HTLV-1
   SV40
4. Gene-mediated transformation
   HTLV-1 *tax* gene
   SV40 large T-antigen gene
5. Antigen-specific stimulation of T cells
6. Stromal cell-dependent cell lines
   B cells
   T cells
7. Hybridomas
   B cells
   T cells

Table 2. Human cell lines used in immunological research

Cell line	Cell type	Tissue of origin	Refs	Growth characteristics	Positive CD markers	Comments
CEM	Lymphoblastoid (T cell)	Peripheral blood of patient with acute lymphoblastic leukemia	4	RPMI 1640 + 20%FCS	CD4,5,7,w17,29, 30,31,38,44,45,46, 47,50,53,58,71	
HUT102	Lymphoblastoid (T cell)	Peripheral blood of patient with mycosis fungoides	5,6	RPMI 1640 + 10%FCS Growth rate increases with IL-2	CD4,9,25,29,30, w49d,51,63,70,71	HTLV-1 producer
Jurkat	Lymphoblastoid (T cell)	Peripheral blood of patient with acute lymphoblastic leukemia	7, 8	RPMI 1640 + 10%FCS	CD1,2,3,5,7,9, 11a,28,29,31, 38,45,58,59,w60	Lectin-induced IL-2 secretion
MOLT-4	Lymphoblastoid (T cell)	Peripheral blood of patient with acute lymphoblastic leukemia	9	RPMI 1640 + 10%FCS Doubling time 24 h	CD1,5,7,11a,18,29, 31,38,43,44,45,45R, 46,47,48,w49d,50, 53,58,59,w60,71	Lectin-induced IL-2 secretion
Daudi	Lymphoblastoid (B cell)	Burkitt lymphoma	10–12	RPMI 1640 + 20%FCS Grows 11-fold in 10 days	CD10,19,20,21,22, 29,37,38,40,45,45R, 46,47,48,53,54,55, 71,72,w75,77,w78	EBNA (+)[a] VCA (+)[b]
Raji	Lymphoblastoid (B cell)	Burkitt lymphoma	13, 14	RPMI 1640 + 10%FCS Grows 5-fold in 5 days	CD10,11a,18,19,20, 20,21,22,29,37,38, 40,45,45R,46,47,48, w49d,53,54,58,63,70, 71,74,w75,76,77,w78	EBNA(+) No endogenous viruses detected
HL-60	Promyelocyte	Peripheral blood of patient with acute promyelocytic leukemia	15, 16	RPMI 1640 + 20%FCS Grows 7-fold in 7 days	CD13,15,w17,29,33, 43,44,45,47,63,71	Differentiate to granulocyte with retinoic acid
K-562	Erythroblastoid	Pleural effusion of patient with chronic myelogenous leukemia	17, 18	RPMI 1640 + 20%FCS Doubling time 28 h	CD9,15,29,32,43, 45,46,47,53,54,55, 58,59,63,w65,71	Highly sensitive to NK cells

	Morphology	Origin		Medium	Markers	Uses
U-937	Monocytoid	Pleural effusion of patient with diffuse histiocytic lymphoma	19	RPMI 1640 + 20% FCS Doubling time 24 h	CD4,11a,13,18,29, 31,32,33,36,38, 43,44,45,45R,46,47, w49d,50,54,w60,63, w65,69,71	Differentiate to monocyte with various agents
HeLa	Epithelial-like	Biopsy specimen of patient with cervix carcinoma	20–22	Eagle's MEM[c] + 10% FCS		Useful for studies of cell cycle and gene functions

[a]EBNA, EBV nuclear antigen.
[b]VCA, viral capsid antigen.
[c]MEM, minimum essential medium.

Table 3. Animal cell lines used in immunological research

Cell	Origin	Cell type	Comments	Refs
EL-4	Mouse (C57BL/6)	T lymphoma	Cultured in DMEM + 10% FCS Subline secretes IL-2	23
A20	Mouse (BALB/c)	B lymphoma	Cultured in RPMI 1640 + 10% FCS + 50 μM 2-mercaptoethanol sIgG$^+$, MHC class II$^+$, FcR^{+a} Serves as antigen-presenting cells	24
P388D1	Mouse (DBA/2)	Macrophage	Cultured in DMEM + 10% horse serum Expresses more MHC class II with interferon-γ Subline secretes IL-1	25
WEHI-3	Mouse (BALB/c)	Macrophage	Cultured in DMEM + 10% FCS Secretes IL-3	26, 27
P815	Mouse (DBA/2)	Mastocytoma	Cultured in DMEM + 10% FCS Used as a target cell for CTL	28
F9	Mouse (strain 129)	Epithelial-like	Cultured in DMEM + 15% FCS Originally initiated from a testicular teratocarcinoma Differentiates with specific agents	29, 30
CHO	Chinese hamster	Fibroblast/epithelial-like	Cultured in Ham's F-12 medium + 10% FCS Originally initiated from a biopsy of an ovary Used extensively for the expression of recombinant proteins	31–33
COS-7	Monkey (African green)	Fibroblast-like	Cultured in DMEM + 10% FCS Derived from CV-1 Transformed by an origin-defective SV40 mutant Used as a suitable host for transfection of vectors requiring expression of SV40 large T antigen	34
B95-8	Marmoset	EBV-transformed leukocyte	Cultured in RPMI 1640 + 10% FCS Initiated by exposing marmoset blood leukocytes to EBV Releases high titers of EBV Used as a source of EBV to establish continuous lymphoblastoid cell lines from human donors	35

[a]FcR, Fc receptor.

Table 4. Cytokine-dependent cell lines

Cell line	Cytokine	Comments	Ref.
D10.G4.1	IL-1	AKR/J(H-2^k)-derived helper T-cell clone Conalbumin-specific and MHC-restricted Also alloreactive to I–A^{b,v} Cultured at 10% conventional rat growth factor	36
HT-2-A5E	IL-2	BALB/c(H-2^d)-derived helper T-cell clone Selected for sensitivity to IL-2 Also responds to IL-4 Cultured at 100–200 IU ml^{-1} IL-2	37
CTLL-2	IL-2	C57BL/6(H-2^b)-derived CTL clone Cultured at 0.25–1 nM IL-2	38
FDC-P1	IL-3	Derived from DBA/2(H-2^d) bone marrow cells Also responds to IL-4 Cultured at 10% WEHI-3 conditioned supernatant (as an IL-3 source)	39
CT.4S	IL-4	Derived from CTLL line CT.EV Selected for IL-2 receptor-negative variant after mutagenesis Grows in IL-4 or in > 100 U ml^{-1} IL-2 Cultured at 500 U ml^{-1} IL-4	40
T88-M	IL-5	Derived from DBA/2 bone marrow cells Selected for high-affinity IL-5 receptor Also responds to IL-3 Lymphoid-like; sIgM$^-$, cIgM$^-$, Ly1$^+$, Ly2$^-$, MHC class II$^-$, and IL-2 receptor$^+$ Cultured at 40 pg ml^{-1} IL-5	41

MH-60.BSF2	IL-6	Hybridoma line between IL-6-immunized
		    BALB/c lymphocytes and P3U1 myeloma
		Selected in HAT medium with IL-6
		Dependent on IL-6; doubling time = 20 h
		Does not respond to IL-3, IL-4 or IL-5
		Secretes anti-IL-6 monoclonal antibody
		    which cannot neutralize IL-6 activity
		Cultured at 1 U ml^{-1} IL-6
			42

BMC-MP	IL-7	Derived from MRL-*lpr/lpr*(H-2^k) bone marrow cells Lymphoid line; CD3$^-$4$^-$8$^-$45$^-$w90$^-$, sIgM$^+$ Rearrangement of both TCRγ and IgH J$_H$ Cultured at 2–10 U ml^{-1} rIL-7 Split every 2–3 days	43
BMC-TP	IL-7	Derived from B10-Thy-1.1(H-2^b) bone marrow cells Same as BMC-MP	43

(60–62). EBV can selectively transform human B cells. Characteristics of EBV-transformed lymphoblastoid cell lines (LCL) are shown in *Table 5*. Transforming EBV is released by a marmoset leukocyte line, B95-8 (see *Table 3*). Procedures for EBV transformation of human B cells (59) are summarized in *Table 6*.

On the other hand, HTLV-1 selectively transforms T cells. The transformed T cells have a phenotype of activated helper T cells, namely $CD2^+$, $CD3^+$, $CD4^+$, $CD8^-$, $CD25^+$ and $HLA-DR^+$ (60–62). Although genes responsible for EBV-mediated transformation remain to be investigated, those involved in HTLV-1-mediated transformation have been clarified. Besides the structural genes *gag*, *pol* and *env*, HTLV-1 genome has the unique *pX* region between *env* and the 3' long terminal repeat (63). The *pX* region, especially a *tax* gene within this region, has been shown to encode the functions of HTLV-1 that immortalize a distinct subpopulation of human T cells (64, 65).

Table 5. Characteristics of EBV-transformed lymphoblastoid cell lines (LCL)

Cells to be transformed:	B cells
Characteristics:	Easy to establish
	high frequency (>95%)
	requires short period (<1 month)
	Easy to prepare a huge number of cells
	doubling time = 24 h
	Easy to maintain, cryopreserve and thaw
	Small changes in chromosome numbers
	Do not produce EBV
	Do not form tumors in nude mice
	Do not form colonies in soft agar
	Difficult to clone by limiting dilution method

Table 6. Epstein–Barr virus transformation (59)

Medium	RPMI 1640 with 20% FCS
Procedures	1. Prepare culture supernatant of a virus-producing cell line, B95-8 (see *Table 3*), by growing the line at 3×10^5 ml^{-1} for 5–7 days
	2. Prepare peripheral blood lymphocytes, and suspend at 1×10^6 ml^{-1} in medium supplemented with 200 ng ml^{-1} cyclosporin A
	3. Mix #1 and #2 at 1:4 (vol/vol), and culture the mixture at 1 ml/well at 37°C in a humidified atmosphere of 5% CO_2
	4. Change medium twice a week
	5. LCL are established in a week or later, thereafter split the culture at 1:3–1:5 twice a week (approx. at $3–5 \times 10^5$ cells ml^{-1})
Note	1. Handle EBV/B95-8 carefully on a biohazard bench
	2. Select a batch of FCS prior to use
	3. Cyclosporin A is employed to eliminate CTL which might kill virus-infected B cells (58)

CTL, cytotoxic T lymphocyte.

3. STROMAL CELL-DEPENDENT B-CELL AND T-CELL LINES

Long-term bone marrow cultures (66, 67) are potentially valuable systems for investigating interactions between stromal cells and developing blood-cell precursors (see Chapter 3, *Figure 1*). Recently, several stromal-cell clones which can support the growth and differentiation of hematopoietic cell lineages have been established (68, 69). These stromal-cell lines support B lymphopoiesis under conditions described by Whitlock and Witte (67), while they also support myelopoiesis under conditions described by Dexter *et al.* (66). Thus, such stromal-cell clones can serve as an *in vitro* microenvironment for hematopoiesis. A sample procedure to establish a bone marrow-derived stromal-cell line (68) is shown in *Table 7*, and the characteristics of some established lines are presented in Chapter 4 (*Table 4*). Lined cells are then cloned by a limiting dilution method (see Section 7). Stromal cell-dependent B-cell lines (69) are generated as in *Table 8*.

In order to investigate the thymic microenvironment essential to normal T-cell development, thymus-derived stromal-cell lines have been generated (70–72). These stromal-cell lines are so heterogeneous that they differ not only in morphology but in capability of producing IL-7 and supporting an *in vitro* T-cell differentiation. A sample procedure to establish a thymus-derived stromal-cell line (70) is shown in *Table 9*, and the characteristics of some established lines in Chapter 4 (*Table 5*).

Table 7. Bone marrow-derived stromal-cell lines (68)

Medium: RPMI 1640 supplemented with 5% FCS, 50 µM 2-mercaptoethanol, 50 µg ml^{-1} streptomycin and 50 U ml^{-1} penicillin (67)

1. Start a Dexter culture (66) or a Whitlock–Witte culture (67) until stromal-cell layers develop (for approximately 3 weeks)
2. Harvest the adherent cells using trypsin-EDTA[a]
3. Start a new culture at $3–4 \times 10^5$ cells/flask
4. Split cells when they become confluent over the flask
5. Repeat #2–#4
6. When the proliferation of the cells becomes extremely slow (so-called crisis), decrease the culture volume and increase the concentration of serum in media
7. Select the bulk line that proliferates the best, and continue growing it
8. Test its stromal-cell function for hematopoiesis (67,68)
9. Cloning (see *Table 32*)

[a]EDTA, ethylenediaminetetraacetic acid.

Table 8. Stromal cell-dependent B-cell lines (69)

Medium: as in *Table 7*

1. Establish a stromal-cell line (see *Table 7*)
2. Prepare a monolayer of the stromal-cell line in a T25 flask
3. Culture $1–2 \times 10^6$ bone marrow cells on the stromal-cell layer
4. Change medium twice a week
5. Transfer the bone marrow-derived cells on to a new stromal-cell layer every 3–4 days
6. In 3 months, B-cell lines can be generated
7. Cloning on a stromal-cell layer (see *Table 32*)

Table 9. Thymus-derived stromal-cell lines (70)

Medium: DMEM supplemented with 10% FCS, 50 µM 2-mercaptoethanol and 50 µg ml^{-1} gentamycin

1. Kill a mouse by a cervical dislocation, and remove thymus
2. Prepare two 60-mm dishes with Hank's solution, put the thymus into one of the dishes, rinse gently, and transfer it into the other dish
3. Cut the thymus with scissors into 0.5 mm pieces
4. Put 0.1–0.15 ml of culture medium into each well of a 24-well plate
5. Put 2–3 pieces of the thymus fragment into each well
6. In a week, stromal cells grow around some of the thymus fragment, then put 0.1 ml of prewarmed medium into the wells with positive growth
7. Add 0.1 ml of prewarmed medium twice a week until the total volume reaches 2 ml
8. Change half the medium once a week
9. Test stromal-cell function, and clone positive lines

4. ANTIGEN-SPECIFIC T-CELL LINES

Mature T lymphocytes recognize specific antigens in association with major histocompatibility complexes (MHC) on the surface of antigen-presenting cells (APC), and function as either helper T cells or cytotoxic T lymphocytes (CTL). The experimental problems in T-lymphocyte biology are the low frequency of specific T cells and the small amount of active T-cell products. In order to circumvent these problems, T-cell lines and clones have been isolated and propagated. Two basic strategies have been devised: one is to clonally expand normal immune T cells in an antigen-specific and MHC-restricted manner, and the other is to immortalize normal T cells by fusion with continuously replicating tumor cells, so-called T-cell hybridoma technology (see Section 6). Comparison of the two strategies (76, 116) is given in *Table 10*.

Table 10. T-cell clones and T-cell hybridomas (76,116)

Characteristics	Clones	Hybridomas
Repertoire generated	Biased[a]	Less biased
Antigen specificity	Stable	Unstable (requires repeated recloning)
Optimal antigen concentration for *in vitro* activation	Low[b]	High[b]
Co-stimulatory signals for activation (besides TCR/CD3)	Requires	Does not require
Contamination of APC/feeder cells	Possible	None
Manipulation (e.g. gene trans-fection, mutagenesis, etc.)	Difficult	Easy
Cell number available	$\sim 10^{9c}$	Infinite
Adoptive transfer (*in vivo*)	Possible	Rarely possible
Somatic cell genetics	Impossible	Possible

[a]Repertoire with capability of vigorous proliferation is preferentially selected as T-cell clones.
[b]Varies in functions (116).
[c]With the help of growth factors.

4.1. Murine helper T-cell lines

These are generated by stimulating T cells with a specific antigen. The antigen used includes allogeneic cells, processed antigens expressed by MHC-compatible APC or viral antigens expressed on infected syngeneic host cells. An optimal concentration of the antigen must be determined prior to the stimulation of bulk immune T cells. The procedure (76) is described in *Table 11*. Once the optimal conditions to stimulate immune T cells are determined, antigen-specific T cells are expanded into lines by repeating the cycles of stimulation and 'rest' (73, 74) (*Table 12*). Cloning is carried out thereafter (see Section 7). A periodical rest is important for maintaining functional T-cell lines and clones. Otherwise, continuous stimulation without 'rest' may result in exhaustion and failure of the T-cell lines. On the other hand, IL-2 is helpful in a vigorous proliferation of T cells immediately after an antigen-specific stimulation, and hence is employed in cloning. Continuous presence of IL-2, however, may result in loss of antigen specificity. The CTL lines are also established in a similar way, except that IL-2 is required even during rest (75).

Table 11. Dose–response curve for antigen-specific proliferation of T cells

I.   Medium: RPMI 1640, supplemented with:
          10% fetal calf serum
          2 mM L-glutamine
          50 μM 2-mercaptoethanol
          100 U ml^{-1} penicillin
          0.1 mg ml^{-1} streptomycin

II.   Preparation of antigen-primed lymph node cells
    1. Immunize mice by a subcutaneous injection (e.g. at the base of the tail) with antigen emulsified in adjuvant (e.g. Freund's complete adjuvant, FCA)
    2. On day 8–11, kill the mice by cervical dislocation, remove the regional lymph nodes (para-aortic and inguinal in the case of base of tail injections), and prepare a sterile single-cell suspension

III. Antigen-specific proliferation assay
    1. Wash and resuspend the cells in complete medium at 1×10^6 cells ml^{-1}
    2. Dispense 0.1 ml aliquots of cells into the wells of a 96-well culture plate (1×10^5 cells/well)
    3. Add 0.1 ml of a range of antigen concentrations in triplicate, including medium alone (no antigen)
    4. Incubate for 4–6 days at 37°C in a humidified atmosphere of 5% CO_2 in air, depending on the peak of the proliferative response
    5. During the last 18 h of incubation, add 37×10^3 Bq (1 μCi)/well of [^{3}H]thymidine
    6. Harvest with an automated cell harvester
    7. Measure radioactive incorporation using a scintillation counter
    8. Plot a dose–response curve for the antigen to determine the optimal concentration for *in vitro* stimulation

IV. Note
    1. A control lymph node cell population should be prepared from mice immunized with FCA alone, and a similar experiment should be carried out to check that the antigen used reacts specifically with the cells primed with the antigen *in vivo*
    2. The immunization protocol must be varied to take account of the antigen employed, including the requirement of a booster injection
    3. An antigen-specific proliferation assay requires T cells, antigen and antigen-presenting cells (APC); lymph node cells include both T cells and APC

Table 12. Murine T-cell lines (73,74)

I. *In vivo* and *in vitro* stimulation
    1. Immunize mice, and prepare a single-cell suspension of lymph node cells
    2. Resuspend the cells in complete medium[a], and aliquot at $1-5 \times 10^6$ cells/well in 24-well culture plates
    3. Add antigen at an optimal concentration to achieve a total volume of 2 ml
    4. Incubate for 4–6 days

II. Rest
    1. After the stimulation, harvest the responding T-cell blasts and purify them from dead cells by density-gradient centrifugation
    2. Wash the T cells twice, resuspend in complete medium, and count in a hemocytometer
    3. Prepare a single-cell suspension from spleens of syngeneic mice for use as feeder cells
    4. Wash the spleen cells, irradiate them with 30 Gy (3000 rad) irradiation, and count in a hemocytometer
    5. Plate 4×10^6 irradiated feeder cells with 4×10^5 T-cell blasts per well in a total volume of 2 ml
    6. Incubate for 10 days

III. Stimulation
    1. Remove 1.5 ml of supernatant from each well after the 10-days' rest
    2. Add 1.5 ml of fresh medium containing antigen to give an optimal final concentration and 5×10^6 irradiated syngeneic spleen cells as APC
    3. Incubate for 4 days

IV. Maintenance of T-cell lines
    1. Repeat the cycles of 4-days' rest and 10-days' stimulation to enrich and maintain antigen-specific T-cell lines
    2. Note: experiments should be done by using T cells immediately before stimulation
    3. Note: antigen-specific CTL lines can be obtained in a similar way except by using IL-2 during stimulation and rest

---

[a]Medium: as in *Table 11*.

4.2. Human helper T-cell lines

The procedures for establishing human T-cell lines are essentially similar to those for murine T-cell lines described above. There are, however, some important differences and limitations. For example, human lines are usually established using either ethical immunization (vaccination) or fortuitous priming (infection, transfusion, transplant or autoimmunity). Human lines require repeated donation from the same individual to provide not only T cells (once) but also feeder and antigen-presenting cells (every week for several months). To circumvent the latter donations, several alternatives have been employed:

(i)   select a panel of HLA-DR/DP/DQ-matched individuals;
(ii) transfect the required MHC gene into immortalized APC; or
(iii) establish EBV-transformed autologous B cells (see *Table 6*) from the T-cell donor (76).

5. B-CELL HYBRIDOMAS

5.1. General aspects of hybridoma production

In order to obtain monoclonal antibodies with a desired specificity, a powerful technique for B-cell hybridomas has been established. In 1975, Köhler and Milstein (77) described details of such a method in which myeloma cells were fused with normal spleen cells to produce potentially immortal 'hybrid' cells that secreted homogeneous antibody of exquisite specificity. As shown schematically in *Table 13*, clones of hybrid cells derived in this way inherit the genetic material determining a given antibody specificity from normal plasma cells and the potential for unlimited proliferation from the continuously growing myeloma parent. An outline scheme for the production of monoclonal antibodies is listed in *Table 14*.

5.2. Myelomas

Induction and maintenance

Myelomas are induced in BALB/c mice by injecting intraperitoneally either mineral oil or pristane (2,6,10,14-tetramethylpentadecane) at 2-monthly intervals. Lou/C rats have a high incidence of spontaneous myeloma production. For human hybridomas, [rodent myeloma × human lymphocyte] hetero-fusions used to be employed, but without constant success. Therefore, tumor lines of a human origin have been devised as a fusion partner. These include either myelomas or, more frequently, lymphoblastoid cell lines (LCL) which are derived from EBV-transformed B lymphocytes (see Section 2). These cell lines are rendered suitable for a fusion partner as described below.

Preparation of drug-resistant mutants

In order to select hybridomas after fusion, drug-resistant mutant myelomas have been developed. Most commonly, the myeloma is a mutant lacking in one of two salvage-pathway enzymes required for the synthesis of DNA when *de novo* biosynthesis of DNA is blocked by a folic acid antagonist such as aminopterin. The relationship between the salvage and *de novo*

Table 13. Characteristics of hybridomas

Antibody-producing cells	Capability of	
	producing antibodies with a desired specificity	proliferating continuously
Normal plasma cells	+	−
Myeloma cells	−	+
Hybridoma of the above two	+	+

Table 14. Procedure for the production of monoclonal antibodies

1. Selection of mutant myeloma cells as a fusion partner
2. Preparation of spleen cells from an immunized animal
3. Fusion
4. Selection of the 'hybrid' cells (hybridomas)
5. Screening
6. Cloning
7. Preparation of monoclonal immunoglobulins

pathways of DNA synthesis and the enzymes involved are shown schematically in *Figure 1*. Drugs used to produce the mutant which is deficient in either of the two salvage-pathway enzymes and details of the selection method are described in *Table 15*, together with another drug marker, resistance to ouabain. In practice, many suitable fusion partners are available for rodent systems (*Table 16*) and human systems (*Table 17*). Some human mutant myelomas are double mutants which are HGPRT negative and ouabain resistant. These double mutants also help later selection (see below).

5.3. Normal lymphocytes
Antigen-primed B lymphocytes must be prepared prior to fusion. Procedures for immunization of animals and preparation of the lymphocytes are described elsewhere (Chapters 1 and 4). For details of immunization, see also refs 76, 95, 96.

5.4. Cell fusion
Several fusion methods are employed to fuse together myeloma cells with normal B lymphocytes. The methods fall into one of three categories; *biological, chemical* and *physical*, as listed in *Table 18*. Among them, the chemical method, which uses polyethylene glycol (PEG) as a fusogen, is the most common. A sample fusion protocol using PEG (95) is shown in *Table 19*. A typical frequency for hybridoma generation in rodent fusions using PEG is between eight and ten hybridomas per 10^5 myelomas, whereas these frequencies are much lower in human systems, approximately one hybridoma per 10^7 myelomas (96). Electrofusion, which yields a higher fusion frequency than PEG, is favorable for human hybridomas (101).

Figure 1. Pathways of nucleotide biosynthesis showing enzymatic steps that are altered in mutant cells used as fusion partners. Mutant cells lacking HGPRT (hypoxanthine guanine phosphoribosyl transferase) or TK (thymidine kinase) cannot use the corresponding salvage pathway for nucleotide biosynthesis. Such mutants cannot survive in medium containing aminopterin, which poisons the *de novo* synthesis pathway by inhibiting DHFR (dihydrofolate reductase). Individual mutant cells that have fused with spleen cells and thus contain the HGPRT or TK enzyme can survive in appropriate selective medium (HAT medium) by using the salvage pathway.

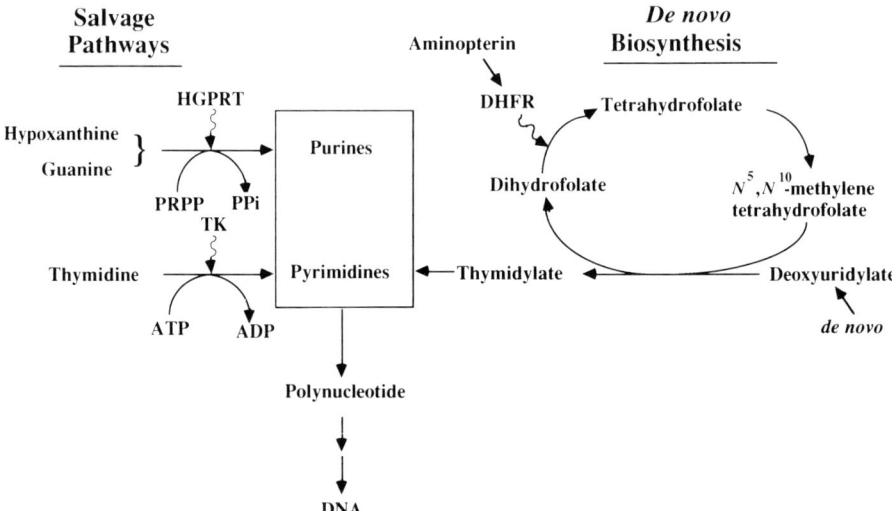

Table 15. Selection of drug-resistant mutant tumor lines

Drug	Phenotype	Selection methods
8-Azaguanine (8-AG) or 6-thioguanine (6-TG)	Deficiency of hypoxanthine guanine phosphoribosyl transferase (HGPRT⁻)	Culture cells in increasing concentrations of 8-AG or 6-TG. HGPRT$^+$ cells incorporate the toxic analog into DNA and are killed. HGPRT is encoded by X chromosomes and only one active copy must be inactivated. HGPRT$^-$ lines, once established, must be cultured in medium containing 0.02 mg ml^{-1} (0.13 mM) 8-AG every 3–6 months to eliminate revertants
5-Bromodeoxyuridine (5-BrdU)	Deficiency of thymidine kinase (TK⁻)	Culture cells in increasing concentrations of 5-BrdU. TK$^+$ cells incorporate the analog into DNA and are killed. TK is an autosomal enzyme and two copies must be inactivated
Ouabain	Resistance to ouabain (oua^R)	Mutagenize cells by γ-irradiation and culture them in increasing concentrations of ouabain over 2 weeks. Ouabain inhibits Na$^+$K$^+$-ATPase

Table 16. Fusion partners for mouse and rat B-cell hybridomas

Cell line[a]	Designation	Origin	Ig production	Ref.
Mouse				
P3-X63-Ag8 (X63)	Myeloma	BALB/c	IgG1(kappa)	77
P3-X63-Ag8.653	Myeloma	[X63]	None	78
P3-X63-Ag8.U1 (P3U1)	Myeloma	[X63]	None	79
P3-NS I/1-Ag4/1 (NS I)	Myeloma	[X63]	Kappa (nonsecretor)	80
NSO/1	Myeloma	[NS I]	None	81
Sp2/0-Ag14	Hybridoma	X63 × BALB/c spl[b]	None	82
Rat				
Y3-Ag1.2.3 (Y3)	Myeloma	Lou/C	Kappa	83
YB2/O	Hybridoma	Y3 × AO spl[b]	None	84

[a]All cell lines listed are HGPRT⁻.
[b]spl, spleen cells.

Table 17. Fusion partners for human × human B-cell hybridomas

Cell line	Designation	Ig production	Drug marker[e]	Ref.
GM1500 6TG-Al 1	LCL	IgG2 (kappa)	HGPRT⁻	85
KR-4	LCL[a]	IgG2 (kappa)	HGPRT⁻, Oua[R]	86
H35.1.1	LCL	IgM (kappa)	HGPRT⁻	87
0467.3	LCL	IgM (lambda)	HGPRT⁻	87
MC/NMS-2	Myeloma	None	HGPRT⁻	88
M/H D-33	Myeloma[b]	None	HGPRT⁻	89
UC729-HF2	LCL	None	HGPRT⁻	90
LTR228	LCL	IgM (kappa)	HGPRT⁻	91
W1-L2-727-HF2-6TG	LCL	IgG (kappa)	HGPRT⁻, Oua[R]	92
TAW-925	LCL[c]		HGPRT⁻, Oua[R]	93
HAB-1	Myeloma[d]	None	HGPRT⁻	94

[a]Derived from GM1500 6TG-Al 1.
[b]Mouse (Ag8.653) × human (pSV2-*neo*[R]-transfected myeloma) heteromyeloma.
[c]Derived from W1-L2, a 6-thioguanine-resistant human B-LCL, by selecting in the presence of 5×10^{-7} M ouabain.
[d]Mouse (Ag8) × human (lymphocytes) heteromyeloma.
[e]For details see *Table 15*.

Table 18. Fusion methods used in hybridoma production

Method	Details
Biological	Several groups of viruses such as HVJ (Sendai virus) (97)
	Rarely used now
	Disadvantages: possible contamination of viral products
Chemical	Polyethylene glycol (PEG) (98, 99)
	Cell membranes are fused together indiscriminately
	Heterokaryons are formed first, the nuclei being also fused at the first cell division
	Each batch of PEG should be tested for good fusion frequencies and low toxicity prior to use
	Relatively cheap method
	For details see *Table 19*
Physical	Electrofusion (100)
	Cells are brought together in a low-amplitude continuous AC electric field, and then subjected to a high-amplitude short pulse
	This brief pulse causes membrane breakdown, and cell fusion results when the membrane breakdown occurs in areas of cell–cell contact
	High fusion frequencies: 10 times more efficient than PEG (101)
	Useful for human hybridomas (101–103)
	Needs a special device

CELLULAR IMMUNOLOGY LABFAX

Table 19. Fusion protocol using polyethylene glycol (PEG) (95)

Medium: serum-free medium, e.g. RPMI 1640

Temperature for fusion: 37°C

PEG: Molecular weight: 1000–4000
      Concentration: 30–50% (vol/vol)

Cells to be fused:
  Myelomas in exponential growth
  Spleen cells from an appropriately immunized animal
  Ratio of spleen cells : myeloma between 1 : 1 and 10 : 1

Sample protocol:
Day 0
  1. Combine myeloma and spleen cells in a tube
  2. Pellet cells and take off supernatant
  3. Suspend cells homogeneously
  (*The following steps 4–8 and 10–12 should be done in a 37°C water bath*)
  4. Add 1 ml of warm PEG dropwise over 1 min by repeating a step of dropping 2 drops and swirling the pellet twice with the tip of a pipette
  5. Swirl gently for 1 min with the tip of a pipette
  6. Add 1 ml of warm medium dropwise over 1 min as #4
  7. Repeat #6
  8. Add 7 ml of warm medium dropwise over 2–3 min by repeating a step of dropping 5 drops and swirling pellet 5 times with the tip of a pipette
  9. Spin for 6 min at 200 *g* to pellet cells
  10. Suspend cells gently in HT medium[a]
  11. Swirl cells gently with the tip of a pipette
  12. Dilute to the equivalent of 0.5–1×10^6 cells ml^{-1}
  13. Put 0.1 ml/well in 96-well flat-bottomed plates
  (As controls, plate out separately spleen cells and myeloma cells that have not been fused, which should both die)

Day 1
Add 0.1 ml of HAT medium[a] to the fusion plates

Days 2, 3, 5 and every 3 days thereafter
Remove 0.1 ml of supernatant and add 0.1 ml of fresh HAT medium[a]

Day 11 and later
Watch plates for growth, and test supernatants of growing wells for desired antibody

[a]For HT and HAT media, see *Tables 20* and *21*.

5.5. Selection of hybridomas
Media used for culture and selection of hybridomas
Culture media used for culture and selection of hybridomas are shown in *Table 20*. Useful stock solutions for the media are listed in *Table 21*. Some of these stock solutions are commercially available (e.g. Sigma).

Table 20. Media used in hybridoma experiments (76, 95)

Media	Contents
Serum-free medium	RPMI 1640 or Dulbecco's modification of Eagle's medium (DMEM)
Complete medium	RPMI or DMEM, supplemented with: 10–20% fetal calf serum (FCS) 2 mM glutamine 100 U ml^{-1} penicillin 0.1 mg ml^{-1} streptomycin 25 mM Hepes
HT medium	Complete medium containing: 1.0×10^{-4} M hypoxanthine (H) 1.6×10^{-5} M thymidine (T)
HAT medium	Complete medium containing: 1.0×10^{-4} M hypoxanthine (H) 4.0×10^{-7} M aminopterin (A) 1.6×10^{-5} M thymidine (T)
HAT/ouabain medium	Complete medium containing: 1.0×10^{-4} M hypoxanthine (H) 4.0×10^{-7} M aminopterin (A) 1.6×10^{-5} M thymidine (T) 1.0×10^{-5} M ouabain

Hepes, N-2-hydroxyethyl piperazine-N'-ethane-sulfonic acid.

HAT selection

In order to select hybridomas out of unfused cells, the cells are cultured in medium containing hypoxanthine, aminopterin and thymidine (HAT selection medium, see *Table 20*) commencing the day after fusion. The basis of the HAT selection is shown in *Table 22*. Since tumor cells employed are HGPRT$^-$, tumor/tumor hybrids as well as unfused tumors die in HAT medium. Unfused lymphocytes and lymphocyte/lymphocyte hybrids die because of no immortality. On the other hand, tumor/lymphocyte hybrids can grow in HAT medium because they inherit immortality from tumor parents and HGPRT from normal lymphocyte parents. The reason why cells are suspended in HT medium immediately after fusion (see *Table 19*) is that HT medium may induce HGPRT activity in normal lymphocytes before adding aminopterin. After initial growth in HAT medium for approximately 10–14 days, the hybridomas are transferred to HT medium for 2–3 weeks, and then to complete hybridoma medium. Do not transfer hybridomas directly from HAT to normal medium as sufficient aminopterin may be carried over to prevent a resumption of *de novo* synthesis of DNA.

HAT/ouabain double selection procedure

This method is used in human fusions where the antibody-producing cells to be fused are EBV-transformed lymphoblastoid cells, rather than normal B lymphocytes. The basis for the selection is shown in *Table 23*. The selection medium used is HAT/ouabain medium (see *Table 20*). The technique improves by 10–100 times the frequency of hybrids obtained (104).

Table 21. Useful stock solutions for fusions

Stock solution	Contents
50 × HT	Hypoxanthine (mol. wt = 136.1) 5×10^{-3} M Thymidine (mol. wt = 242.2) 8×10^{-4} M Dissolve 68 mg hypoxanthine and 19.4 mg thymidine in 100 ml TDW[a] at 50°C Sterilize by membrane filtration and store in 2–5 ml aliquots at -20°C See *Table 20*
50 × A	Aminopterin (mol. wt = 440.4) 2×10^{-5} M (Note: highly toxic and a potent carcinogen aminopterin must be protected from light) Add 0.9 mg aminopterin to 90 ml TDW Add 1 M NaOH until aminopterin dissolves Titrate to pH 7.5 with 1 M HCl Adjust final volume to 100 ml with TDW Sterilize by membrane filtration and store in 2–5 ml aliquots at -20°C See *Table 20*
50 × 8AG	8-Azaguanine (mol. wt = 152.1) 6.6×10^{-3} M (1 mg ml^{-1}) Dissolve 100 mg 8-azaguanine in 20 ml 5% $NaHCO_3$ Add TDW to 100 ml Sterilize by membrane filtration and store in 5 ml aliquots at -20°C See *Table 15*

[a]TDW, twice-distilled water.

Table 22. Mechanism of HAT selection

Cells	HGPRT	Potential for immortality	Growth in HAT medium
Unfused tumor	−	+	−
Unfused lymphocyte	+	−	−
Tumor/tumor fusion	−	+	−
Lymphocyte/lymphocyte fusion	+	−	−
Tumor/lymphocyte fusion	+	+	+

Table 23. Mechanism of HAT/ouabain double selection

Cells[a]	HGPRT	Ouabain	Growth in HAT/ouabain
Myeloma/LCL tumor cell	−	Resistant	−
Antibody-producing lymphocyte	+	Sensitive	−
Hybridoma of the above two	+	Resistant	+

[a]Homologous fusions are avoided for brevity's sake.

Cell Lines, Hybridomas

5.6. Screening

The purpose of screening is to determine which wells contain the desired antibody so as to avoid unnecessary culture of nonproductive cells. The major issue in choosing a screening assay is that the assay must be fast and simple enough to allow screening of many samples and that it must be unequivocal. Therefore, clear-cut discrimination between positives and negatives is often more important than exquisite sensitivity. A screening assay, which is difficult to perform on a large number of samples, can also be applied by testing supernatants pooled from a small number of hybrids, provided the assay sensitivity is high (136). Components of positive pools are then screened individually. A list of techniques which may be appropriate for screening of B-cell hybridomas is given in *Table 24*.

5.7. Culture of hybridoma clones

After identification of positive cultures, antibody-secreting hybridomas must be cloned to ensure that the antibody is homogeneous and monospecific. Details for cloning are described in Section 7. Cloned hybridoma cells must be expanded gradually, from 96-well plates to 24-well plates and then to flasks. Large amounts of hybridoma-derived antibody can be prepared either *in vitro* or *in vivo* (*Table 25*). For purification, isolation, enzymic modification and further characterization of monoclonal antibodies, see refs 76, 106.

Table 24. Screening procedures for B-cell hybridomas (76, 96)

Screening procedure	Comments
Enzyme-linked immunosorbent assay (ELISA)	Measures antibodies of 1–100 μg ml^{-1} Many modifications
Radioimmunoassay (RIA)	Solid-phase RIA is similar to ELISA, but less convenient Competitive RIA is more sensitive
Immunodiffusion	Measures precipitation of soluble antigen by antibody Less sensitive than ELISA or RIA, although much cheaper
Immunofluorescence	Measures antibodies for a cell surface antigen Uses either fluorescence microscopy or a cytofluorimeter
Immunoblotting	Measures antibodies for a soluble antigen blotted on to nitrocellulose
Histochemistry	Measures antibodies for tissue antigens Second antibodies are either fluorescence- or enzyme-labeled
Indirect agglutination	Measures IgM antibodies capable of agglutinating red cells or latex particles coated with antigen Uses second antibodies to detect IgG antibodies
Miscellaneous	Cytotoxicity assay (dye-exclusion method or ^{51}Cr-release assay) Inhibition of cell-mediated lysis Inhibition of receptor activation Inhibition of cytokine activity, and so on

CELLULAR IMMUNOLOGY LABFAX

Table 25. Preparation of monoclonal antibody from hybridomas

Method	Comments
In vitro	Culture hybridomas in flasks Can yield 1–20 µg ml^{-1} immunoglobulin For reference to large-scale production, see ref. 105
In vivo	Inject hybridomas into the peritoneal cavities of syngeneic mice, and antibodies can be recovered from the peritoneal fluid and serum Can yield 1–20 mg ml^{-1} (1000 times more than *in vitro*) (83) Day −14: IP injection of 0.5 ml pristane into syngeneic mice Day 0: IP injection of 10^6–10^7 hybridoma cells Day 14: Drain off ascites fluid, and clarify it by centrifugation Later: Repeat the drainage for the lifetime of the mice, and finally also bleed the mice to prepare serum Note: Ascites fluid should be drained off before the mice increase in weight by 20%. Do not maintain a hybridoma by serial passage in mice because of the risk of accumulating nonsecreting cells (76)

5.8. Application of monoclonal antibodies

Monoclonal antibodies potentially show no undesirable cross-reactivity and have very high titers. Furthermore, they can be prepared easily and reproducibly in large quantities. These advantages allow many experiments that were not possible or practical before (136). Monoclonal antibodies have also been applied to clinical use (107–113). These are listed in *Table 26*.

Table 26. Application of monoclonal antibodies

Research	Immunofluorescence of cell-surface markers; e.g. CD markers Immunohistochemistry Cell separation by panning or killing Immunochemical separation; e.g. affinity chromatography Study of the structure of antigen–antibody complexes by X-ray diffraction Study of antibody diversity Study of antibody idiotype
Clinical	Assays for circulating hormones or toxins by ELISA or RIA Diagnostic imaging (107) Therapeutic uses in cancer (108–110), transplantation (111) or autoimmune diseases (112, 113) For therapeutic uses, genetically engineered antibodies expressing a human constant region together with a murine variable region (114, 115) have been devised

Cell Lines, Hybridomas

6. T-CELL HYBRIDOMAS

The methods used to generate T-cell hybridomas are essentially the same as for B-cell hybridomas. Outline procedures for T-cell hybridomas are listed in *Table 27*. In this section, only points which differ from those for B-cell hybridomas are described.

6.1. T-lymphoma lines

As in B-cell hybridomas, normal T cells are fused to homologous tumor partners because of the instability in interspecies hybrids. *Tables 28* and *29* summarize the rodent and human T-lymphoma lines, respectively, that have served as a fusion partner. The most common is murine AKR thymoma BW5147.G.1.4 (BW5147) which is HGPRT⁻ and hence HAT sensitive (117). As immunoglobulin-nonsecretors are preferred for a fusion partner in B-cell hybridomas, T-cell receptor-negative mutant lines have also been generated. So far, a variant of BW5147, BW1100.129.237, which is lacking in both α- and β-chains of the T-cell receptor, is reported to give a good fusion efficiency (118). Most of the fusion partners used are drug-marked, and hence HAT and HAT/ouabain are popular methods for selection. Since thymidine, which is present in HAT medium, has been shown to inhibit the growth of lymphoid cells, other selection methods which do not employ HAT medium have also been devised (see below).

Table 27. Procedure for T-cell hybridomas

---

1. Selection of a mutant T-cell tumor line as a fusion partner
2. Enrichment of specific T lymphocytes derived from an immunized host
3. Fusion
4. Selection of the 'hybrid' cells (hybridomas)
5. Expansion (24-well plates)[a]
6. Screening
7. Cloning
8. Culture and isolation of lymphokines, TCR and other products

---

[a]One of the major differences from B-hybridoma procedures.

Table 28. Fusion partners for mouse and rat T-cell hybridomas

Cell line	Designation	Origin	Drug marker[a]	Selection	Ref.
Mouse					
BW5147.G.1.4	Thymoma	AKR/J	HGPRT⁻	HAT	117
BW1100.129.237	Thymoma	[BW5147]	HGPRT⁻	HAT	118
AKR1.G.1.Oua^R.1.26	Thymoma	AKR/J	HGPRT⁻, Oua^R	HAT/ouabain	119
EL4.BU	T lymphoma	C57BL/6	TK⁻	HAT	120, 121
EL4.BU.1.Oua^R.1.1	T lymphoma	[EL4.BU]	TK⁻, Oua^R	HAT/ouabain	120
Rat					
C58(NT)D.1.G. Oua^R.1	T lymphoma	W/Fu	HGPRT⁻, Oua^R	HAT/ouabain	122

[a]For details see *Table 15*.

Table 29. Fusion partners of human × human T-cell hybridomas

Cell line	Designation	Drug marker[a]	Selection	Ref.
CEM-T15	Leukemia	HGPRT⁻	HAT	123
CEM-AG^R	Leukemia	HGPRT⁻	HAT	124
J3R7	Leukemia	HGPRT⁻	AH[b]	125
CEM-6	Leukemia	HGPRT⁻	HAT	126
Jurkat-6TG-3	Leukemia	HGPRT⁻	HAT	127
SH9	Leukemia	HGPRT⁻	HAT	128
D1R11	Leukemia	TK⁻	HAT	129
CEM	Leukemia	–	EA[c]	130
Molt-4 or Jurkat	Leukemia	–	Soft agar	131

[a]For details see *Table 15*.
[b]AH, azaserine and hypoxanthine, which select by the same mechanism as HAT.
[c]EA, emetine and actinomycin D (for details see text).

6.2. Normal T lymphocytes

Since there are very few specific T cells in a population of immune lymphocytes, it is crucial to enrich those T cells prior to fusion. The procedures are listed in *Table 30*. Among them, antigen-specific helper T cells are enriched in the same way as for normal T-cell lines (*Table 12*). Although the nature of the cell type that actually fuses is not clear, activation of the immune T lymphocytes appears to select cells that are favored for the fusion, although there is no direct evidence for this. Thus, some of the procedures described above may serve to enrich cells that are not only antigen specific but amenable for fusion.

6.3. Fusion and selection

Fusions for T-cell hybridomas are carried out exactly in the same way as in B-cell hybridomas (*Table 19*). HAT selection and HAT/ouabain double selection are also the same. In T-cell hybridomas, however, two additional selection strategies have been devised: emetine/actinomycin D (130) and a soft-agar manipulation (131). In these methods, tumor partners need not be drug-marked; in other words, these methods may be applied to any tumor-cell line.

Emetine/actinomycin D selection
This method was originally applied to a human T-leukemia line, CEM, which is not drug-marked (130). Tumor cells are treated with emetine and actinomycin D before fusion. Emetine inhibits protein synthesis in mammalian cells at the *trans*-location that involves

Table 30. Enrichment of specific T cells

Procedure	T cells
Restimulation with antigen and APC	Antigen-specific helper T cells
Restimulation with targets and interleukin-2	Antigen-specific cytotoxic T cells
Panning by antigen-coated plastic dishes	Antigen-specific suppressor T cells
Rosetting by antigen-coated erythrocytes	Antigen-specific suppressor T cells
FACS after staining with anti-I-J antibodies	I-J⁺ murine suppressor T cells

FACS, fluorescence-activated cell sorter.

movement of mRNA along a ribosome, and actinomycin D inhibits RNA synthesis irreversibly. Therefore, tumor cells treated with these drugs die of ribosomal depletion, unless they fuse with normal untreated cells and are provided with the devices required for replication. Thus, only normal–tumor hybrids are permitted to grow in normal culture medium. Although this method is applicable to any tumor-cell line, the optimal concentrations of the treatment must be carefully determined for each line. This method appears to yield higher hybridoma frequencies than HAT (10^{-4} vs. 10^{-6}).

Soft-agar selection
An alternative strategy is based on the ability of T-cell hybridomas to form colonies in soft agar, whereas neither fusion parent forms colonies. Originally, antigen-specific normal human T cells were fused with nonmutagenized human T-lymphoblastoid cell lines such as MOLT-4 or Jurkat, and a mixture of the fused and unfused cells were cultured in a Petri dish containing 0.3% soft agar for 10–60 days. Only hybridoma cells formed colonies, enabling the selection (131).

6.4. Screening
This is one of the major differences from procedures for B-cell hybridomas. Since T hybridomas themselves, not culture supernatants, are required for screening, hybridomas must be expanded enough prior to screening. In most assay systems, T hybridomas need to be expanded to 24-well plates. Enrichment procedures mentioned above can also serve as screening assays as long as they are fast and simple enough to allow screening of many samples and are unequivocal. Some of the procedures are given in *Table 31*.

Table 31. Screening for T-cell hybridomas

T-cell hybridoma	Screening procedure
Helper T-cell hybridoma	Lymphokine-producing capability in an antigen-specific and MHC-restricted manner
Cytotoxic T-cell hybridoma	Standard CTL assay using an appropriate target
Suppressor T-cell hybridoma	Screen by their antigen-binding capacity, and then select by their suppressor activity

7. CLONING
7.1. Definition and purpose
When cell cultures are initiated, even with prior enrichment of certain cell types, the bulk culture represents the cellular progeny of many different cells. A homogeneous population can only be produced following 'cloning' of individual cells. The term 'cloning' in cell culture is defined as the initiation of a cell line from a single progenitor; that is, cloned cells are all genetically identical. This process is used to establish clones of various cell types that have been described in this chapter. The purpose of cloning is not only to select a desirable population but also to eliminate undesirable or unstable populations. For instance, in antibody-secreting hybridomas, it is necessary to repeat the cloning in order to ensure that nonproducers, arising as spontaneous variants, do not outgrow the antibody-secreting hybridomas.

7.2. Procedures
Several methods are available for cloning cells (2, 132). These are summarized in *Table 32*.

Table 32. Cloning methods (2, 96, 132)

Method	Cells
Limiting dilutions	Most cell types; often used for cloning T-cell clones and T- and B-hybridomas (for details see text)
Cloning in semisolid media	Hybridomas, tumor cells and bone marrow stem cells Plate out cells at low densities into semisolid media in which single cells grow into discrete colonies Pick off each colony using sterile Pasteur pipettes and transfer to 96-well microplates Media used include bacto-agar, Noble agar, agarose or methylcellulose
Micromanipulation	Adherent cells Uses cloning plates or cloning rings
Cell sorter	Uses a continuous-flow cytofluorimeter (flow cytometer) (for details see ref. 76)

Cloning by limiting dilutions

Cells are seeded into wells at very low cell densities, namely 1, 0.5 and 0.25 cells/well in 96-well plates, together with feeder cells at 1×10^5–5×10^5 cells per well. For hybridomas or T-cell lines, these may be mouse or rat thymocytes, spleen cells or peritoneal exudate cells (133–135). The feeder cells must be irradiated at 30–40 Gy or treated with mitomycin C to prevent any growth in culture, and need not be histocompatible. Positive cultures at the lowest cell density employed may be assumed to be clones, and are subjected to further analysis. Cloning should be carried out at least twice. The reason is as follows: the initial distribution of the cells seeded in each well follows Poisson statistics. The probability of each well receiving r cells is given by a Poisson formula:

$$F_m(r) = \frac{(m^r)(e^{-m})}{r!}$$

Equation 1

where m is the mean number of the cells per well and e is the base of the natural logarithms. Therefore, the probabilities of each well receiving no cells and one cell are:

$$F_m(0) = \frac{(m^0)(e^{-m})}{0!} = e^{-m},$$

Equation 2

and

$$F_m(1) = \frac{(m^1)(e^{-m})}{1!} = me^{-m},$$

Equation 3

respectively. Therefore, the probability of each well receiving two or more cells is

$$1 - \{F_m(0) + F_m(1)\} = 1 - (1 + m)e^{-m}.$$

Equation 4

Table 33 gives individual values of expected probabilities, calculated using Equations 2–4, for values of m between 0.1 and 1.0. For example, when m is 0.2 and 0.5 cells/well, the probability of each well receiving two or more cells is 0.018 and 0.090, respectively. Thus, a significant proportion of wells receive two or more cells and hence cannot initiate a true clone. Therefore, cloning must be repeated to ensure the homogeneity of the cells.

Table 36. Continued

Cell line	Appears in	ATCC
EL-4	*Table 3*	TIB 39
A20	*Table 3*	TIB 208
P388D1	*Table 3*	CCL 46
P388D1 (IL-1 high producer)	*Table 3*	TIB 63
P815	*Table 3*	TIB 64
F9	*Table 3*	CRL 1720
CHO	*Table 3*	CCL 61
COS-7	*Table 3*	CRL 1651
B95-8	*Table 3*	CRL 1612
D10.G4.1	*Table 4*	TIB 224
HT-2-A5E	*Table 4*	CRL 1841
CTLL-2	*Table 4*	TIB 214
P3-X63-Ag8(X63)	*Table 16*	TIB 9
P3-X63-Ag8.653	*Table 16*	CRL 1580
P3-X63-Ag8.U1	*Table 16*	CRL 1597
P3-NS I/1-Ag4/1(NS1)	*Table 16*	TIB 18
Sp2/0-Ag14	*Table 16*	CRL 1581
Y3-Ag1.2.3(Y3)	*Table 16*	CRL 1631
YB2/O	*Table 16*	CRL 1662
LTR228	*Table 17*	HB 8502
BW5147.G.1.4	*Table 28*	TIB 48
AKR1.G.1.Oua^{R}.1.26	*Table 28*	TIB 232
EL4.BU	*Table 28*	TIB 40
C58(NT)D.1.G.Oua^{R}.1	*Table 28*	TIB 236
CEM-AG^{R}	*Table 29*	CRL 8081

9. REFERENCES

1. Jakoby, W.B. and Pasten, I.H. (1988) *Cell Culture: Methods in Enzymology.* Academic Press, San Diego, Vol. 58.

2. Freshney, R.I. (1987) *Culture of Animal Cells: A Manual of Basic Technique (2nd edn).* Alan R. Liss, New York.

3. Klaus, G.G.B. (1987) *Lymphocytes: A Practical Approach.* IRL Press, Oxford.

4. Foley, G.E., Lazarus, H., Farber, S., Uzman, B.G., Boone, B.A. and McCarthy, R.E. (1965) *Cancer,* **18,** 522.

5. Gazder, A.F., Garney, D.N., Bunn, P.A., Russel, E.K., Jaffe, E.S., Schechter, G.P. and Guccion, J.G. (1980) *Blood,* **55,** 409.

6. Poiesz, B.J., Ruscetti, F.W., Gazder, A.F., Bunn, P.A., Minna, J.D. and Gallo, R.C. (1980) *Proc. Natl. Acad. Sci. USA,* **77,** 7415.

7. Schneider, U., Schwenk, H.U. and Bornkamm, G. (1977) *Int. J. Cancer,* **19,** 621.

8. Gillis, S. and Watson, J. (1980) *J. Exp. Med.,* **152,** 1709.

9. Minowada, J., Ohmura, T. and Moore, G.E. (1972) *J. Natl. Cancer Inst.,* **49,** 891.

10. Klein, E. and Klein, G. (1968) *Cancer Res.,* **28,** 1300.

11. Huber, C., Sundstrom, C., Nilsson, K. and Wigzell, H. (1976) *Clin. Exp. Immunol.,* **25,** 367.

12. Nilsson, K., Giovanella, B.C., Stehlin, J.S. and Klein, G. (1977) *Int. J. Cancer,* **19,** 337.

13. Pulvertoft, R.J.V. (1964) *Lancet,* **i,** 238.

14. Epstein, M.A., Achong, B.G., Barr, Y.M., Zajac, B., Henle, G. and Henle, W. (1966) *J. Natl. Cancer Inst.,* **37,** 547.

15. Collins, S.J., Gallo, R.C. and Gallagher, R.E. (1977) *Nature*, **270**, 347.

16. Gallagher, R., Collins, S., Trujillo, J., McCredie, K., Ahearn, M., Tsai, S., Metzgar, R., Aulakh, G., Ting, R., Ruscetti, F. and Gallo, R. (1979) *Blood*, **45**, 713.

17. Lozzio, C.B. and Lozzio B.B. (1975) *Blood*, **45**, 321.

18. Lozzio, B.B. and Lozzio, C.B. (1979) *Leukemia Res.*, **3**, 363.

19. Sundstrom, C. and Nilsson, K. (1976) *Int. J. Cancer*, **17**, 565.

20. Gey, G.O., Coffman, W.D. and Kubicek, M.T. (1952) *Cancer Res.*, **12**, 264.

21. Scherer, W.F., Syverton, J.T. and Gey, G.O. (1953) *J. Exp. Med.*, **97**, 695.

22. Scherer, W.F. and Hoogasian, A.C. (1954) *Proc. Soc. Exp. Biol. Med.*, **87**, 480.

23. Gorer, P.A. *et al.* (1950) *Br. J. Cancer*, **4**, 372.

24. Kim, K.J., Kanellopoulos-Langevin, C., Merwin, R.M., Sachus, D.H. and Asofsky, R. (1979) *J. Immunol.*, **122**, 549.

25. Dawe, C.J. and Potter, M. (1957) *Am. J. Pathol.*, **33**, 603.

26. Ralph, P. and Nakoinz, I. (1977) *J. Immunol.*, **119**, 950.

27. Moore, M.A.S. (1982) *J. Cell Physiol.*, **1** (Suppl.), 53.

28. Plant, M., Lichtenstein, L.M., Gillespie, E. and Henney, C.S. (1973) *J. Immunol.*, **111**, 389.

29. Bernstine, E.G., Hooper, M.L., Grandchamp, S. and Ephrussi, B. (1973) *Proc. Natl. Acad. Sci. USA* **70**, 3899.

30. Strickland, S., Smith, K.K. and Marotti, K.R. (1980) *Cell*, **21**, 347.

31. Puck, T.T., Cieciura S.J. and Robinson, A. (1958) *J. Exp. Med.*, **108**, 945.

32. Kao, F.-T., Johnson, R.T. and Puck, T.T. (1969) *Science*, **164**, 312.

33. Kao, F.-T., Chasin, L. and Puck, T.T. (1969) *Proc. Natl. Acad. Sci. USA*, **64**, 1284.

34. Gluzman,, Y. (1981) *Cell*, **23**, 175.

35. Miller, G. and Lipman, M. (1973) *Proc. Natl. Acad. Sci. USA*, **70**, 190.

36. Kaye, J. and Janeway, C.A. (1984) *Lymphokine Res.*, **3**, 175.

37. Watson, J. (1979) *J. Exp. Med.*, **150**, 1510.

38. Gillis, S. and Smith, K.A. (1977) *Nature*, **268**, 154.

39. Dexter, T.M., Garland, J., Scott, D., Scolnick, E. and Metcalf, D. (1980) *J. Exp. Med.*, **152**, 1036.

40. Hu-Li, J., Ohara, J., Watson C., Tsang, W. and Paul W.E. (1989) *J. Immunol.*, **142**, 800.

41. Tominaga, A., Mita, S., Kikuchi, Y., Yasumichi, H., Takatsu, K., Nishikawa, S.-I. and Ogawa, M. (1989) *Growth Factors,* **1**, 135.

42. Matsuda, T., Hirano, T. and Kishimoto, T. (1988) *Eur. J. Immunol.*, **18**, 951.

43. Takai, Y., Sakata, T., Iwagami, S., Tai, X.-G., Kita, Y., Hamaoka, T., Sakaguchi, N., Yamagishi, H., Tsuruta, Y., Teraoka, H. and Fujiwara, H. (1992) *J. Immunol.*, **148**, 1329.

44. Salzman, N.P. (1986) *The Papovaviridae.* Plenum Press, New York, Vol. 1.

45. Klein, G. (1987) *Advances in Viral Oncology.* Raven Press, New York, Vol. 6.

46. Graham, F.L., van der Eb, A.J. and Heijnecker, H.L. (1974) *Nature,* **251**, 687.

47. Wigler, M., Silverstein, S., Lee, L.S., Pellicer, A., Cheng, Y.-C. and Axel, R. (1977) *Cell*, **11**, 222.

48. Wong, T.K. and Neumann, E. (1982) *Biochem. Biophys. Res. Commun.*, **107**, 584.

49. Capecchi, M.R. (1980) *Cell*, **22**, 479.

50. Anderson, W.F., Killos, L., Sanders-Haigh, L., Kretschmer, P.J. and Diacumakos, E.G. (1980) *Proc. Natl. Acad. Sci. USA*, **77**, 5399.

51. Gluzman, Y., Frisque, R.J. and Sambrook, J. (1980) *Cold Spring Harbor Symp. Quant. Biol.,* **44**, 293.

51. Gluzman, Y., Sambrook, J. and Frisque, R.J. (1980) *Proc. Natl. Acad. Sci. USA*, **77**, 3898.

53. Tsuyama, N., Miura, M., Kitahira, M., Ishibashi, S. and Ide, T. (1991) *Cell Struct. Funct.*, **16**, 55.

54. Hronis, T.S., Steinberg, M.L., Defendi, V. and Sun, T.-T. (1984) *Cancer Res.*, **44**, 5797.

55. Reddel, R.R., Ke, Y., Gerwin, B.I., McMenamin, M.G., Lechner, J.F., Su, L.T., Brash, D.E., Park, J.-B., Rhim, J.S. and Harris, C.C. (1988) *Cancer Res.*, **48**, 1904.

56. Goto, M., Okamoto, M., Sasano, M., Nishizawa, K., Aotsuka, S., Yamaguchi, N., Obinata, M. and Ikeda, K. (1991) *Clin. Exp. Immunol.*, **86**, 387.

57. Katsuki, T. and Hinuma, Y. (1975) *Int. J. Cancer*, **15**, 203.

58. Bird, A.G., McLachlan, S.M. and Britton, S. (1981) *Nature*, **289**, 300.

59. Hashimoto, T. *et al.* (1983) *Hum. Genet.*, **63**, 75.

60. Miyoshi, I., Kubonishi, I., Yoshimoto, S.,

Akagi, T., Ohtsuki, Y., Shiraishi, Y., Nagata, K. and Hinuma, Y. (1981) *Nature*, **294**, 770.

61. Yamamoto, N., Okada, M., Koyanagi, Y., Kannagi, M. and Hinuma, Y. (1982) *Science*, **217**, 737.

62. Popovic, M., Lange-Wantzin, G., Sarin, P.S., Mann, D. and Gallo, R.C. (1983) *Proc. Natl. Acad. Sci. USA*, **80**, 5402.

63. Seiki, M., Hattori, S., Hirayama, Y. and Yoshida, M. (1983) *Proc. Natl. Acad. Sci. USA*, **80**, 3618.

64. Grassmann, R., Dengler, C., Muller-Fleckenstein, I., Fleckenstein, B., McGuire, K., Dokhelar, M.-C., Sodroski, J.G. and Haseltine, W.A. (1989) *Proc. Natl. Acad. Sci. USA*, **86**, 3351.

65. Akagi, T. and Shimotohno, K. (1993) *J. Virol.*, **67**, 1211.

66. Dexter, T.M., Allen, T.D. and Lajtha, T.G. (1977) *J. Cell. Physiol.*, **91**, 335.

67. Whitlock, C.A. and Witte, O.N. (1982) *Proc. Natl. Acad. Sci. USA*, **79**, 3608.

68. Ogawa, M., Nishikawa, S., Ikuta, K., Yamamura, F., Naito, M., Takahashi, K. and Nishikawa, S.-I. (1988) *EMBO J.*, **7**, 1337.

69. Nishikawa, S.-I., Ogawa, M., Nishikawa, S., Kunisada, T. and Kodama, H. (1988) *Eur. J. Immunol.*, **18**, 1767.

70. Ogata, M., *et al.* (1989) *J. Lymphocyte Biol.*, **45**, 49.

71. Palacios, R., Studer, S., Samaridis, J. and Pelkonen, J. (1989) *EMBO J.*, **8**, 4053.

72. Watanabe, Y., Mazda, O., Aiba, Y.-I., Iwai, K., Gyotoku, J.-I., Ideyama, S., Miyazaki, J. and Katsura, Y. (1992) *Cell. Immunol.*, **142**, 385.

73. Kimoto, M. and Fathman, C.G. (1980) *J. Exp. Med.*, **152**, 759.

74. Fathman, C.G. and Fitch, F.W. (1983) *Isolation, Characterization, and Utilization of T Lymphocyte Clones.* Academic Press, New York.

75. Baker, P.E., Gillis, S. and Smith, K.A. (1979) *J. Exp. Med.*, **149**, 273.

76. Hudson, L. and Hay, F.C. (1989) *Practical Immunology.* Blackwell Scientific Publications, Oxford.

77. Köhler, G. and Milstein, C. (1975) *Nature*, **256**, 495.

78. Kearney, J.F., Radbruch, A., Liesegang, B. and Rajewsky, K. (1979) *J. Immunol.*, **123**, 1548.

79. Yelton, D.E., Diamond, B.A., Kwan, S.P. and Scharff, M.D. (1978) *Curr. Top. Microbiol. Immunol.*, **81**, 1.

80. Köhler, G. and Milstein, C. (1976) *Eur. J. Immunol.*, **6**, 511.

81. Galfre, G. and Milstein, C. (1981) *Methods Enzymol.*, **73**, 3.

82. Shulman, M., Wilde, C.D. and Kohler, G. (1978) *Nature*, **276**, 269.

83. Galfre, G., Milstein, C. and Wright, B. (1979) *Nature*, **277**, 131.

84. Kilmartin, J.V., Wright, B. and Milstein, C. (1982) *J. Cell Biol.*, **93**, 576.

85. Croce, C.M., Linnenbach, A., Hall, W., Stteplewski, Z. and Koprowski, H. (1980) *Nature*, **288**, 488.

86. Kozbor, D., Lagarde, A. and Roder, J.C. (1982) *Proc. Natl. Acad. Sci. USA*, **79**, 6651.

87. Chiorrazzi, N., Wasserman, R.L. and Kunkel, H.G. (1982) *J. Exp. Med.*, **156**, 930.

88. Ritts, R.E., Ruiz-Arguelles, A., Weyl, K.G., Bradley, A.W., Weihmeir, B., Jacobsen, D.J. and Strehlo, B.L. (1983) *Int. J. Cancer*, **31**, 133.

89. Teng, N.N.H., Lam, K.S., Riera, F.C. and Kaplan, H.S. (1983) *Proc. Natl. Acad. Sci. USA*, **80**, 7308.

90. Abrams, P.G., Knost, J.A., Clarke, G., Wilburn, S., Oldham, R.K. and Foon, K.A. (1983) *J. Immunol.*, **131**, 1201.

91. Larrick, J.W., Truitt, K.E., Raubitschek, A.A., Senyk, G. and Wang, J.C.N. (1983) *Proc. Natl. Acad. Sci. USA*, **80**, 6376.

92. Emanuel, D., Golg, J., Colacino, J., Lopez, C. and Hammerling, U. (1984) *J. Immunol.*, **133**, 2202.

93. Ichimori, Y., Harada, K., Hitotsumachi, S. and Tsukamoto, K. (1987) *Biochem. Biophys. Res. Commun.*, **142**, 805.

94. Faller, G., Vollmers, H.P., Weiglein, I., Marx, A., Zink, C., Pfaff, M. and Muller-Hermelink, H.K. (1990) *Brit. J. Cancer*, **62**, 595.

95. Oi, V.T. and Herzenberg, L.A. (1980) in *Selected Methods in Cellular Immunology* (B.B. Mishel and S.M. Shiigi, eds). Freeman, San Francisco, p. 351.

96. Dawson, M. (1992) in *Cell Culture Labfax* (M. Butler and M. Dawson, eds). BIOS Scientific Publishers, Oxford, pp. 59, 75.

97. Spear, P.G. (1987) in *Cell Fusion* (A.E. Sowers ed.). Plenum Press, New York, p. 3.

98. Pontecorvo, G. (1975) *Somatic Cell Genet.*, **1**, 397.

99. Davidson, R.L. and Gerald, P.S. (1976) *Somatic Cell Genet.*, **2**, 165.

100. Bates, G.W., Saunders, J.A. and Sowers, A.E. (1987) in *Cell Fusion* (A.E. Sowers, ed.). Plenum Press, New York, p. 367.

101. Karsten, U., Papsdorf, G., Roloff, G., Stolley, P., Abel, H., Walther, I. and Weiss, H. (1985) *Eur. J. Cancer Clin. Oncol.*, **21**, 733.

102. Bischoff, R., Eisert, R.M., Schedel, I., Vienken, J. and Zimmermann, U. (1982) *FEBS Lett.*, **147**, 64.

103. Lo, M.M.S., Tsong, T.Y., Conrad, M.D., Strittmatter, S.M., Hester, L.D. and Snyder, S.H. (1984) *Nature*, **310**, 792.

104. Roder, J.C., Cole, S.P.C. and Kozbor, D. (1986) *Methods Enzymol.*, **121**, 140.

105. Spier, R.E., Griffiths, J.B., Stephenne, J. and Crooy, P.J. (1989) *Advances in Animal Cell Biology and Technology for Bioprocesses*. Butterworths, London.

106. Goding, J.W. (1986) *Monoclonal Antibodies: Principles and Practice (2nd edn)*. Academic Press, San Diego.

107. Goldenberg, D.M. (1987) *J. Cancer Res. Clin. Oncol.*, **113**, 203.

108. Sears, H.F., Herlyn, D., Streplewski, Z. and Koprowski, H. (1984) *J. Biol. Response Mod.*, **3**, 138.

109. Frankel, A.E., Houston, L.L. and Issel, B.F. (1986) *Ann. Rev. Med.*, **37**, 125.

110. Miller, R.A., Maloney, D.G., Warnke, R. and Levy, R. (1982) *New Engl. J. Med.*, **306**, 517.

111. Ortho Multicenter Transplant Study Group (1988) *New Engl. J. Med.*, **313**, 337.

112. Mackay, I.R. and Rowley, M.J. (1988) *Postgrad. Med. J.*, **64**, 522.

113. Shizuru, J.A., Alters, S.E. and Fathman, C.G. (1992) *Immunol. Rev.*, **129**, 105.

114. Riethmuller, G., Rieber, E.P., Kiefersauer, S., Prinz, J., van der Lubbe, P., Meiser, B., Breedveld, F., Eisenberg, J., Kruger, K., Deusch, K., Sanders, M. and Reiter, C. (1992) *Immunol. Rev.*, **129**, 81.

115. Riechmann, L., Clark, M., Waldmann, H. and Winter, G. (1988) *Nature*, **332**, 323.

116. Ozaki, S., Duram, S.K., Muegge, K., York-Jolley, J. and Berzofsky, J.A. (1988) *J. Immunol.*, **141**, 71.

117. Hyman, R. and Stallings, V. (1974) *J. Natl. Cancer Inst.*, **52**, 429.

118. White, J., Blackman, M., Bill, J., Kappler, J., Marrack, P., Gold, D.P. and Born, W. (1989) *J. Immunol.*, **143**, 1822.

119. Hyman, R., Cunningham, K. and Stallings, V. (1980) *Immunogenetics*, **10**, 261.

120. Hyman, R. and Stallings, V. (1976) *Immunogenetics*, **3**, 75.

121. Mohit, B. and Fan, K. (1971) *Science*, **171**, 75.

122. Hyman, R. and Trowbridge, I. (1981) *Immunogenetics*, **12**, 511.

123. Irigoyen, O., Rizzolo, P.V., Thomas, Y., Rogozinski, L. and Chess, L. (1981) *J. Exp. Med.*, **154**, 1827.

124. Okada, M., Yoshimura, N., Kaieda, T., Yamamura, Y. and Kishimoto, T. (1981) *Proc. Natl. Acad. Sci. USA*, **78**, 7717.

125. Foung, S.K.H., Sasaki, D.T., Grumet, C. and Engleman, E.G. (1982) *Proc. Natl. Acad. Sci. USA*, **79**, 7484.

126. Butler, J.L., Muraguchi, A., Lane, H.C. and Fauci, A.S. (1983) *J. Exp. Med.*, **157**, 60.

127. DeFreitas, E.C., Vella, S., Linnenbach, A., Zmijewski, C., Koprowski, H. and Croce, C.M. (1982) *Proc. Natl. Acad. Sci. USA*, **79**, 6646.

128. Le, J., Vilcek, J., Saxinger, C. and Prensky, W. (1982) *Proc. Natl. Acad. Sci. USA*, **79**, 7857.

129. Grillot-Courvalin, C. and Brouet, J.-C. (1981) *Nature*, **292**, 844.

130. Kobayashi, Y., Asada, M., Higuchi, M. and Osawa, T. (1982) *J. Immunol.*, **128**, 2714.

131. Platsoucas, C.D., Calvelli, T.A. and Kunicka, J.A. (1987) *Hybridoma*, **6**, 589.

132. McCullough, K.C. and Speir, R.E. (1990) *Monoclonal Antibodies in Biotechnology: Theoretical and Practical Aspects*. Cambridge University Press, Cambridge.

133. Lernhardt, W., Anderson, J., Coutinho, A. and Melchero, F. (1978) *Exp. Cell Res.*, **111**, 309.

134. Galfre, G. and Milstein, C. (1981) *Methods Enzymol.*, **73**, 3.

135. Hengartner, H., Luzzati, A.L. and Schreiner, M. (1978) *Curr. Top. Microbiol. Immunol.*, **81**, 92.

136. Berzofsky, J.A., Epstein, S.L. and Berkower, I.J. (1989) in *Fundamental Immunology (2nd edn)* (W.E. Paul, ed.). Raven Press, New York, p. 315.

137. Hay, R., Caputo, J., Chen, T.R., Macy, M., McClintock, P. and Reid, Y.V. (1992) *ATCC Catalogue of Cell Lines and Hybridomas (7th edn)*. ATCC, Rockville.

CHAPTER 6
ASSAYS OF CELLULAR ACTIVITY
J. McCulloch and S.J. Martin

In the immune system, as elsewhere in the body, cells become specialized to perform specific functions. Expression of effector function by a particular cell type generally arises as a result of interaction between its cell-surface receptors (membrane-bound immunoglobulins, T-cell receptors, complement receptors, cytokine receptors, etc.) and the appropriate ligand. Interactions of the correct specificity and affinity are followed by biochemical signals which communicate this event to the interior of the cell and initiate the appropriate cellular response. The cells of the immune system are among the most diverse cells in the body in terms of the range of functions they perform, from the scavenging macrophage which hunts down, engulfs and digests a wide range of prey, to the highly specialized T cell which will only respond to one particular MHC–peptide combination. In response to this challenge, immunologists have devised an equally diverse range of assays with which to measure these different functions.

1. MEASUREMENT OF EARLY ACTIVATION

One of the earliest measurable consequences of interaction between the surface receptors on a cell and their appropriate ligand is a rapid (within 30 sec) increase in intracellular ionized calcium $[Ca^{2+}]_i$ levels. This is accompanied by other biochemical reactions, such as the breakdown of phosphoinositides, predominantly mediated by phospholipase C (PLC), which leads to the generation of at least two active biochemical messengers, diacylglycerol (DAG) and inositol 1,4,5-triphosphate $[Ins(1,4,5)P_3]$. These and other signaling events also lead to the activation of a variety of protein kinases which phosphorylate proteins at either serine, threonine or tyrosine residues, thus changing the activity of these proteins. *Table 1* details a selection of assays that can be used as early indicators of cellular activation.

Measurement of increases in $[Ca^{2+}]_i$ levels is probably the most commonly used assay of early lymphocyte activation, and several detailed protocols exist for measuring such responses in either B or T cells by flow cytometry (1, 2). The up-regulation of specific cell-surface ligands (such as MHC class II or IL-2 receptor) upon activation is sometimes used as an indicator of cellular activation and can also be assessed by flow cytometry.

2. PHAGOCYTOSIS AND MICROBICIDAL ACTIVITY

Phagocytosis, the process whereby foreign (and also senescent self) material is ingested, is a relatively nonspecific immune reaction performed mainly by two classes of phagocytes: polymorphonuclear leukocytes (PMNs; primarily neutrophils) and mononuclear phagocytes (monocytes and macrophages). The ability to endocytose and digest foreign organisms/particles is an important property of phagocytes, both as a first line of defence against invading micro-organisms and also as a first step in the pathway leading to intracellular processing and subsequent presentation of components of the phagocytosed particle to T cells, thus triggering the adaptive immune response. Although PMNs and monocytes/macrophages are not the only cells capable of phagocytosis, these cells are particularly adapted to this task and are the focus of this discussion (see refs 5, 6 for more details).

Table 1. Assays of early activation

Assay	Method	Comment	Ref.
Increases in intracellular Ca^{2+} $[Ca^{2+}]_i$	Purified cells are labeled with a calcium-binding fluorescent dye. Cells are then activated and assessed for increases in $[Ca^{2+}]_i$ either at the single-cell level (flow cytometry), or at the population level (spectrofluorimetry)	Calcium-binding dyes available for flow cytometry are Indo-1, or where the cytometer is not equipped with a UV light source, Fluo-3. Indo-1 can also be used for spectrofluorimetry, as can Fura-2 or Quin-2 A Ca^{2+} ionophore such as ionomycin should be used as a positive control	1,2
Phosphoinositide breakdown	Cells are labeled with [^{3}H]inositol, treated with the activating stimulus, and are then analyzed for [^{3}H]inositol phosphate (InsP) formation by Dowex ion-exchange chromatography or HPLC	Dowex chromatography will resolve inositol monophosphate ($InsP_1$), $InsP_2$ and $InsP_3$, but not isomers of these HPLC is required to resolve $Ins(1,4,5)P_3$ from $Ins(1,3,4)P_3$	3
Protein phosphorylation analysis	Cells are metabolically labeled with [^{32}P]orthophosphate, lysed with detergent, and proteins separated by SDS-PAGE. Labeled proteins are then visualized by autoradiography. This procedure is carried out before and after activation in order to assess changes in the protein phosphorylation state	Further analysis can be used to determine whether serine, threonine or tyrosine residues are being phosphorylated, thereby indicating the category of protein kinase responsible	4

Abbreviations: HPLC, high-pressure liquid chromatography; UV, ultraviolet.

As a rule, phagocytes do not require antibody or complement as opsonins to enable them to recognize and engulf their targets (being in possession of an array of receptors which recognize a diverse set of ligands), but operate much more efficiently in their presence. Attachment of the opsonized particle to the phagocyte cell-surface receptors (Fc receptors, complement receptors, etc.) triggers the formation of pseudopods which surround the foreign particle in a zipper-like fashion (via receptor–opsonin interactions). When the surrounding pseudopods meet, they fuse together, forming a phagocytic vacuole, or phagosome, which then fuses with lysosomes in the phagocyte which discharge a battery of enzymes and antibacterial toxic oxygen species into the phagosome, thus killing and initiating digestion of the foreign particle. There are several *in vitro* assays available for measuring phagocytic capability, as listed in *Table 2*.

Table 2. Phagocytosis assays

Assay	Method	Comments	Ref.
Uptake of ^{14}C-radiolabeled *Staphylococcus aureus*	A defined amount of heat-killed, radiolabeled *S. aureus* bacteria are added to the phagocyte suspension in the presence of 10% autologous (or pooled AB as a second choice) serum and incubated under gentle rotation. Phagocytosis is stopped at defined times, phagocytes are pelleted by centrifugation, washed to remove bound noningested bacteria, and the amount of radioactivity associated with the cell pellets is expressed as a percentage of the total radioactivity associated with the same amount of radiolabeled bacteria as was introduced into the assay	Normal values for human neutrophils are 45–55% phagocytosis at a bacteria:neutrophil ratio of 10:1. This decreases as the ratio of bacteria to neutrophils increases The wash steps need to be carefully adhered to in order to minimize falsely high readings due to adherent but noningested bacteria Main drawback of this assay is the requirement for radiolabeled bacteria	7
Plating assay	Phagocytes are incubated under gentle rotation in a 1:1 ratio with bacteria in the presence of serum. At specified times, samples are withdrawn, the phagocytes pelleted by centrifugation, and the supernatants (containing the live uningested bacteria) are plated in serial 10-fold dilutions on agar plates. After an appropriate period of incubation the colonies are counted. The decrease in extracellular micro-organisms over time can then be calculated	Can be used with many strains of bacteria Some laboratories may not wish to grow bacterial cultures due to possibility of contamination of cell lines	8
Morphological assessment	Live or heat-killed bacteria are incubated with phagocytes as described in the plating assay. Samples of the cell suspension are withdrawn at defined times, cytospin preparations are made, then fixed and stained with Giemsa. The percentage of cells which have ingested bacteria can then be directly assessed under microscopy Yeasts such as *Candida albicans* can be substituted for bacteria	A minimum of 200 cells should be counted. The mean number of bacteria phagocytosed per cell can also be calculated (counting only cells which have phagocytosed) Cells which have bound but not ingested micro-organisms should be rigorously excluded from the count.	8,9

Assays

Table 2. Continued

Assay	Method	Comments	Ref.
Nitroblue tetrazolium (NBT) uptake	NBT dye is added to a neutrophil suspension in the presence of plasma and endotoxin as a source of neutrophil stimulant. Stimulated neutrophils incorporate the dye into phagosomes. The dye is reduced upon lysosomal fusion to form insoluble blue crystals of formazan. Cytospins are then made of the cell suspension	The percentage of phagocytic cells may be counted directly by microscopy. This assay is also a measure of hexose monophosphate shunt activation (oxidative-dependent microbicidal pathway)	9

In PMNs, the interaction between opsonized micro-organisms and the phagocyte membrane, apart from leading to the uptake of the organism by phagocytosis as discussed above, also produces a respiratory burst which results in the production of several oxygen-containing species with microbicidal (microbe-killing) activity. These include, the superoxide anion (O_2^-), singlet oxygen ($'O_2$) and hydroxyl radicals (OH·). The phagocyte is protected from the toxic effects of the reactive oxygen species it produces by the enzyme superoxide dismutase (SOD) and the glutathione redox cycle.

In addition to this oxygen-dependent microbicidal system there is also oxygen-independent microbicidal activity (e.g. lysozyme, lactoferrin, cationic proteins, and the low pH within the lysosome) within PMNs. Very few micro-organisims can survive the combined assault of these potent antimicrobials, but there are one or two notable exceptions (e.g. some mycobacteria and *Salmonella* species).

Although similar enzymes to those in PMNs have also been described in monocytes and macrophages, the exact mechanisms whereby these cells kill ingested micro-organisms are still unclear (10).

Assays used for the determination of microbicidal activity in phagocytes are described in *Table 3*.

Table 3. Assays of microbicidal activity

Assay	Method	Comments	Ref.
Plating method utilizing lysostaphin	Neutrophils are incubated with live staphylococci under rotation in the presence of serum. At a set time, lysostaphin is added to all tubes (to kill extracellular bacteria) and incubation is continued for defined times (to allow intracellular killing to proceed). The various neutrophil samples are then hypotonically lysed to release ingested bacteria, which are then plated at various dilutions on agar to allow growth of bacteria which have survived ingestion	Count colonies on plates with less than 500 colonies % survival is calculated against a neutrophil-free control tube (100% survival) Normal values are 93–98% killing in 60–90 min for human neutrophils	8
Differential centrifugation method	Where lysostaphin cannot be used (i.e. with bacteria other than staphylococci) the uningested bacteria have to be removed before plating by differential centrifugation. The other details are similar to the lysostaphin protocol		7

Assays

3. CHEMOTAXIS AND CHEMOKINESIS

A variety of substances (chemoattractants) can stimulate directional (chemotactic) and nondirectional (chemokinetic) locomotor activity in immune cells. Immune-cell migration is important for mobilizing phagocytes to inflammatory sites, as well as facilitating interactions between antigen-presenting cells and T and B lymphocytes.

In *chemotaxis*, the direction of locomotion is determined by the direction and magnitude of the chemoattractant, which is usually in the form of a concentration gradient. In leukocytes, chemotaxis is always positive, i.e. towards the gradient source (11).

In *chemokinesis*, the speed of locomotion and/or the frequency of turning is determined by the magnitude of the stimulus, but locomotor activity is random, i.e. although an individual cell in a population may move in a consistent direction, the cells of the population move in directions that are random to one another. More precise definitions of these terms can be found elsewhere (11, 12).

Chemoattractants are diverse substances, ranging from breakdown products of bacterial cell walls, to fragments of complement proteins, cytokines and neuropeptides. Gradients of these substances are 'sensed' via specific cell-surface receptors on the responding cells and translated into locomotor activity by as yet incompletely understood mechanisms (13). Some of the many substances with known chemotactic/chemokinetic activity are listed in *Table 4*.

Table 4. Leukocyte chemoattractants

Neutrophil	Monocyte/macrophage	Lymphocyte
C5a	C5a	Interleukin-2 (IL-2)
C5a des Arg	C5a des Arg	Interferon-β (IFN-β)
Formyl-Met-Leu-Phe (FMLP)	FMLP	Anti-CD4
Leukotriene-B$_4$ (LTB$_4$)	LTB$_4$	Anti-CD3
Platelet activating factor (PAF)	PAF	Cyclic guanine monophosphate (cGMP)
Substance P	Substance P	Anti-immunoglobulin (anti-Ig)
Interleukin-8 (IL-8)	Tumor necrosis factor-α (TNFα)	IL-8
Met-enkephalin	Tumor necrosis factor-β (TNFβ)	
β-endorphin	Transforming growth factor-β (TGFβ)	
	Met-enkephalin	
	β-endorphin	

Neutrophils are the most mobile immune cells and are therefore usually the first to arrive at a site of inflammation, followed by monocytes and then lymphocytes. Both neutrophils and monocytes will readily migrate on two-dimensional surfaces *in vitro*, whereas lymphocytes migrate poorly under these conditions. It should also be noted that immune cell populations are heterogeneous in their responses to chemoattractants, for example, studies have shown that only 50% of peripheral blood neutrophils, and even fewer monocytes, are chemotactically responsive *in vitro*.

Several methods are available for the study of locomotor activity of leukocytes during exposure to chemoattractants. These assays fall into two categories:

(i) Population assays, where a large population of cells is allowed to move (generally through a porous membrane or under agarose) in the presence or absence of a chemotactic stimulus. After an arbitrary time interval the assay is stopped and the distribution of the cell population is examined.

(ii) Single-cell assays, which involve the use of video or time-lapse microscopy to observe the behavior of single cells in terms of their speed, direction of movement, etc., in gradients or different concentrations of attractants.

Both assay types have their strengths and weaknesses, but for the investigator who is not likely to want to conduct such investigations on a frequent basis, the population assays are probably the simplest and less technically demanding (in terms of specialized equipment, etc.) to perform. *Table 5* outlines some of the available population assay methods, details on the single-cell assay methods can be found elsewhere (14).

4. PROLIFERATION

Upon stimulation by antigen in the presence of the appropriate cytokine growth factors, resting lymphocytes undergo a period of cell division before differentiating to effector or memory cells. This proliferative phase can be used to study stimulation of B and T lymphocytes in two ways:

(i) as a measure of responsiveness to a specific antigen (see Chapter 5, *Table 11*), or

(ii) as a general measure of immunocompetence (mitogenesis assays).

Some general guidelines for lymphocyte proliferation assays are outlined in *Table 6*. The reader is advised to consult source references 19–21 for detailed protocols.

When creating a proliferation assay to measure responses to specific antigen, pilot studies should be performed to optimize conditions for the particular antigen/cell type under investigation. Special attention should be given to the selection of fetal calf or human serum for these assays as there is often considerable batch-to-batch variation. Consequently, samples from several batches should be tested to ensure low background proliferation and high mitogen-stimulated proliferative responses. This is vital for the establishment of a sensitive and specific assay system. It is also important to carry out several replicates of each test and control well (6–12 ideally) since lymphocyte responses of low precursor frequency (see Section 9) can produce large variations between supposedly identical wells under the conditions commonly used in such assays.

Polyclonal stimulation assays are used to measure overall immune potential. A variety of mitogenic stimulants can be used to drive lymphocyte proliferation in a polyclonal fashion. Some examples are shown in Chapter 4, *Table 15*.

5. CYTOKINE PRODUCTION

Cytokines are a diverse group of protein mediators produced during adaptive and innate immune reponses. They act upon, and are produced by, a variety of cell types, not solely

Table 5. Chemotaxis/chemokinesis assays

Assay	Method	Comments	Ref.
Micropore filter assay	Two chambers, separated by a porous filter, are filled with cells (top chamber) and chemoattractant (bottom) or control buffer. The attractant diffuses up through the filter, thus setting up a concentration gradient, which the cells respond to by migrating towards this through the pores in the filter. The assay is stopped by fixing the filter, followed by staining the cells within and scoring the assay by microscopy	The experiment is scored in various ways, the most common of which is by estimating the distance traveled by the leading front of cells (two or more cells), or counting the number of cells which have passed a fixed point in the filter Different filter pore sizes are required, depending on the cell type being used in the assay. For neutrophils this is ~ 3 µm and for monocytes ~ 8–12 µm	14–16
Agarose assay	Gelatin-coated glass slides are overlaid with a 1% agarose solution. Three holes are punched in the agarose equidistant to each other, into which is pipetted the cell suspension (middle), with the chemoattractant and a buffer control in the wells at either side. After a 2 h incubation at 37°C the slides are fixed, the agarose removed and the cells are stained	The distance of migration of the leading front of cells towards the attractant is compared with their distance of migration towards the control buffer Time-lapse films of cells migrating under the agarose can also be taken	14,17
Orientation assay	Uses a Zigmond chamber (a Perspex slide cut with two 1 mm deep × 5 mm wide wells separated by a 1 mm wide bridge). One well is filled with chemoattractant and the other with buffer. A glass coverslip, to which neutrophils (or other cells) have been allowed to attach, is placed cell-side down on to the chamber, thus creating by capillary action a concentration gradient across the bridge separating the two wells. The cells attached to the coverslip are therefore in contact with this gradient. The chamber is incubated for 30–40 min at 37°C, after which the cells are examined under phase contrast for their direction of orientation in the gradient	Simple and rapid assay to perform The ideal gap between the coverslip and the bridge is 10 µm Orientation is scored by counting 100–200 cells and assessing the percentage polarized either towards or away from the gradient of attractant. Unpolarized cells and cells oriented at 90° to the gradient are disregarded	14,18

CELLULAR IMMUNOLOGY LABFAX

Table 6. General criteria for proliferation assays

Parameter	Comment
Cell number per well	1×10^5
Well type	Flat- or round-bottomed
Serum	Human cells: 10% AB serum or 10% autologous serum Mouse cells: 5–10% FCS
Medium	RPMI 1640 with 2 mM L-glutamine, 100 U ml^{-1} penicillin, 100 μg ml^{-1} streptomycin sulfate $\pm$ 50 μM 2-mercaptoethanol (mouse cells generally)
Culture volume/well	200 μl
Culture period	1–7 days, 37°C, 5% CO_2, with humidity
[^{3}H]thymidine per well	0.1–1.0 μCi (37×10^2–37×10^3 Bq)
[^{3}H]thymidine pulse time	6–18 h
Accessory cells/well	2:1–4:1 ratio of effector cells to mitomycin C- or γ-radiation-inactivated (see Chapter 4, *Table 11*) autologous mononuclear cells or EBV-transformed B cells, if using purified T cells
Harvesting samples	Use automatic cell harvester to collect cells on to glass-fiber mats, measure [^{3}H]thymidine incorporation by liquid scintillography
Calculation of results	Calculate arithmetic mean c.p.m. for experimental replicates and control replicates, express results as (experimental-control) c.p.m. or as stimulation indices (experimental c.p.m./control c.p.m.)

cells of the immune system. Many cytokines exhibit functional redundancy, i.e. different cytokines can induce the same effects *in vitro*. It remains to be seen whether this truly reflects their role *in vivo*, but this can cause confusion in bioassays of cytokine release. Hence, when assaying cytokines in body fluids or culture supernatants it is preferable to use a dual approach:

(i) use of specific monoclonal antibodies (mAbs) to quantitate the cytokine of interest (ELISA); and
(ii) assaying the ability of the test sample to support the growth of a cytokine-dependent cell line (bioassay).

Table 7 lists a selection of cell lines/types commonly used to measure cytokines in biological fluids.

Table 7. Cytokine bioassays

Cytokine	Cell line/type	Suitability	Comments
IL-1α,β	Thymocytes	Murine	Co-stimulation with PHA
	D10.G4.1	Murine	Co-stimulation with anti-CD3
IL-2	CTLL-2	Murine and human	Use with blocking anti-IL-4 mAb in murine case
IL-3	FDC-P2, FD.C/1, 32D, MC/9	Murine	Lines have absolute IL-3 dependence
IL-4	CT.4S	Murine	Specific for IL-4
IL-5	BCL$_1$	Murine and human	Check specificity with anti-IL-5 Ab (especially in human IL-5 assays)
IL-6	B9	Murine and human	Block with anti-IL-4 Ab in murine assay
IL-7	IxN/2bx	Murine and human	Line has absolute dependency on IL-7
IL-8	Neutrophils	Murine and human	Compare chemotaxis $\pm$ anti-IL-8 blocking Ab
IL-9	M-07e	Murine and human	Use $\pm$ anti-IL-9 Ab (both cases), also $\pm$ anti-GM-CSF and anti-SCF Abs in human case
IL-10	LB2-1 T$_H$1 cells	Murine and human	Assay supernatants for IFN-γ release, may have to add anti-IL-4 and anti-TGFβ Abs in murine case
IL-11	T10	Murine and human	Compare $\pm$ anti-IL-11 and anti-IL-6 blocking Abs
IL-12	PHA Blasts	Murine and human	Compare with IL-12-depleted samples and/or Ab-immobilized IL-12 (human)
IFN-γ	WEHI-3	Murine	Class II MHC expression $\pm$ sample $\pm$ anti-IFN-γ Ab
	COLO-205	Human	
TNFα,β	L929	Murine and human	Use $\pm$ anti-TNFα,β blocking Abs

SCF, stem cell factor.

In most cases cytokine bioassays measure proliferative responses (as assessed by [^{3}H]-thymidine incorporation, see Section 4) of the cytokine-dependent cell line in response to serial dilutions of the test sample. These assays are calibrated by stimulating the cells with defined amounts of recombinant or standard cytokines, the results of which are used to construct a calibration curve from which the test sample values are extrapolated (e.g. see ref. 22). Some cell lines can also be induced to proliferate by unrelated cytokines to the cytokine being assayed, hence specificity should always be checked by blocking with antibodies specific for the cytokine of interest and its functional mimic. A few cytokine assay cell lines with absolute specific dependency are available and in these cases the use of blocking antibodies is less crucial. The reader should also note that for several cytokines (often human) there are no generally available dependent cell lines. Here one has to rely on sensitive antigen-capture ELISA systems which are commercially available (but generally expensive).

6. ANTIBODY SECRETION

B lymphocytes function primarily to secrete immunoglobulins (Ig). Accordingly, the most relevant measure of B-cell activity is detection of secreted Ig. The researcher can employ several methods to detect secreted Ig (see ref. 19 for practical details). Hemolytic plaque assays and enzyme-linked immunospot (ELISPOT) can be used to analyze Ig secretion by *individual* B cells *in vitro*, while quantitation of Ig secreted by B-cell *populations* is best achieved using an ELISA-based method. Finally, determination of Ig secretion profiles *in vivo* can be achieved by reliance upon intracytoplasmic Ig measurement of B cells in histological sections. These methods are summarized in *Table 8*.

Table 8. Assays for Ig secretion

Assay	Comment
Direct hemolytic plaque assay (HPA)	Used to enumerate Ig-secreting B cells. Antigen-coated sheep red blood cells (SRBCs) are used as indicators. Secreted specific Ig binds to indicator SRBCs and causes them to lyse in the presence of C'. Restricted mainly to IgM. Can subsequently analyze plaque-forming cells
Indirect HPA	As above, but can also be used to analyze other Ig isotypes. Achieved by use of anti-isotype serum which fixes C' well. Hence, can simultaneously analyze antigen specificity and isotype. May suffer from lack of sensitivity and specificity
Reverse HPA	As above but uses protein A-coated SRBCs. Anti-isotype serum binds secreted Ig and also has strong C'-fixing activity and affinity for protein A. This variant is solely for determination of secreted Ig isotype
ELISPOT	Used to enumerate Ig-secreting B cells, and to assess their specificity and isotype. B cells are cultured in antigen-coated plates. Detection of secreted Ig is made using anti-Ig-enzyme conjugates. Zones of specific Ig release around B cells then appear as color spots. This method is sensitive and straightforward but no analysis of the plaque-forming cell can be made after the assay
ELISA	For the detection of secreted Ig in B-cell cultures or body fluids. Here used for detecting specific antibody, to quantitate the amount of secreted Ig and its isotype (Ig capture assay)
Intracytoplasmic Ig measurement	To determine the isotype and/or specificity of B cells in a histological section or cytospin preparation. Can use antigen or anti-Ig directly conjugated to FITC (direct assay) or to biotin (indirect assay). The indirect approach allows better appreciation of histology and specific staining

7. CYTOTOXICITY

Cytotoxicity, either self-MHC-restricted or natural, is a central immune function in the limitation and clearance of intracellular pathogens. Antigen-specific, self-MHC-restricted cytotoxicity is mediated by cytotoxic T lymphocytes (CTLs). Most usually $CD3^+CD4^-CD8^+$ cells, these effectors recognize an endogenously processed antigenic peptide in association with class I MHC. In addition, CTLs can effectively lyse cells bearing allogeneic (different individuals of the same species) MHC (alloreactivity), as measured in the mixed lymphocyte reaction (MLR). Natural cytotoxicity is mediated by large granular lymphocytes (LGL), primarily of the $CD3^-CD16^+CD56^+$ phenotype. It is not yet known how LGLs recognize their target cells. LGLs can lyse targets in various ways:

(i) natural killer (NK) cells can lyse a wide variety of syngeneic, allogeneic and even xenogeneic targets without prior stimulation;
(ii) lymphokine-activated killer (LAK) cells have similar lytic potential but require *in vitro* expansion with IL-2 before reaching effector function; and
(iii) antibody-dependent cellular cytotoxic (ADCC) or killer (K) cells are LGLs which lyse IgG-coated targets after triggering by FcRIII (CD16) receptor occupancy. *Table 9* summarizes the main characteristics of these various forms of cellular cytotoxicity.

Assays for cytotoxicity are many and varied, and some of the common features are described below (see ref. 23 for more details). Typically, cells are cultured at various effector to target (E:T) ratios with a constant number of ^{51}Cr-labeled targets. After the appropriate incubation period, supernatant samples are taken from each well. The effectiveness of killing is reflected in the proportion of ^{51}Cr released by lysed targets. As controls, spontaneous release (SR) is

Table 9. Features of cellular cytotoxicity assays

Feature	CTL-Ag	CTL-allo	NK	LAK	ADCC
In vitro stimulation	+	±	−	−	−
In vitro expansion with Ag + IL-2	+	±	−	−	−
In vitro expansion with IL-2 only	−	−	−	+	−
$CD3^+$ $CD4^-$ $CD8^+$	+	+	−	−	−
$CD3^-$ $CD16^+$ $CD56^+$	−	−	+	+	+
Self-MHC-restricted cytolysis	+	−	−	−	−
Allogeneic cytolysis	−	+	+	+	+
Xenogeneic cytolysis	−	−	+	+	+
6–18 h assay	+	+	−	−	−
≥ 4 h assay	−	−	+	+	+
IgG-opsonization required	−	−	−	−	+

Note: CTL-Ag = antigen-specific, MHC-restricted CTL assay.
  CTL-allo = alloreactive CTL assay (MLR).

determined by culturing targets in medium alone, whereas total possible release (TR) is recorded by lysing targets with a detergent such as Triton X-100. After counting γ activity in each sample (c.p.m. or d.p.m.), the results are expressed as either percentage specific lysis or lytic units (LU). Specific lysis takes account of the experimental release (ER) in comparison to controls and is calculated as follows:

$$[(ER - SR)/(TR - SR)] \times 100.$$

Lytic units (LU) are obtained by plotting percentage specific lysis vs. log of effector cell number. The region of the curve which equates to approximately 30% specific lysis should be straight. In this case the effector cell number at this point is 1 LU. Express the number of LU per 10^6 cells, and mention the percentage specific lysis extrapolation value.

8. HELPER FUNCTION

T cell help for Ig production involves two principal factors; T cell–B cell contact and lymphokine secretion. Cell contact permits presentation of an antigenic peptide in the context of class II MHC by the B cell, recognition of this complex by the TCR on the T cell, and accessory signals from an array of adhesion molecule interactions on both cells. Although direct cellular juxtaposition may be critical for the initiation of specific Ig production from resting B cells, not all of the factors involved in this process have been established. Therefore in order to measure T helper (T_H) cell activity it is more practical to quantitate B-cell growth and differentiation factors released by the T_H cell. Supernatants from stimulated T_H-cell cultures can also be used to drive Ig synthesis by polyclonally activated B-cell cultures. This section briefly outlines approaches that can be employed to measure T_H activity in either way.

8.1. T_H lymphokine secretion

Various lymphokines have been shown to play a role in help for Ig production, notably IL-2, IL-4, IL-5 and IL-6 (24). Supernatants from cultures of T_H cells stimulated with γ-irradiated or mitomycin C-treated APCs plus antigen can be assayed via capture ELISA or cytokine bioassay (see Section 5) in order to quantitate these lymphokines.

8.2. Supernatant helper function

T_H lines are stimulated as above and the supernatants kept. A B-cell culture is then stimulated with suboptimal amounts of anti-Ig or LPS (see Section 4), a range of dilutions of supernatant from the T_H-cell culture are added to this and Ig production is assessed by capture ELISA (see Section 6).

9. LIMITING DILUTION

Limiting-dilution analysis (LDA) can be used to estimate the frequency of cells in a population which are capable of responding to an activation signal (precursor frequency) either by proliferation (or IL-2 release), antibody secretion or generation of cytotoxic T-cell activity. In previous sections, assays were described for measuring these cellular functions in bulk cultures (Sections 5–8) but these methods do not give any indication of the actual number of cells participating in the response. LDA is a relatively simple yet powerful tool for investigating both T- and B-lymphocyte repertoires.

In general, it can be said that the higher the precursor frequency of a lymphocyte clone with a particular epitope specificity, the more potent the immune response will be to this epitope. Therefore, lymphocyte clones of low precursor frequency will frequently be missed in conventional bulk culture assays (2×10^5 cells/well) unless a large number of replicate wells are used.

ASSAYS OF CELLULAR ACTIVITY

Several methods exist for the measurement of precursor frequencies, depending on the effector cell function one wishes to measure, and the reader is referred to several detailed protocols which have been published elsewhere (25–27). Whichever protocol is used, the basic technique involves setting up many replicate cultures (24–96) of the cell population containing the responder cells to be measured, at several cell concentrations (as a general starting point, concentrations of 5, 4, 3, 2, 1, 0.75, 0.5 and 0.25×10^4 cells/well can be used). Each well also contains a fixed number (5×10^4) of feeder cells (e.g. irradiated PBMC). Parallel cultures are set up in the presence and absence of the antigen (or target cell) of interest and the cultures are incubated for a period of time (3–10 days, again depending on the assay type and the responder cell being measured), after which the proportion of non-responding or negative wells is calculated for each dilution of the responder cell population (since a positive well may be the product of one or more cells).

The fraction of negative wells (converted to its negative log) is then plotted against the responder cell concentration (*Figure 1*). Statistical methods such as regression analysis or the least squares method are then used to fit a straight line to this graph, and using the zero term of the Poisson equation:

$$F_0 = e^{-u}$$

where F_0 is the fraction of negative wells, e the base of the natural logarithim and u the average number of precursors per well. Where u is 1, we have:

$$F_0 = e^{-1} = 1/e = 0.37.$$

We can predict that when 37% (0.37) of the test wells are negative, there is on average one precursor cell per well. Thus the frequency of precursors specific for a given antigen in a cell population can be extrapolated directly from the graph, as shown (*Figure 1*). Several programs exist for carrying out this type of analysis (see ref. 25, for example).

10. APOPTOSIS

Assays for the measurement of cell death may seem out of place in a section on cellular activity. However, apoptosis has several important features which sets it apart from the passive mode

Figure 1. Semilog plot of cell input vs. the fraction of nonresponder culture wells. The frequency of responder cells can be interpolated from $F_0 = 0.37$. From the plot we can estimate the frequency of responders as 1 in 2×10^4 cells.

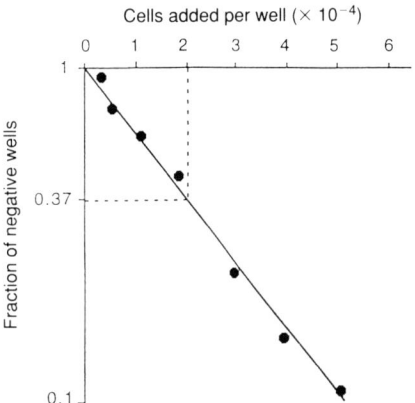

of cell death, necrosis. Apoptosis is often referred to as programmed cell death (PCD), which is not strictly correct since not all cells undergoing PCD exhibit the characteristic features of apoptosis. Apoptosis is commonly observed when death is a desirable event; during tissue turnover, deletion of autoreactive thymocytes, and NK or CTL killing, for example (28). Necrosis, on the other hand, is generally the result of a major cellular insult which provokes severe membrane damage and rapidly leads to uncontrolled cell swelling followed by lysis. Cells undergoing apoptosis often, but not always, require ongoing RNA and/or protein synthesis and possess intact and functional organelles. Thus, apoptotic cell death can be loosely described as an active mode of cell death. Therefore, when measuring cell death it is important to be able to discriminate between cells dying via apoptosis or necrosis. *Table 10* summarizes the main differences between apoptotic and necrotic cells and how these can be assayed. Further details of dyes used in the measurement of apoptosis can be found in Chapter 4 (*Table 8*).

Apoptosis is most reliably quantitated on the basis of cellular morphology. This can be done either on cytospun cell preparations (for single cell suspensions) or tissue sections, stained with hematoxylin and eosin or a similar differential stain. Under light microscopy, the most obvious feature distinguishing apoptotic cells is their highly condensed (intensely stained) and fragmented nuclei (*Figure 2*).

Table 10. Features of apoptosis and necrosis

Feature	Apoptosis	Necrosis	Assay method	Ref.
Loss of plasma membrane integrity	Slow (hours)	Rapid (minutes)	Vital dye (trypan blue, propidium iodide) exclusion assays on hemocytometer or flow cytometer	29,30
Cell body	Initially shrunken, then fragmented (apoptotic bodies)	Initially swollen, then ruptured	Phase-contrast microscopy, LM or EM	31
Cytoplasm	Condensed	Swollen	Density gradient centrifugation	32
Organelles	Intact	Disrupted	EM, or specific functional assay (e.g. MTT)	33,34
Nucleus	Initially condensed, then fragmented	Swollen	LM on H&E-stained cell preparations, or EM in the case of small cells (e.g. resting lymphocytes)	31,33,34
Chromatin	Highly condensed (pyknosis)	Flocculated	LM or EM	31,33
DNA	Cleaved into 200 bp (internucleosomal) multiples	Nonspecific pattern of degradation	Electrophoresis of DNA in 1–2% agarose or DNA content analysis on flow cytometer	30,35

Abbreviations: LM, light microscopy; EM, electron microscopy; MTT, 3-(4,5-dimethylthiazol-2-yl)-2,5-diphenyltetrazolium bromide; H&E, hematoxylin and eosin.

Assays

Figure 2. Morphology of human leukemia HL-60 cells undergoing apoptosis. Note the condensed (darkly staining) and fragmented nature of the nucleus in several cells, as well as the presence of small cell fragments containing pieces of condensed chromatin (termed apoptotic bodies). These features are typical of apoptotic cell death.

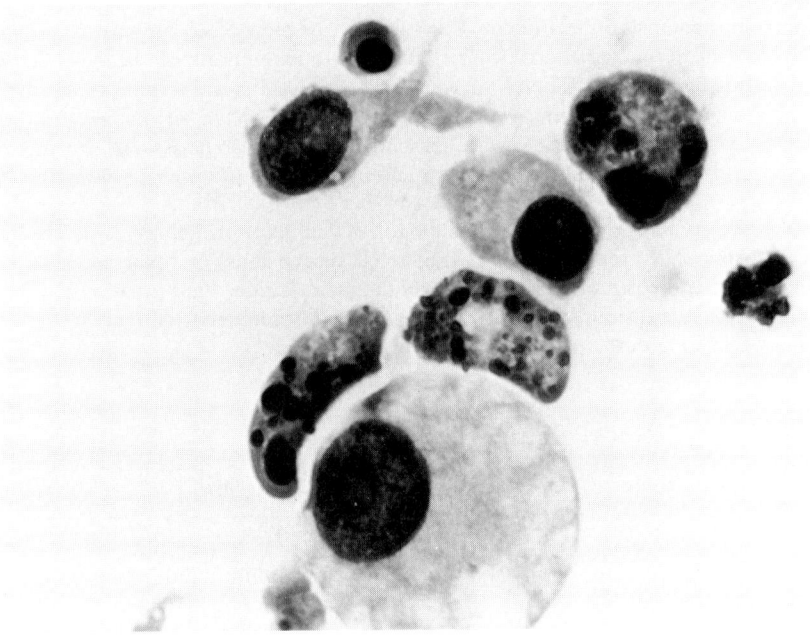

Since morphological assessment can be both subjective and tedious, investigators are currently searching for biochemical parameters with which to measure apoptosis in a more objective and quantitative manner. To date, most methods of this type rely upon detecting the DNA fragmentation that occurs during apoptosis. *Table 11* summarizes some of the methods in use.

Table 11. Biochemical assays for the measurement of apoptosis

Method	Parameter measured	Comments	Ref.
Diphenylamine binding (Burton method)	DNA fragmentation	Reliable but fairly long and tedious	30,36
[³H]thymidine release	DNA fragmentation	Rapid, important to do several replicates (4–6), but only works with cycling cells	30
Nick-translation in conjunction with flow cytometry	DNA strand breakages	Highly quantitative but technically quite complex. May get false positives due to detection of single-strand breaks	37
Hoechst 33342 uptake	Discrete increase in plasma membrane permeability	Rapid and highly quantitative but requires specialized flow cytometry equipment	38
Cell cycle analysis	Decrease in DNA content as assessed by propidium iodide binding in conjunction with flow cytometry	Rapid, quantitative, demonstrated to work on many cell types. Important to pre-incubate fixed cells in PBS before staining	39,40

11. REFERENCES

1. Mond, J.J. and Brunswick, M. (1991) in *Current Protocols in Immunology* (J.E. Coligan, A.M. Kruisbeek, D.H. Margulies, E.M. Shevach and W. Strober, eds). John Wiley & Sons, New York, p. 3.9.1.

2. June, C.H. and Rabinovitch, P. S. (1991) in *Current Protocols in Immunology* (J.E. Coligan, A.M. Kruisbeek, D.H. Margulies, E.M. Shevach and W. Strober, eds). John Wiley & Sons, New York, p. 5.5.1.

3. Imboden, J.B., Shoback, D. M. and Inokuchi, S. (1991) in *Current Protocols in Immunology* (J.E. Coligan, A.M. Kruisbeek, D.H. Margulies, E.M. Shevach and W. Strober, eds). John Wiley & Sons, New York, p. 11.01.

4. Siegel, J. N. (1991) in *Current Protocols in Immunology* (J.E. Coligan, A.M. Kruisbeek, D.H. Margulies, E.M. Shevach and W. Strober, eds). John Wiley & Sons, New York, p. 11.2.1.

5. Verhoef, J. (1992) in *Encyclopedia of Immunology* (I.M. Roitt and P.J. Delves, eds). Academic Press, London, p. 1220.

6. Roos, D. (1992) in *Encyclopedia of Immunology* (I.M. Roitt and P.J. Delves, eds). Academic Press, London, p. 1163.

7. Clark, R.A. and Nauseef, W.M. (1991) in *Current Protocols in Immunology* (J.E. Coligan, A.M. Kruisbeek, D.H. Margulies, E.M. Shevach and W. Strober, eds). John Wiley & Sons, New York, p. 7.23.4.

8. Leijh, P.C.J., Van Furth, R. and Van Zwet, T.L. (1986) in *Handbook of Experimental Immunology* (D.M. Weir, L.A. Herzenberg, C. Blackwell and L.A. Herzenberg, eds). Blackwell Scientific Publications, Oxford, Vol. 2, p. 46.1.

9. Hudson, L. and Hay, F.C. (1989) in *Practical Immunology (3rd edn)*. Blackwell Scientific Publications, Oxford, p. 187.

10. Van Furth, R. (1992) in *Encyclopedia of Immunology* (I.M. Roitt and P.J. Delves, eds). Academic Press, London, p. 1092.

11. Wilkinson, P.C. (1992) in *Encyclopedia of Immunology* (I.M. Roitt and P.J. Delves, eds). Academic Press, London, p. 329.

12. Keller, H.U., Wilkinson, P.C., Abercombie, M., Becker, E. L., Hirsch, J.G., Miller, M.E., Ramsey, W.S. and Zigmond, S.H. (1977) *Clin. Exp. Immunol.*, **27**, 377.

13. Harvath, L. (1992) in *Encyclopedia of Immunology* (I.M. Roitt and P.J. Delves, eds). Academic Press, London, p. 1097.

14. Wilkinson, P.C. (1986) in *Handbook of Experimental Immunology* (D.M. Weir, L.A. Herzenberg, C. Blackwell and L.A. Herzenberg,

eds). Blackwell Scientific Publications, Oxford, Vol. 2, p.51.1.

15. Leonard, E.J., Sylvester, I. and Yoshima, T. (1991) in *Current Protocols in Immunology* (J.E. Coligan, A.M. Kruisbeek, D.H. Margulies, E.M. Shevach and W. Strober, eds). John Wiley & Sons, New York, p. 6.12.3.

16. Boyden, S.V. (1962) *J. Exp. Med.*, **115**, 453.

17. Wilkinson, P.C. (1972) *Chemotaxis and Inflammation*. Churchill-Livingstone, Edinburgh.

18. Zigmond, S.H. (1977) *J. Cell Biol.*, **75**, 606.

19. Mond, J.J. and Brunswick, M. (1991) in *Current Protocols in Immunology* (J.E. Coligan, A.M. Kruisbeek, D.H. Margulies, E.M. Shevach and W. Strober, eds). John Wiley & Sons, New York, p. 3.8.1.

20. Kruisbeek A.M. and Shevach E. (1991) in *Current Protocols in Immunology* (J.E. Coligan, A.M. Kruisbeek, D.H. Margulies, E.M. Shevach and W. Strober, eds). John Wiley & Sons, New York, p.3.12.1.

21. James, S.P. (1991) in *Current Protocols in Immunology* (J.E. Coligan, A.M. Kruisbeek, D.H. Margulies, E.M. Shevach and W. Strober, eds). John Wiley & Sons, New York, p. 7.10.1.

22. Bottomly, K., Davis, L. S. and Lipsky, P.E. (1991) in *Current Protocols in Immunology* (J.E. Coligan, A.M. Kruisbeek, D.H. Margulies, E.M. Shevach and W. Strober, eds). John Wiley & Sons, New York, p. 6.3.1.

23. Wunderlich, J. and Shearer, G. (1991) in *Current Protocols in Immunology* (J.E. Coligan, A.M. Kruisbeek, D.H. Margulies, E.M. Shevach and W. Strober, eds). John Wiley & Sons, New York, p. 3.11.1.

24. Fitch, F.W. (1992) in *Encyclopedia of Immunology* (I.M. Roitt and P.J. Delves, eds). Academic Press, London, p. 656.

25. Waldmann, H., Cobbold, S. and Lefkovits, I. (1987) in *Lymphocytes: a Practical Approach* (G.G.B. Klaus, ed.). IRL Press, Oxford, p. 163.

26. Miller, R.A. (1991) in *Current Protocols in Immunology* (J.E. Coligan, A.M. Kruisbeek, D.H. Margulies, E.M. Shevach and W. Strober, eds). John Wiley & Sons, New York, p. 3.15.1.

27. Sharrock, C.E.M., Kaminski, E. and Man S. (1990) *Immunol. Today*, **11**, 281.

28. Duvall E. and Wyllie, A.H. (1986) *Immunol. Today*, **7**, 115.

29. Lennon, S.V., Martin, S.J. and Cotter, T.G. (1991) *Cell Prolif.*, **24**, 203.

30. Duke, R.C. and Cohen, J.J. (1991) in *Current Protocols in Immunology* (J.E. Coligan, A.M. Kruisbeek, D.H. Margulies, E.M. Shevach and W. Strober, eds). John Wiley & Sons, New York, p.3.17.1.

31. Martin, S.J. and Cotter, T.G. (1991) *Int. J. Radiat. Biol.*, **59**, 1001.

32. Martin, S.J., Lennon, S.V., Bonham, A. and Cotter, T.G., (1990) *J. Immunol.*, **145**, 1859.

33. Wyllie, A.H. (1981) in *Cell Death in Biology and Pathology* (I.D. Bowen and R.A. Lockshin, eds). Chapman and Hall, London, p.9.

34. Shi, Y., Szalay, M.G., Paskar, L., Boyer, M., Singh, B. and Green, D.R. (1990) *J. Immunol.*, **144**, 3326.

35. Wyllie, A.H. (1980) *Nature*, **284**, 555.

36. Burton, K. (1956) *Biochem. J.*, **62**, 315.

37. Meyaard, L., Otto, S.A., Jonker, R.R., Mijnster, M.J., Keet, R.P. and Miedema, F. (1992) *Science*, **257**, 217.

38. Ormerod, M.G., Sun, X.-M., Snowden, R.T., Davies, R., Fearnhead, H. and Cohen, G.M. (1993) *Cytometry*, **14**, 595.

39. Nicoletti, I., Migliorati, G., Pagliacci, M.C., Grignani, F. and Riccardi, C. (1991) *J. Immun. Meth.*, **139**, 271.

40. Garvey, B.A., Telford, W.G., King, L.E. and Fraker, P.J. (1993) *Immunology*, **79**, 270.

Assays

CHAPTER 7

CELL-SURFACE ANTIGENS
P.J. Delves

1. INTRODUCTION

The cell surface is covered with protein molecules which are held in the membrane by hydrophobic transmembrane segments or glycosyl-phosphatidylinositol (GPI) anchors. Antigens found on cell surfaces comprise not only those encoded by the cell itself but also the products of intracellular parasites (e.g. glycoproteins of enveloped viruses). Furthermore, soluble ligands may be bound to receptor structures on the cell membrane, e.g. IgE bound to FcεR1 on mast cells, or lectin-like molecules bound to cell-surface carbohydrate structures. The molecules which form an integral part of the cell surface may be proteins, glycoproteins or glycolipids and, broadly, subserve one of three major functions: adhesion, antigen recognition or receptors for soluble mediators. However, many cell-surface molecules possess more than one function, e.g. molecules involved in cell–cell or cell–extracellular matrix adhesion can also themselves be involved in signal transduction.

While certain molecules, such as MHC class I encoded gene products, are extremely broad in their cellular distribution, others are highly restricted, e.g. TCR molecules are found exclusively on the surface of T lymphocytes. The expression of many cell-surface molecules is not only cell-lineage dependent but is also influenced by parameters such as stage of cell development ('differentiation antigens') or activation state ('activation antigens'). Many cell-surface antigens occur as allelic variants, MHC gene products being the extreme example. The genetic recombination and nucleotide additions (and somatic hypermutation for immunoglobulin genes) involved in the production of lymphocyte antigen receptors create further diversity at the surface of lymphocytes, as does alternative splicing (e.g. CD45 isoforms), and heterogeneous post-translational modifications (e.g. glycoforms of IgG). Although the number of any given molecule per cell is influenced by cell cycle and activation state, different cell-surface antigens are generally expressed at characteristic levels. For example, in the case of thymocytes, it has been estimated that there are approximately 10^6 molecules of CD90, 10^5 molecules of CD43 and 10^4 molecules of CD44 per cell (1).

The cell-surface antigens of leukocytes have been intensively studied due to ease of access to this cellular compartment. International workshops convened every few years assign a 'cluster of differentiation' (CD) nomenclature based on reactivity with groups of mAbs. Validation of CD antigens occurs when the cDNA is cloned and sequenced and then expressed to confirm its reactivity with the relevant mAbs. Antigens that remain to be fully characterized are given provisional workshop designations by the use of a 'w' before the assigned number, e.g CDw12. The definitive reference works for the leukocyte cell-surface antigens are the *Leucocyte Typing* books published after each workshop, the most recent of which, *Leucocyte Typing V*, reports the results of the meeting held in November 1993 (2). In addition, a leukocyte differentiation antigen database (LDAD) has been compiled by Stephen Shaw. The LDAD program runs on IBM PCs (and via emulation on Macintosh) and may be downloaded freely by anonymous ftp from balrog.nci.nih.gov. Other excellent and comprehensive reviews dealing with leukocyte antigens have also been published recently (3,4). For details of cell-surface antigens of rat, guinea-pig, hamster, dog, rabbit, birds, toad, pigs, cattle and sheep, the reader is directed to ref. 5.

Information relating to flow cytometric analysis of cell-surface antigens can be found in Chapter 4. The major characteristics of many of the known leukocyte cell-surface antigens are presented in the tables which follow this preamble. Although these lists are by no means exhaustive, they do cover most of the better-characterized antigens described in the literature. The biochemical characteristics are cataloged in *Table 1*, whereas *Table 2* deals with function and distribution. The reader should note that many of these molecules have several ligands, many of which may yet remain to be determined. The cell types listed under primary distribution are meant as a guide to the published distributions of these molecules on cells of the immune system; in many cases only a limited number of cell types has been examined. Functional aspects of most of these molecules also remain to be fully explored.

Table 1. Leukocyte cell-surface antigens – biochemical characteristics

Antigen	MW (kDa)	Chromosome		Membr. (#)	Glyco- N/O	Comments	Ref.
		Human	Mouse				
CD1a	49	1q22-q23	3	TM-1	4/−	Noncovalently linked to β_2m; similar structure to MHC class I but does not show significant polymorphism	6
CD1b	45	1q22-q23	3	TM-1	4/−		
CD1c	43	1q22-q23	3	TM-1	4/−		
CD2	45-58	1p13	3	TM-1	3/−	CD2R neo-epitope expressed on the CD2 molecule after cellular activation	7
CD3γ	21-28	11q23	9	TM-1	2/−	ζ and η alt. spl. from same gene; disulfide-linked ζ homodimer or ζη heterodimer noncovalently associated with CD3 γ,δ,ε and TCR αβ or γδ. ζ also associates with CD16	8
CD3δ	20-26	11q23	9	TM-1	2/−		
CD3ε	20-25	11q23	9	TM-1	−/−		
ζ	16	1q22-q25	1	TM-1	−/−		
η	22	1q22-q25	1	TM-1	−/−		
CD4	55-59	12pter-p12	6	TM-1	2/−		9
CD5	67	11q13	19	TM-1	2/?	Co-precipitates with TCR/CD3 complex	10
CD6	100-130			TM-1	8/+		11
CD7	40	17q25		TM-1	2/?	Mouse has three potential N-linked glycosylation sites	12
CD8α	32-34	2p12	6	TM-1	−/+	Disulfide-linked αβ heterodimers and αα homodimers; also soluble forms by alt. spl.	13
CD8β	32-34	2p12	6	TM-1	1/+		
CD9	22-27	12p13		TM (4)	1/+		14
CD10	100-110	3q21-q27		TM-II	6/?		15
CD11a	180	16p13.1-p11	12	TM-1	12/?	Noncovalent dimer with CD18	16
CD11b	170	16p13.1-p11	12	TM-1	19/?	Noncovalent dimer with CD18	
CD11c	150	16p13.1-p11	12	TM-1	8/?	Noncovalent dimer with CD18	

Cell-surface Antigens

Table 1. Continued

Antigen	MW (kDa)	Chromosome Human	Chromosome Mouse	Membr. (#)	Glyco- N/O	Comments	Ref.
CDw12	90–120						17
CD13	150–160/chain	15q25		TM-II	11/+	Noncovalent homodimer	15
CD14	52–55	5q23–q31		GPI	4/?	+ Soluble forms	18
CD15	–	–	–	–	Antigen	On glycoproteins and glycolipids (e.g. CD11/CD18 integrins). CD15s is sialyl-CD15	19
CD15s	–	–	–	–	Antigen		
CD16	50–65	1q23–q24	1	TM-I	5/?	In man TM (FcγRIIIa) and GPI (FcγRIIIb) isoforms encoded by two linked genes; TM form noncovalently associated with a γ chain (also used by FcεR1) and ζ chain (also used by CD3) disulfide-linked homo- or heterodimers; also, soluble isoforms by proteolytic cleavage	20
CD16b	48	1q23–q24		GP1	5/?		
CDw17	–	–		–	Antigen	Glycosphingolipid	21
CD18	95	21q22.3		TM-I	6/?	Noncovalent dimers with CD11a,b,c	22
CD19	90–95			TM-I	5/?	Noncovalent complex with CD21, CD81 and Leu-13	23
CD20	33,35,37	11q12–q13.1	19	TM (4)	–/–	Isoforms due to different degrees of phosphorylation	24
CD21	145	1q32	1	TM-I	11–12/?	Two isoforms by alt. spl, also soluble form; complexes noncovalently with CD19, CD81 and Leu-13	23
CD22	130 α	19q13.1	7	TM-I	10/–	Non-covalently associated αβ heterodimer. α and β alt. spl. from same gene	25
	140 β	–		TM-I	11/–		

CD	MW	Chromosome		TM		Comments	Ref
CD23	45–50	19p13.3	8	TM-II	1/?	Mouse has two potential N-linked glycosylation sites; two alt. spl. forms, FcεRIIa/FcεRIIb, differ in cytoplasmic region; high-affinity receptor is trimer or tetramer; also soluble forms	26
CD24	35–52			GPI	2/?		27
CD25	55	10p15–p14		TM-I	2/+	See: IL-2R	28
CD26	110–120	11pter–p11.2		TM-II	8/?	May exist as noncovalent trimer consisting of CD26 homodimer plus CD45	15
CD27	45–55/chain	12p13	6	TM-I	1/+	Disulfide-linked homodimer; also soluble form	29
CD28	44/chain	2q33	1	TM-I	5/?	Disulfide-linked homodimer	30
CD29	130	10p11.2		TM-I	12/?	Noncovalently associated with integrin α-subunits (CD49a–f)	31
CD30	105–120	1p36		TM-I	2/?	+ Soluble form	32
CD31	130–140			TM-I	9/?	Isoforms by alt. spl.	33
CD32	40	1q23–q24	1	TM-I	2/–	FcγRIIa, b and c forms encoded by three closely linked genes, show further variation by alt. spl. including soluble forms	20
CD33	67	19q13.3		TM-I	5/–		34
CD34	105–120	1q12–qter	1	TM-I	9/+		35
CD35	160–260	1q32	1	TM-I	20/?	Several cell surface and soluble forms	36
CD36	85–88	7		TM (2)	10/+		37
CD37	40–52	19p13–q13.4	7	TM (4)	3/–		38
CD38	45	4		TM-II	4/?		39
CD39	70–100				+/–		40

Cell-surface Antigens

CELL-SURFACE ANTIGENS

Table 1. Continued

Antigen	MW (kDa)	Chromosome Human	Chromosome Mouse	Membr. (#)	Glyco- N/O	Comments	Ref.
CD40	44–50	20	2	TM-I	2/–		41
CD41	123 α	17q21.32		–	4/?	Post-translationally cleaved into αβ which are disulfide-linked; noncovalently associated with CD61	42
	23 β	–		TM-I	1/?		
CD42a	17–23			TM-I	1/?	Tetramer of CD42a and CD42d noncovalently associated with disulfide-linked CD42b–CD42c αβ heterodimer	43
CD42b	135–145	17pter-p12		TM-I	4/+		
CD42c	22–25	22		TM-I	1/?		
CD42d	80–85			TM-1	8/–		
CD43	95–135	16p11.2	7	TM-I	1/+	Range of glycoforms, also proteolytically cleaved soluble forms	44
CD44	85–250	11p13	2	TM-I	7/+	Several variants by alt. spl.; chondroitin sulfate linked at several sites. CD44R restricted epitope	45
CD45	170–240	1q31–q32	1	TM-I	11–16/+	Several variants by alt. spl.; may be noncovalently associated with CD26	46
CD45RA	205,220	1q31–q32	1	TM-I	11–16/+		
CD45RB	190,205,220	1q31–q32	1	TM-I	11–16/+		
CD45RO	180	1q31–q32	1	TM-I	11–16/+		
CD46	51–68	1q32		TM-I	3/+	Variants by alt. spl.	47
CD47	47–52	3q13.1–2		TM (5)	+/?	Associated with CD51/CD61 VNR	48
CD48	40–47	1q21.3–q22	1	GPI	6/?		49

CD	MW (kDa)	Chromosome	Exons	Structure		Comments	Ref
CD49a	210	5		TM-I	24/?	CD49 molecules are noncovalently associated with integrin β1 subunit (CD29) to form VLA1-6; αβ-subunits post-translationally cleaved and disulfide-linked	50
CD49b	160–165	5q23-q31		TM-I	10/?		
CD49c	110 α	17		—	}13/?}		
	30 β	—					
CD49d	150 uncleaved	2q31-q32	9	TM-I	11/?		
	80 α			—	6/?		
	70 β	—		TM-I	5/?		
CD49e	135 α	12q11-q13		—	14/?		
	25 β	—		TM-I	—/?		
CD49f	125 α	2		—	8/?		
	30 β	—		TM-I	2/?		
CD50	116–140			TM-I	15/?		51
CD51	125 α	2q31-q32		—	10/?	Post-translationally cleaved, αβ disulfide-linked, noncovalently associated with CD61	52
	24 β	—		TM-I	3/?		
CD52	21–28			GPI	1/?		53
CD53	32–42	1p31-p12	3	TM (4)	2/?		38
CD54	85–110	19p13.2	9	TM-I	8/?	+ Soluble form	54
CD55	64–75	1q32		GPI	1/+	+ Soluble forms	55
CD56	140	11q23-q24	9	TM-I	6/?	Isoforms by alt. spl., yielding further variants by post-translational modifications; other isoforms by alt. spl. not on leukocytes (e.g. 180 kDa on neural tissue)	56
	120	11q23-q24	9	GPI	6/?		
CD57	—	—	—	—	Antigen	Present on several different molecules	57
CD58	55, 70	1p13		GPI/TM-I	6/?	TM or GPI isoforms by alt. spl. + soluble form	58
CD59	18–20	11p14-p13		GPI	1/?		59
CDw60	—	—	—	—	Antigen		60

Table 1. Continued

Antigen	MW (kDa)	Chromosome		Membr. (#)	Glyco- N/O	Comments	Ref.
		Human	Mouse				
CD61	105–110	17q21.32		TM-I	6/?	Noncovalently associated with CD41 to give GPIIb–IIIa complex or with CD51 to give vitronectin receptor	52
CD62E	107–115	1q12-qter		TM-I	11/?	Glycoforms and soluble form	61
CD62L	74–100	1q23-q25		TM-I	7/–	+ Soluble form	
CD62P	140–150	1q21-q24		TM-I	12/?	? Oligomeric. Also soluble form by alt. spl.	
CD63	53	12q12-q13		TM (4)	3/?		62
CD64	72–75	1q21.2–q21.3	3	TM-I	7/?	FcγRIa, b and c forms encoded by three closely linked genes. Also soluble isoforms	20
CDw65	–	–	–	–	Antigen		63
CD66a	160–190	19q13.1–q13.2		TM-I	20/?	Members of NCA subgroup of CEA molecules. Isoforms by alt. spl. described for most of these molecules	64
CD66b	95–110			GPI	11/?		
CD66c	90			GPI	+/?		
CD66d	30			TM-I	2/?		
CD66e	180–200			GPI	+/?		
CD68	110			TM-I	9/?	? Isoforms by alt. spl. Also soluble form	65
CD69	34/28	12p13-p12	8	TM-II	1/?	Disulfide-linked 'homodimer' of two glycoforms	66
CD70	55,75,95, 110,170	19p13		TM-II			67
CD71	95/chain	3q26.2-qter		TM-II	3/+	Disulfide-linked homodimer. Also soluble form	68
CD72	42/chain	9p	4	TM-II	1/?	Disulfide-linked homodimer	69

CD	MW	Chromosome		Anchor		Comments	Ref
CD73	69–74	6q14–q21		GPI	4/?	Disulfide-linked homodimer	15
CD74	43,41,35,33	5q32	18	TM-II	2–4/+	Two initiation codons + alt. spl. yield four isoforms. Mostly intracellular but some on cell surface	70
CDw75	–	–		–	Antigen	Sialic acid-dependent determinant	71
CDw76	–	–		–	Antigen	Sialic acid-dependent determinant	72
CD77	–	–		–	Antigen	Neutral glycosphingolipid	73
CDw78	–	–		–	Antigen		74
CD79a	32–45	17q23		TM-I	6/?	Disulfide-linked CD79a/CD79b heterodimer noncovalently associated with mIg. CD79a smaller in mice due to only two potential N-linked glycosylation sites. Truncated isoform of CD79b ('Ig-γ')	75
CD79b	36–38			TM-I	3/?		
CD80	60	3q13.3–q21		TM-I	8/?		76
CD81	22–26	11p	7	TM (4)	–/–	Noncovalent complex with CD19, CD21 and Leu-13	77
CD82	50–53						78
CD83	40–45			TM	3/?		79
CDw84	73						80
CD85	83–120						80
CD86	80						80
CD87	50–56			GPI	5/?		81
CD88	40	19q13.3		TM (7)	1/?		82
CD89	50–70	19q13.4		TM-I	5/?		83
CDw90	19–35	11q22.3–q23	9	GPI	3/–	Noncovalently associated with Sca-1	84

Cell-surface Antigens

Table 1. Continued

Antigen	MW (kDa)	Chromosome		Membr. (#)	Glyco- N/O	Comments	Ref.
		Human	Mouse				
CD91	α500 β85	12q13–q14		– TM-I	52/?	Post-translationally cleaved into αβ subunits. α-subunit noncovalently linked to TM β-subunit	85
CDw92	70						2
CD93	120						2
CD94	43/chain					Disulfide-linked homodimer	86
CD95	42–43	10q24.1	19	TM-I	2/?		87
CD96	160			TM-I	15/+		88
CD97	74,80,89						2
CD98	80–85 40	11q12–q22 11q12–q22		TM-II	4/? –/–	Disulfide-linked heterodimer	89
CD99	32	Xpter-p22.32 Ypter-p11.2		TM-I	–/+	Pseudoautosomal. CD99R restricted epitope	90
CD100	150/chain				+/+	Disulfide-linked homodimer	91
CDw101	140						92
CD102	55–65	17q23–q25		TM-I	6/?		93
CD103	150			TM-I		Heterodimer with integrin β^7	94
CD104	220			TM-I	5/?		95
CD105	95/chain			TM-II	5/+	Disulfide-linked homodimer. Variants by alt. spl.	96
CD106	90, 110	1p32–p31	3	GPI/ TM-I	5–6/?	Isoforms by alt. spl. + soluble form	97

CD107a	90–120	13q34		TM-I	17–20/+	Bear bulky poly-N-acetyllactosamines	98
CD107b	95–120	Xq24–q25		TM-I	16–17/+		99
CD108	80						100
CD109	150, 170						101
CD115	150	5q33.2–q33.3	18	TM-I	11/?		
CDw116	70–85	Xp22.32 Yp11.3	19	TM-I	11/?	Pseudoautosomal α-chain of GM-CSFR non-covalently associates with β-subunit common to IL-3R and IL-5R. α-chain constitutes low aff. receptor, αβ high aff. receptor. + soluble form of α-chain by ? alt. spl.	102
CD117	145	4q11–q22	5	TM-I	10/?		103
CDw119	90–100	6q23–q24	10	TM-I	5/?	+ Soluble form	104
CD120a	50–60	12p13.2	6	TM-I	3/–	+ Soluble forms by proteolytic cleavage	105
CD120b	75–85	1p36.3–p36.2	4	TM-I	2/?		
CDw121a	80	2q12	1	TM-I	6/?	IL-1R: two types encoded by linked genes. Soluble form of CDw121b by proteolytic cleavage	106
CDw121b	60–70	2q12–q22	1	TM-I	5/?		
CDw122	75	22q11.2–12	15	TM-I	4/?	See IL-2R	28
CDw124	130–150	16p12.1–11.2	7	TM-I	6/?	IL-4R. Associates with a γ-chain shared with IL-2R and IL-7R. + Soluble form by alt. spl.	107
CD126	80	1q21	3	TM-I	5/?	IL-6R α-chain. Low aff. receptor. Noncovalent heterodimer with CDw130 forms high aff. receptor. Soluble form by proteolytic cleavage.	108
CDw127	70–80	5p13	15	TM-I	5/?	IL-7R. Associates with a γ-chain shared with IL-2R and IL-4R. + Soluble form by alt. spl.	109
CDw128	44–58	2q34–q35		TM(7)	4/?	IL-8RI	110
CDw128	67–70	2q34–q35		TM(7)	3/?	IL-8RII	

Table 1. Continued

Antigen	MW (kDa)	Chromosome Human	Chromosome Mouse	Membr. (#)	Glyco- N/O	Comments	Ref.
CDw130	130			TM-I	10/?	IL-6R, IL-11R and LIF-R β-chain.	108
2B4	66		1	TM			111
4-1BB	30–40			TM-I	2/?	Monomeric and dimeric forms	112
114-A10	150			TM-I	3/?	8 Glucosaminoglycan attachment sites	113
A15	36			TM (4)	4/?		114
APA	130–160/chain			TM-II	9/?	Disulfide-linked homodimer. Molecular weight variants are glycoforms	15
B7-2	60–100			TM-I	8/?		115
C-CAM	105			TM-I	16/?	C-CAM1 and C-CAM2 isoforms differ in cytoplasmic domain	116
CD40L	32–39	Xq26.3-q27.1	X	TM	+/?		117
CLA	Multiple			TM (2)	10/?	Isoforms by alt. spl. and glycoforms	118
CMRF35	25–50			TM-I	2/+		119
CTLA-4	26–27/chain	2q33	1	TM-I	1/?	Disulfide-linked homodimer	120
EPO-R	66	19p13.2	9	TM-1	1/–		121
FcεRI α	45–65	1q23	1	TM-I	6/?	Disulfide-linked γ homodimer associated noncovalently with an α- and β-chain. γ-chain also associates with CD16	122
β	32–33	1q23	1	TM (4)	2/?		
γ	7–12	1q23	1	TM-I	–/–		
flk-2				TM-I	9/?		123

fMLP-R	55–70	19q13.3		TM (7)	3/?		124
G-CSFR	130–150	1p35–p34.3	4	TM-I	9/?	Two isoforms with different length cytoplasmic regions and also soluble form by ?alt. spl.	125
gp42	40–45			GPI	3/?		126
gp49	49			TM-I	3/–		127
IgH	IgM 70	14q32.33	12	TM-I	5/–	Diverse disulfide-linked tetramers of two identical polypeptide heavy chains and two identical polypeptide light chains. Can have additional V region glycosylation. Multiple C region glycoforms. Soluble forms by alt. spl., GPI-linked isoform of IgD	128
	IgD 50–63	14q32.33	12	TM-I	7/–		
Igκ	25–28	2p12	6	–	–/–		
Igλ	25–28	22q11.2	16	–	–/–		
IL2-R α	55	10p15–p14	15	TM-I	2/+	α-chain (CD25) is low affinity IL-2R. β (CDw122)/γ dimer is int. aff. IL-2R, αβγ trimer is high aff. IL-2R. Chains noncovalently associated. Also soluble forms. γ-chain also used by IL-4R and IL-7R	28
β	75	22q13		TM-I	4/?		
γ	64	Xq13.1		TM-I	6/?		
IL-3R α	70	Yp13.3, Xp22.3		TM-I	6/?	Specific pseudoautosomal α-chain + noncovalently linked β-chain common to IL-5 and GM-CSF receptors. α-chain constitutes low aff. receptor, αβ high aff. receptor	102
β	120–140	22q13.1	15	TM-I	3/?		
IL-5R α	55–60	3p26	6	TM-I	6/?	Specific α-chain + noncovalently linked β-chain common to IL-3 and GM-CSF receptors. α-chain constitutes low aff. receptor, αβ high aff. receptor; soluble form by alt. spl.	129
β	120–140	22q13.1	15	TM-I	3/?		
IL-9R	64						130
IL-10R	151						131
IL-11R						Noncovalent heterodimer with CDw130	130

Table 1. Continued

Antigen	MW (kDa)	Chromosome Human	Chromosome Mouse	Membr. (#)	Glyco- N/O	Comments	Ref.
IL-12R	110–135						132
IL-13R							133
Integrin α^7	120–180			TM-I			95
Integrin β^7	120	12q13.13	15	TM-I	8/?		95
LAG-3	50–70	12p13.3		TM-I	4/?		134
LDL-R	160	19p13.2	9	TM-I	5/+		135
Leu-13						Noncovalent complex with CD19, CD21 and CD81	136
LIF-R	190	5p13-p12	15	TM-I	20/?	Noncovalent heterodimer with CDw130. Soluble isoform in mouse by alt. spl.	137
ltk	100	15q13-q21		TM-I	3/?		138
Ly-9	100		1	TM-I	8/?		139
Ly-49	44		6	TM-II	3/?	Disulfide-linked homodimer. Several isoforms	140
M6	42–54			TM-I	3/?		141
M130	130			TM-I	11/?	Isoforms by alt. spl.	142
Mac-2	29–35			–	–/?	No membrane anchor, presumably bound to other surface molecules. Isoforms by alt. spl.	143
MAFA-G63	28–40						144
Mannose-R	175–190			TM-I	8/?		145

	M_r (kDa)			Anchor	Exons/Introns	Comments	Reference
M-ASGP-BP	42			TM-II	2/?	Functionally active form is a homohexamer or homooctomer	146
MDR1	170	7q21.1	5	TM (12)	3/?		147
MHC I α	44	6p21.3	12	TM-I	1/—	α-chain highly polymorphic, non-covalently associated with nonpolymorphic β_2-microglobulin	148
β2m	12	15q21–q22.2	2	—	—/—		
MHC II α	32–35	6p21.3	17	TM-I	2/—	Highly polymorphic, noncovalently associated αβ heterodimer	149
β	28–32	6p21.3	17	TM-I	1/—		
MKW	52			TM	—/—		150
MPL	65–90	1p34	4	TM-I	4/?	Two isoforms, MPLP and MPLK, differ in length of cytoplasmic region	151
MRC OX-2	40–50		3	TM-I	6/—		152
MRC OX-40	47–51			TM-I	1/?		153
NB1	58–64			GPI	+/?		154
NK1.1	30–39/chain		6	TM-II		Disulfide-linked homodimer	140
NKG2	35–40?	12		TM-II	3/?	Identified by cDNA. Prob. disulfide-linked homodimer. Four or five linked genes	155
PAF-R	39–50	1		TM (7)	2/?		156
PC-1	115–120			TM-II	10/?	Disulfide-linked homodimer	157
R2	40–50	11p12		TM (4)	3/?		38
RT6.1	24–35		7	GPI	1/?		158
6.2	25–28		7	GPI	—/—		
Sca-1	12–18		15	GPI	—/?	Several isoforms encoded by linked genes. Noncovalently associated with CDw90	159

Table 1. Continued

Antigen	MW (kDa)	Chromosome		Membr. (#)	Glyco-N/O	Comments	Ref.
		Human	Mouse				
Scav R	77	8p22		TM-I	7/?	Trimers, ? homo- , ? hetero-	160
SN8	α 49 β 40			TM-I		Disulfide-linked αβ heterodimer	161
Syndecan	80–120			TM-I	1/+	Bears five glycosaminoglycans	162
TCRα	40–45	14q11.2	14	TM-I	5/−	Diverse disulfide-linked αβ heterodimers. 'Pre-TCR'	8
TCRβ	38–45	7q35	6	TM-I	2/−	may consist of β-chain homodimers. Various soluble forms of TCR described but significance controversial	
TCRγ	45–60	7p15	13	TM-I	4/−	Diverse disulfide-linked and noncovalently linked γδ	8
TCRδ	40–45	14q11.2	14	TM-I	2/−	heterodimer isoforms	
TGFβ-R	53–300			TM-I	3-6/?	Several isoforms. Type III has up to five glycosaminoglycan chains attached. Also soluble form	163
TSA-1	12–13			GPI			164

MW: Typical molecular weight obtained on an SDS-PAGE gel run under reducing conditions.
Membr., membrane association. TM-I, type-I (C-terminus cytoplasmic) transmembrane; TM-II, type-II (N-terminus cytoplasmic) transmembrane. For molecules which pass through the membrane more than once, the number (#) of transmembrane regions is given in parentheses. GPI, glycosyl-phosphatidylinositol anchor. −, associated with a membrane-bound chain.
Glyco-N, number of consensus sequences for nitrogen-linked glycosylation; Glyco-O, oxygen-linked glycosylation (+, yes; −, no; ?, not known); Glyco-Antigen, carbohydrate antigen.
alt. spl., alternative splicing.

Table 2. Leukocyte cell-surface antigens – functions and distribution

Antigen	Synonyms	Ligand(s)	Primary distribution	Function
CD1a	T6, Ly-38 (mouse)	Peptides?	Cort. Thy, DC, NK	Ag presentation?
CD1b		Peptides?	Cort. Thy, DC	Ag presentation?
CD1c		Peptides?	Cort. Thy, DC, B-sub, NK	Ag presentation?
CD2	Sheep erythrocyte (SRBC) receptor, LFA-2, T11, Leu-5, Tp50, Ly-37 (mouse), MRC OX54 (rat)	CD48, CD58, CD59, SRBC	Thy, T, NK-sub, BM, Mono, mouse, B, rat Macro	Adhesion, signaling
CD3γ	CD3 complex, T3	—	Thy, T	TCR signal transduction
CD3δ			Thy, T	
CD3ε			Thy, T	
ζ			Thy, T, NK	
η			Thy, T	
CD4	T4, Leu-3, L3T4 (mouse), Ly-4 (mouse), W3/25 (rat), MRC OX35 (rat)	MHC class II, HIV-1, HIV-2	Thy, T-sub, DC, human and rat Mono, human and rat Macro	Adhesion, signaling. Accessory molecule for TCR–MHC class II interaction
CD5	T1, Leu-1, Ly-1 (mouse), Lyt-1 (mouse), LyA (mouse), MRC OX19 (rat)	CD72	Thy, T, B-sub	Signaling
CD6	T12		Thy, T-sub, B-sub	? Signaling
CD7	gp40		Thy, T-sub, NK, pluripotent hematopoietic cells	
CD8α	T8, Leu-2, Ly-2/Ly-3 (mouse), Lyt-2/Lyt-3 (mouse), LyB/LyC (mouse), MRC OX8 (rat)	MHC class I	Thy, T-sub, NK-sub, rat NK	Adhesion, signaling. Accessory molecule for TCR–MHC class I interaction
CD8β				

Cell-surface Antigens

Table 2. Continued

Antigen	Synonyms	Ligand(s)	Primary distribution	Function
CD9	Motility-related protein-1 (MRP-1), p24		Pre-B, Mono, Pt, Eosino, Baso, T-act, Endo, BM	Adhesion, signaling
CD10	Common acute lymphoblastic leukemia antigen (CALLA), enkephalinase, neutral endopeptidase (NEP), metalloendopeptidase	Peptides, f-Met-Leu-Phe (f-MLP)	Pre-B, Pre-T, Neutro, B blasts, BM stroma	Zinc metalloprotease
CD11a	Leukocyte function-associated antigen-1 (LFA-1), LeuCAMa (CD11a/CD18), integrin α^L subunit, Ly-15 (mouse), WT.1 (rat)	CD50 (ICAM-3), CD54 (ICAM-1), CD102 (ICAM-2)	T, B, Neutro, Mono, Macro, NK, BM	Adhesion
CD11b	Mac-1, CR3, LeuCAMb, integrin α^M subunit, Mo-1, OKM-1 (CD11b/CD18), Ly-40 (mouse), MRC OX41 (rat)	C3bi, fibrinogen, CD54, factor X	Mono, Neutro, NK, T-sub, Macro-sub, FDC	Adhesion, complement receptor
CD11c	CR4, integrin α^x subunit, p150 (p150,95), LeuCAMc (CD11c/CD18)	C3bi, fibrinogen	Macro, Mono, Neutro, NK, B-act, T-act	Adhesion, complement receptor
CDw12	p90–120		Mono, Neutro	
CD13	Aminopeptidase N (APN), MY7 antigen, coronavirus receptor	Peptides, coronavirus	Mono, Baso, Eosino, Neutro, Endo, BM stroma	Zinc metalloprotease

CD	Other names	Ligand / associated molecules	Cellular distribution	Function
CD14	LPS–BP receptor	LPS, LPS–LPS binding protein (LBP) complex, LPS–Septin complex	Mono, Macro-sub, LC, FDC, Neutro, human B-sub	? Clearance of LBP-opsonized Gram −ve bacteria. Signaling
CD15	Lewis^x (Le^x), 3-fucosyl-N-acetyl-lactosamine (FAL)	Homotypic	Neutro, Eosino, Mono-sub	Adhesion
CD15s	Sialyl-Lewis^x	CD62E	NK, Neutro, Mono	Adhesion
CD16a CD16b	FcγRIIIa, Leu-11 FcγRIIIb	Human: IgG1 = IgG3 > > > IgG2, IgG4. Mouse: IgG3 > IgG2a > IgG2b > >IgG1	CD16a TM isoform: NK, Macro, Mono-sub, Mono-act, T-sub, FDC, mouse Neutro CD16b GPI isoform: human Neutro, human Eosino-act	Low affinity receptor for complexed IgG. Signaling
CDw17	Lactosylceramide		Mono, Pt, Neutro	? Adhesion
CD18	Integrin β_2 subunit, WT.3 (rat)	CD54, CD102, C3bi	Leuko (see CD11)	Adhesion
CD19	B4		B, FDC	Antigen co-receptor, signaling
CD20	B1, Ly-44 (mouse)		B, FDC	? Signaling
CD21	CR2, Epstein–Barr virus (EBV) receptor, B2, HB5	iC3b, C3d, C3dg, EBV, CD23, IFNα	Thy-sub, B-mat, FDC	Complement receptor, adhesion, signaling
CD22	B-lymphocyte cell adhesion molecule (BL-CAM), Lyb-8 (mouse)	Sialylated glycoproteins, CD75, CD45RO	B-sub	Adhesion, signaling
CD23	FcεRII, BLAST-2, B6, Leu-20, Ly-42 (mouse)	IgE, CD21	FcεRIIa: B FcεRIIb: Mono-act, Eosino, FDC-sub, T, Pt, LC, NK	Low affinity receptor for IgE. Adhesion, signaling. FcεRIIb form not found in mice, and in man it is dependent on IL-4 for expression

Table 2. Continued

Antigen	Synonyms	Ligand(s)	Primary distribution	Function
CD24	BA-1, heat-stable antigen (HSA) (mouse), J11d (mouse), B2A2 (mouse), M1/69 (mouse), Ly-52 (mouse)		B, Neutro, FDC-sub, Thy (mouse), BM (mouse)	? Signaling, ? adhesion
CD25	IL-2Rα subunit, Tac, p55, Ly-43 (mouse), MRC OX39 (rat)	IL-2	T-act, B-act, Mono-act	Cytokine receptor subunit, signaling
CD26	Dipeptidyl aminopeptidase IV (DPP4), thymocyte-activating molecule (THAM) (mouse), MRC OX61 (rat)	Peptides, collagen, fibronectin	Thy, T-sub, Macro, Endo	Protease, adhesion, ? signaling
CD27		CD70 (CD27 ligand)	Medullary Thy, T-act, B-act, PC	Adhesion, ? signaling
CD28	Tp44	CD80 (B7-1), B7-2	Thy, T-sub, B-act, PC	Signaling
CD29	Very late antigen (VLA) β-subunit, platelet glycoprotein (Pt-GP) IIα, integrin β¹-subunit, fibronectin receptor (FNR) β-subunit	Fibronectin, collagen, laminin, vascular cell adhesion molecule-1 (VCAM-1)	Broad	Adhesion, signaling
CD30	Ki-1, Ber-H2 antigen	CD30 ligand (CD30L)	T-act, B-act, Endo	
CD31	Platelet-endothelial cell-adhesion molecule-1 (PECAM-1), endoCAM	Homophilic (and heterophilic)	Pt, Mono, Neutro, Eosino, T-sub, Endo, BM	Adhesion
CD32	FcγRII, Ly-17 (mouse), Lym-20 (mouse)	Human: IgG3≥IgG1≥IgG2, IgG4 Mouse: IgG2a = IgG2b≥IgG1	Macro, Mono, Neutro, B, Eosino, Endo, Baso, Pt, LC, FDC	Low-affinity receptor for complexed IgG. Signaling

CD	Other names	Ligand	Distribution	Function
CD33	MY9 antigen, gp67		Mono, Neutro, BM (myeloid precursors)	? Adhesion, ? signaling
CD34	gp105–120		Immature hematopoietic cells, Endo	
CD35	CR1	C3b, iC3b, C3c, C4b	Neutro, Eosino, Mono, T-sub, B, NK-sub, RBC, FDC	Complement receptor, adhesion
CD36	Pt-GPIV, GPIIb, OKM5 antigen	Thrombospondin, collagen, sequestrin, *Plasmodium falciparum*-infected RBC	Mono, Macro, Pt, small vessel endo, BM	Adhesion
CD37			B-mat, T-sub, Neutro, Mono, BM	
CD38	T10		Thy, Pre-B, T-act, B-act, PC, FDC, NK-sub, Neutro, Mono	
CD39			NK-act, T-act, B-act, Neutro, Mono, Macro, LC, FDC	? Adhesion
CD40		CD40 ligand (CD40L, TNF-related activation protein (TRAP), gp39, T-BAM)	B, thymic epith, FDC, Mono, Pt	Signaling
CD41	Pt-GPIIb, integrin α^{IIb}	Fibrinogen, fibronectin, vitronectin, von Willebrand factor, thrombospondin	Pt, Megakaryo, Mono	Platelet aggregation
CD42a	Pt-GPIX	von Willebrand factor, thrombin	Pt, Megakaryo, Mono	Adhesion
CD42b	Pt-GPIbα, glycocalicin	von Willebrand factor, thrombin	Pt, Megakaryo, Endo, Mono, Neutro, B-act	Adhesion

Table 2. Continued

Antigen	Synonyms	Ligand(s)	Primary distribution	Function
CD42c	Pt-GPIbβ	von Willebrand factor, thrombin	Pt, Megakaryo, Endo	Adhesion
CD42d	Pt-GPV	von Willebrand factor, thrombin	Pt, Megakaryo, Mono	Adhesion
CD43	Leukosialin, Sialophorin, Ly-48 (mouse), W3/13 (rat)	CD54 (ICAM-1)	Thy, T, NK, Macro, Mono, Neutro, PC, Pt, B-act, BM, DC	Adhesion, ? signaling
CD44	Phagocyte glycoprotein-1 (Pgp-1), extracellular matrix receptor-III (ECMR-III), HERMES, H-CAM, HUTCH-1, Ly-24 (mouse), MRC OX50 (rat)	Hyaluronate, collagen, fibronectin, MIP-1β chemokine	Broad	Lymphocyte adhesion to HEV, ? signaling
CD45	Leukocyte common antigen (LCA), B220, T200, Ly-5 (mouse), MRC OX1 (rat)	Phosphotyrosine	Leuko	Phosphatase, signaling
CD45RA	Restricted LCA	Phosphotyrosine	T-sub, B, Mono, BM	
CD45RB	Restricted LCA	Phosphotyrosine	T-sub, B, Neutro, Mono, NK, BM	
CD45RO	Restricted LCA	Phosphotyrosine, CD22	T-sub, B, Neutro, Mono, NK, Pt, BM	
CD46	Membrane cofactor protein (MCP), measles virus receptor	C3b, C4b, measles virus	Broad	Cofactor for cleavage of C3b and C4b
CD47	gp42, neurophilin, integrin-associated protein (IAP), OA3, GR63, ID8		Broad	Adhesion, signaling
CD48	Blast-1, BCM1 antigen (mouse),	CD2	Leuko	Adhesion, ? signaling

	Sgp-60 (mouse), MRC OX-45 (rat)			
CD49a	VLA-α1, integrin α1 subunit	Laminin, collagen	T-act, Mono, NK	Adhesion, signaling
CD49b	VLA-α2, integrin α2 subunit, Pt-GPIα, ECMR-I	Laminin, collagen	Leuko, Pt, Endo	Adhesion, signaling
CD49c	VLA-α3, integrin α3 subunit, ECMR-II	Fibronectin, laminin, collagen, invasin, epiligrin	B, FDC, Endo, NK	Adhesion, signaling
CD49d	VLA-α4, integrin α4 subunit, LPAM-α1	VCAM-1, fibronectin	Mono, T, B, Thy, Pt, NK, Eosino, FDC	Adhesion, signaling
CD49e	VLA-α5, integrin α5 subunit, FNR α-chain	Fibronectin	Broad	Adhesion, signaling
CD49f	Pt-GPIc, ECMR-IV, VLA-α6, integrin α6 subunit	Laminin, invasin	Pt, Mono, T, Thy, FDC, Endo, NK	Adhesion, signaling
CD50	ICAM-3, ICAM-R	CD11a/CD18 (LFA-1)	Leuko	Adhesion
CD51	Integrin-αv subunit, vitronectin receptor (VNR) α-chain	Vitronectin, fibrinogen, von Willebrand factor, thrombospondin, fibronectin, osteopontin, collagen	Pt, B-sub, Endo, Mono, Macro, Megakaryo	Adhesion
CD52	Campath-1 antigen		Leuko	
CD53	MRC OX44 (rat)		Leuko, BM, Pt	
CD54	Intercellular adhesion molecule-1 (ICAM-1), MALA-2, rhinovirus receptor, Ly-47 (mouse), 1A29 (rat)	CD11a/CD18 (LFA-1), rhinovirus, CD43, CD11b/CD18 (Mac-1), Plasmodium falciparum–infected RBC	Broad	Adhesion, ? signaling
CD55	Decay accelerating factor (DAF)	C3b, C4b, C3 convertases	Broad	Adhesion, ? signaling. Prevents formation of $\overline{C4b2a}$ and $\overline{C3bBb}$

Table 2. Continued

Antigen	Synonyms	Ligand(s)	Primary distribution	Function
CD56	Neural cell adhesion molecule (NCAM) isoform, NKH-1, Leu-19	Homophilic, heparin sulfate, heparin	NK, T-sub(act)	Adhesion
CD57	HNK-1, Leu-7		NK, T-sub, B-sub, Mono-sub	Adhesion
CD58	LFA-3, H19, Fib75	CD2	Broad	Adhesion
CD59	TAP, protectin, homologous restriction factor (HRF) 20, membrane inhibitor of reactive lysis (MIRL), MACIF	CD2	Broad	Adhesion, ? signaling. Binds C8 and C9 and thereby inhibits MAC assembly
CDw60	GD3, NeuAc2-8NeuAc2-3Galβ1-4, UM4D4		T-sub, Mono, Pt	? Signaling
CD61	Pt-GPIIIa, VNR β-chain, integrin β³ subunit	Fibrinogen, fibronectin, vitronectin, von Willebrand factor, ? thrombospondin, ? vitronectin, ? osteopontin, ? collagen	Mono, Macro, Megakaryo, Pt, Endo	Adhesion
CD62E	E-selectin, endothelial leukocyte adhesion molecule-1 (ELAM-1), LECAM-2	CD15s, CD62L, CDw65, CLA, fucosylated N-acetyl lactosamine	Endo-act	Adhesion
CD62L	L-selectin, leukocyte adhesion molecule-1 (LAM-1), LECAM-1, MEL-14, Leu-8, lymph node homing receptor (LHR), Ly-22 (mouse), TQ-1, DREG-56	GlyCAM-1, CD34	Leuko	Adhesion

CD	Other names	Ligand	Cellular distribution	Function
CD62P	P-selectin, GMP140, PADGEM, LECAM-3		Pt-act, Endo, Megakaryo	Adhesion
CD63	MLA1, PTLGP40, ME491, neuroglandular antigen, LIMP	Sialylated fucosylated lactosaminoglycans (e.g. CD15 s, CDw65)	Broad	Adhesion
CD64	FcγRI	Human: IgG3 > IgG1 > IgG4 > > IgG2 Mouse: IgG2a = IgG3 > > > IgG1, IgG2b	Macro, Mono, Neutro-act, Eosino-act	High-affinity receptor for IgG. Signaling
CDw65	Ceramide dodecasaccharide 4c, VIM-2	CD62E, CD62P	Neutro, Eosino	
CD66a	Biliary glycoprotein-1 (BGP-1) NCA-160	Homotypic	Neutro	Adhesion, ? signaling
CD66b	CD67, p100, CGM6, NCA-95		Neutro, Eosino	Adhesion, ? signaling
CD66c	NCA-90	Homotypic	Neutro	Adhesion
CD66d	CGM1		Neutro	Adhesion, ? signaling
CD66e	Carcinoembryonic antigen (CEA)	Homotypic	Neutro	Adhesion
CD68	gp 110, macrosialin (mouse)		Macro, Mono, Neutro, Baso, large lympho	
CD69	Activation inducer molecule (AIM), EA-1, MLR-3, Leu-23, VEA		Thy, T-act, B-act, Macro, NK-act, Pt	Signaling
CD70	CD27 ligand	CD27	T-act, B-act	? Signaling
CD71	Transferrin receptor, T9, MRC OX26 (rat)	Transferrin	Macro, all proliferating cells, Endo, FDC	? Signaling, iron uptake
CD72	Lyb-2 (mouse)	CD5	B, Macro	? Signaling
CD73	Ecto-5′-nucleotidase (E5N)	Nucleoside monophosphate	Thy, B-sub, T-sub, Endo-sub, FDC-sub	Enzyme, ? signaling, ? adhesion

Cell-surface Antigens

Table 2. Continued

Antigen	Synonyms	Ligand(s)	Primary distribution	Function
CD74	Invariant chain (Ii)	Intracellular MHC class II	T-act(sub), B, Mono, Macro	Regulates MHC class II folding, transport and peptide binding
CDw75		CD22	B-mat, T-sub	Adhesion
CDw76			B-mat, T-sub, Endo-sub	
CD77	Globotriacylceramide (Gb3), Burkitt's lymphoma-associated antigen (BLA), Pᵏ blood group		B-sub, FDC, Endo	
CDw78	Ba, Leu-21		B, Macro	
CD79a	Ig-α, MB-1 (mouse), Ly-54 (mouse)	—	B	Co-receptor, signaling
CD79b	Ig-β, B29 (mouse)			
CD80	B7, B7-1, BB1, Ly-58 (mouse)	CD28, CTLA-4	T-act, B-act, DC, Mono, Macro, LC	Signaling
CD81	Target of an anti-proliferative antibody-1 (TAPA-1)	Homotypic	Broad	Adhesion, signaling
CD82	GR15, R2.1A4, 4F9		Broad	
CD83	HB15		Interdigitating reticulum cells, LC, T-act, B-act	
CDw84	GR6		T-sub, B, NK, Mono, Pt	
CD85	GR4		B, Mono	
CD86	GR65		B, Mono	
CD87	Urokinase-type plasminogen activator receptor (u-PAR)	Urokinase-type plasminogen activator (u-PA)	Mono, Neutro, Eosino, Endo	Receptor

CD	Other names	Ligand	Cellular expression	Function
CD88	Complement component C5a receptor (C5aR), GR10	C5a	Mono, Macro, Neutro, Mast cells, Eosino	Complement receptor
CD89	FcαR, IgA receptor	IgA	T-sub, B-sub, Mono, Macro, Neutro, Eosino	Receptor for IgA
CDw90	Thy-1, Theta, MRC OX7 (rat)		Human prothymocytes, mouse Thy, mouse T, mouse stem cells, mouse NK	Signaling, ? adhesion
CD91	α2-macroglobulin-R, low density lipoprotein receptor-related protein (LRP)	Protease–antiprotease complexes, plasma lipoproteins	Mono, Macro	Receptor
CDw92	GR9		Broad	
CD93	GR11		Mono, Neutro, Endo	
CD94	Kp43		NK, T-sub	? Signaling
CD95	Fas, Apo-1		B-act, T-act	Apoptotic signaling
CD96	Tactile ('T-cell activation, increased late expression')		T-act, NK-act	? Adhesion, ? signaling
CD97	GR1, BL-KDD/F12		Broad	? Signaling
CD98	4F2, RL-388 (mouse)		T-act, B-act, NK-act, Mono, hematopoietic progenitors	
CD99	MIC2, E2, FMC29, 12E7, HuLy-m6		Thy, T, B, Mono	
CD100	BB18, A8, GR3		T, B-act, NK, Neutro, Mono, Pt	? Signaling
CDw101	BB27, BA27, GR14		T, NK, Mono, Neutro, Eosino, Endo	? Signaling

Table 2. Continued

Antigen	Synonyms	Ligand(s)	Primary distribution	Function
CD102	ICAM-2	CD11a/CD18 (LFA-1)	Endo, T-sub, Mono, DC	Adhesion, co-stimulator for T cell activation
CD103	HML-1 α-chain, M290 (mouse)	Laminin?	Intra-epithelial T	Adhesion
CD104	Integrin β^4 subunit		Thy, Endo-sub	Adhesion
CD105	Endoglin, GR7, β-glycan (rat)	TGF-β1, TGF-β3	BM-sub, Macro-act, Endo	Adhesion
CD106	VCAM-1, INCAM-110	Integrins α^4/β^1 (CD49d/CD29, VLA-4) and α^4/β^7	Endo-act, Macro, B, DC	Adhesion
CD107a CD107b	Lysosomal membrane glycoproteins-1 and 2 (LAMP-1 and LAMP-2)	CD62E	NK, Mono, B, Pt-act, Endo	Adhesion
CDw108	GR2		T, NK, B-act, Endo	Adhesion
CDw109	GR56, 8A3, 7D1		Mono, Pt, Endo	
CD115	Macrophage colony-stimulating factor receptor (M-CSFR), c-fms, CSF-1 receptor	M-CSF (CSF-1)	Macro, Mono and their precursors	Cytokine receptor, signaling
CDw116	Granulocyte macrophage colony-stimulating factor receptor (GM-CSFR), CSF-2R	GM-CSF	Macro, Mono, Neutro, Eosino and their precursors, Endo	Cytokine receptor, signaling
CD117	SLF/HGF-R, stem cell factor receptor, c-kit	Hematopoietic growth factor (HGF) / steel factor (SLF) / mast cell growth factor (MGF) / stem cell factor (SCF) / kit ligand (KL)	BM (most precursor cells), Thy, mast cells	Cytokine receptor, signaling

CDw119	Interferon γ receptor (IFNγR)	IFNγ	Broad	Cytokine receptor, signaling
CD120a	Tumor necrosis factor receptor-I	TNFα, TNFβ (lymphotoxin)	Broad	Cytokine receptor, signaling
CD120b	Tumor necrosis factor receptor-II	TNFα, TNFβ (lymphotoxin)	Broad	Cytokine receptor, signaling
CDw121a	IL-1 receptor type I	IL-1α, IL-1β, IL-1R antagonist	Thy, T, Endo	Cytokine receptor, signaling
CDw121b	IL-1 receptor type II	IL-1α, IL-1β, IL-1R antagonist	B, Macro, Mono	Cytokine receptor, signaling
CD122	IL-2 receptor β, p75	IL-2	T-act, NK	Cytokine receptor, signaling
CDw124	IL-4 receptor (IL-4R)	IL-4	T, B, Macro, Endo	Cytokine receptor, signaling
CD126	IL-6 receptor (IL-6R)	IL-6	Broad	Cytokine receptor, signaling
CDw127	IL-7 receptor (IL-7R)	IL-7	Thy, T, Pro-B, Pre-B, BM (lymphoid precursors), Mono	Cytokine receptor, signaling
CDw128	IL-8 receptor (IL-8R I, IL-8R II)	RI: IL-8. RII: IL-8, GROα/MGSA, MIP-2, NAP-2	T-sub, Mono, Neutro, Baso	Cytokine receptor, signaling
CDw130	IL-6 receptor β (IL-6Rβ), gp130-SIG	IL-6	Broad	Cytokine receptor, signaling
2B4			**Mouse** NK, T-act	Adhesion, ? signaling
4-1BB		Fibronectin, vitronectin, laminin, collagen VI	**Mouse** Thy, T-act	? Signaling
114-A10			**Mouse** myeloid progenitors	
A15			Immature T	

Cell-surface Antigens

Table 2. Continued

Antigen	Synonyms	Ligand(s)	Primary distribution	Function
APA	Aminopeptidase A, gp160, 6C3 (mouse), BP-1 (mouse), Ly-51 (mouse)	Peptides	Early B, thymic cortical epith-sub, BM stroma, Endo	Metalloprotease
B7-2	B70, GL1	CD28, CTLA-4	Monocytes, dendritic cells, B-act, T-act, NK-act	Signaling
C-CAM	Cell-CAM105	Homophilic, calmodulin	Endo, Pt-act, Neutro, Mono	ATPase, adhesion
CD40L	CD40 ligand, TNF-related activation protein (TRAP), gp39, T-BAM	CD40	T-act (CD4$^+$)	Signaling
CLA	Cutaneous lymphocyte-associated antigen, HECA-452	CD62E	Skin memory T	Adhesion
CMRF35			T-sub, B-sub, Mono, Neutro	
CTLA-4	Ly-56 (mouse)	CD80, B7-2, B7-3	T-act	? Signaling, ? adhesion
EPO-R	Erythropoietin receptor	Erythropoietin	Erythroblasts	Cytokine receptor
FcεRI	High-affinity IgE receptor	IgE	Mast cells, Baso	Receptor for IgE, crosslinking leads to signaling for degranulation
flk-2	Fetal liver kinase-2		**Mouse** BM, immature Thy	? Signaling
fMLP-R	Formyl-Met-Leu-Phe receptor	fMet-Leu-Phe (f-MLP)	Mono, Neutro	Chemotaxis
G-CSFR	Granulocyte colony-stimulating factor receptor, CSF-3R	G-CSF	Granulocyte precursors, Neutro, Endo	Cytokine receptor, signaling

		Antigen	Mouse NK-act / Mouse mast cells	
gp42			**Mouse** NK-act	
gp49			**Mouse** mast cells	
IgH κ λ	B-cell antigen receptor, mIg	Antigen	B	Antigen receptor
IL-2R γ	IL-2 receptor γ-chain	IL-2	T-act, B-act, Mono-act.	Cytokine receptor, signaling
IL-3R	IL-3 receptor	IL-3	BM (most precursor cells)	Cytokine receptor, signaling
IL-5R	IL-5 receptor	IL-5	Eosino, Baso, Mouse B	Cytokine receptor, signaling
IL-9R	IL-9 receptor	IL-9	Mast cells, Megakaryo, Thy, T-sub	Cytokine receptor, signaling
IL-10R	IL-10 receptor	IL-10	Macro, Mono, T, B, mast cells	Cytokine receptor, signaling
IL-11R	IL-11 receptor	IL-11	B, BM, Megakaryo	Cytokine receptor, signaling
IL-12R	IL-12 receptor	IL-12	T, NK	Cytokine receptor, signaling
IL-13R	IL-13 receptor	IL-13	Macro, Mono, B	Cytokine receptor, signaling
Integrin α^7	H36	Mucosal vascular addressin, laminin	Mucosal lymphocytes, T-act	Adhesion
Integrin β^7	LPAM-1 β-chain (with α^4[CDw49d]), HML-1 β-chain, M290 β-chain	MadCAM-1 (for $\alpha^4\beta^7$ integrin), VCAM-1, fibronectin	T-sub	Adhesion

Cell-surface Antigens

Table 2. Continued

Antigen	Synonyms	Ligand(s)	Primary distribution	Function
LAG-3	Lymphocyte activation gene 3 product		T-act, NK-act	
LDL-R	Low-density lipoprotein receptor	LDL	T, B, Macro, Mono	Receptor
Leu-13			T, B, FDC, Endo	Signaling
LIF-R	Leukemia inhibitory factor receptor	LIF	Broad	Cytokine receptor
ltk	Leukocyte tyrosine kinase		B	Enzyme
Ly-9	T100		**Mouse** Thy, T, B, BM	
Ly-49	A1, YE1/48	MHC class I	**Mouse** Thy, T, NK-sub	Signaling
M6	Basigin, CE9, HT7, neurothelin, MRC OX-47 (rat)		Thy, T-act, Mono, Neutro, BM, Endo	? Signaling
M130	Ki-M8, Ber-Mac3, GHI/61, SM4		Macro, Mono	? Receptor
Mac-2	Carbohydrate-binding protein (CBP)-35 (mouse), L34 (mouse), RL-29 (mouse), HL-29 (mouse), HL-31 (mouse), IgE-binding protein (IgE-BP) (rat)	Galactose (e.g. on IgE, laminin, etc.)	Macro, DC, Neutro	Lectin
MAFA-G63	Mast-cell function-associated antigen-G63		**Rat** mast cells	Signaling
Mannose-R	Mannose receptor	Mannose	Macro	Adhesion for phagocytosis
M-ASGP-BP	Macrophage asialoglycoprotein binding protein	Gal/GalNAc-terminated glycoproteins	**Rat** Macro	Receptor
MDR1	Multidrug resistance-1, P-glycoprotein (P-gp)	Lipophilic molecules	BM, T-sub, Mono	

MHC I	HLA-A, B, C (human), H2-K, D, L (mouse), RT1A (rat)	Peptides, TCR, CD8	Most nucleated cells	Present Ag (usually endogenous) to TCR
MHC II	HLA-DP, DQ, DR (human), I-A, I-E (mouse), RT1B, RT1D (rat)	Peptides, TCR, CD4	B, Mono, Macro, DC, myeloid and erythroid precursors	Present Ag (usually exogenous) to TCR
MKW			Mono, B-sub	
MPL	c-mpl, v-mpl		? Broad at low level	? Receptor, signaling
MRC OX-2			B, rat Thy, rat FDC, rat Endo	
MRC OX-40			Rat T-act	
NB1			Neutro	
NK1.1	Member of NKR-P1 family	? Carbohydrate	Mouse Neutro, NK, Thy, T-sub	Signaling
NKG2		? Carbohydrate	NK, T-sub	? Receptor
PAF-R	Platelet-activating factor receptor	PAF	Eosino, Mono, Macro Neutro, Pt	Receptor
PC-1	Plasma cell antigen-1, Ly-41 (mouse)		T, PC	Enzyme with both nucleotide pyrophosphatase and alkaline phosphodiesterase-1 activities
R2	4F9, C33		T-sub, B-act	
RT6.1 6.2	ART-2, AgF, Pta		Rat T	? Adhesion, ? signaling

Table 2. Continued

Antigen	Synonyms	Ligand(s)	Primary distribution	Function
Sca-1	Stem cell antigen-1, TAP, MALA-1, DAG, Ly-6		**Mouse** Thy, T, B, BM, Neutro, Mono	? Signaling
Scav-R SN8	Scavenger receptor	Polyanionic molecules	Macro, Endo B-sub	Receptor
Syndecan		Collagen, fibronectin, thrombospondin	Pre-B, immature B, PC (not found on mature B), Endo	Adhesion
TCRα TCRβ	αβ T-cell receptor, TCR-2	MHC-peptide	Thy, T-sub	Receptor for processed antigen
TCRγ TCRδ	γδ T-cell receptor, TCR-1	MHC-peptide	Thy, T-sub	Receptor for processed antigen
TGFβ-R	Transforming growth factor receptors I–VI. β-Glycan (TGFβ-RIII)	TGF β family	Broad	Cytokine receptor, signaling
TSA-1	Thymic shared antigen-1		Cortical Thy, B, thymic medullary epith	

Abbreviations used: act, activated; B, B lymphocytes; Baso, basophils; BM, bone marrow hematopoietic progenitors; Cort Thy, cortical thymocytes; DC, dendritic cells; Eosino, eosinophils; Epith, epithelium; FDC, follicular dendritic cell; LC, Langerhan cells; Leuko, leukocytes; Macro, macrophages; mat, mature; Megakaryo, megakaryocytes; Mono, monocytes; Neutro, Neutrophils; NK, natural killer cells; PC, plasma cells; Pt, platelets; RBC, red blood cells; sub, subpopulation; T, T lymphocytes.

Where species differences occur these are noted.

Mouse or **rat** (bold) signifies molecule described in these species and human homolog not yet established.

2. REFERENCES

1. Williams, A.F. and Barclay, A.N. (1986) *Handbook of Experimental Immunology* (D.M. Weir, L.A. Herzenberg, C. Blackwell and L.A. Herzenberg, eds). Blackwell Scientific Publications, Oxford, Vol. 1, pp. 22.1–22.24.

2. Schlossman, S.F., Boumsell, L., Gilks, W., Harlan, J.M., Kishimoto, T., Morimoto, C., Ritz, J., Shaw, S., Silverstein, R.L., Springer, T.A., Tedder, T.F. and Todd, R.F. (eds) (1994) *Leucocyte Typing V: White Cell Differentiation Antigens*. Oxford University Press, Oxford.

3. Barclay, A.N., Beyers, A.D., Birkeland, M.L., Brown, M.H., Davis, S.J., Somoza, C. and Williams, A.F. (1992) *The Leukocyte Antigen FactsBook*. Academic Press, London.

4. Pigott, R. and Power, C. (1993) *The Adhesion Molecule FactsBook*. Academic Press, London.

5. Miyasaka, M. and Trnka, Z. (1988) *Differentiation Antigens in Lymphohemopoietic Tissues*. Marcel Dekker, New York.

6. Calibi, F. and Bradbury, A. (1991) *Tissue Antigens*, 37, 1.

7. Kaplan, A.J., Chavin, K.D., Yagita, H., Sandrin, M.S., Qin, L.-H., Lin, J., Lindenmayer, G. and Bromberg, J.S. (1993) *J. Immunol.*, 151, 4022.

8. Malissen, B. and Schmitt-Verhulst, A.M. (1993) *Curr. Opin. Immunol.*, 5, 324.

9. Ledbetter, J.A., Deans, J.P., Aruffo, A., Grosmaire, L.S., Kanner, S.B., Bolen, J.B. and Schieven, G.L. (1993) *Curr. Opin. Immunol.*, 5, 334.

10. Osman, N., Ley, S.C. and Crumpton, M.J. (1992) *Eur. J. Immunol.*, 22, 2995.

11. Wee, S., Schieven, G.L., Kirihara, J.M., Tsu, T.T., Ledbetter, J.A. and Aruffo, A. (1993) *J. Exp. Med.*, 177, 219.

12. Yoshikawa, K., Seto, M., Ueda, R., Obata, Y., Fukatsu, H., Segawa, A. and Takahashi, T. (1993) *Immunogenetics*, 37, 114.

13. Miceli, M.C. and Parnes, J.R. (1993) *Adv. Immunol.*, 53, 59.

14. Lanza, F., Wolf, D., Fox, C.F., Kieffer, N., Seyer, J.M., Fried, V.A., Coughlin, S.R., Phillips, D.R. and Jennings, L.K. (1991) *J. Biol. Chem.*, 266, 10638.

15. Shipp, M.A. and Look, A.T. (1993) *Blood*, 82, 1052.

16. Berton, G., Laudanna, C., Sorio, C. and Rossi, F. (1992) *J. Cell. Biol.*, 116, 1007.

17. van der Schoot, C.E., Daams, M., von dem Borne, A.E.G. Kr., Scubitz, K.M., Skubitz, A.P.N., van Agthoven, A., Brailly, H., Romagné, F., Lanini, S., Kniep, B., Civin, C.I., Fackler, M.J., Chorvath, B., Duraj, J., Bazil, V., Horejsí, V., Hildreth, J., Hyman, J. and Tetteroo, P.A.T. (1989) *Leukocyte Typing*, IV, 868.

18. Labeta, M.O., Durieux, J.-J., Fernandez, N., Herrmann, R. and Ferrara, P. (1993) *Eur. J. Immunol.*, 23, 2144.

19. Stocks, S.C., Albrechtsen, M. and Kerr, M.A. (1990) *Biochem. J.*, 268, 275.

20. van de Winkel, J.G.J. and Capel, P.J.A. (1993) *Immunol. Today*, 14, 215.

21. Symington, F.W. (1989) *J. Immunol.*, 142, 2784.

22. Larson, R.S. and Springer, T.A. (1990) *Immunol. Rev.*, 114, 181.

23. Fearon, D.T. (1993) *Curr. Opin. Immunol.*, 5, 341.

24. Tedder, T.F., Klejman, G., Schlossman, S.F. and Saito, H. (1989) *J. Immunol.*, 142, 2560.

25. Wilson, G.L., Najfeld, V., Kozlow, E., Menniger, J., Ward, D. and Kehrl, J.H. (1993) *J. Immunol.*, 150, 5013.

26. Dierks, S.E., Bartlett, W.C., Edmeades, R.L., Gould, H.J., Rao, M. and Conrad, D.H. (1993) *J. Immunol.*, 150, 2372.

27. Nielsen, P.J., Eichmann, K., Köhler, G. and Iglesias, A. (1993) *Int. Immunol.*, 5, 1355.

28. Minami, Y., Kono, T., Miyazaki, T. and Taniguchi, T. (1993) *Ann. Rev. Immunol.*, 11, 245.

29. Gravestein, L.A., Blom, B., Nolten, L.A., de Vries, E., van der Horst, G., Ossendorp, F., Borst, J. and Loenen, W.A. (1993) *Eur. J. Immunol.*, 23, 943.

30. Linsley, P.S. and Ledbetter, J.A. (1993) *Ann. Rev. Immunol.*, 11, 91.

31. Hemler, M.E. (1990) *Ann. Rev. Immunol.*, 8, 365.

32. Durkop, H., Latza, U., Hummel, M., Eitelbach, F., Seed, B. and Stein, H. (1992) *Cell*, 68, 421.

33. Zehnder, J.L., Hirai, K., Shatsky, M., McGregor, J.L., Levitt, L.J. and Leung, L.L. (1992) *J. Biol. Chem.*, 267, 5243.

34. Drexler, H.G., Thiel, E. and Ludwig, W.D. (1991) *Leukemia*, 5, 637.

35. Baumhueter, S., Singer, M.S., Henzel, W., Hemmerich, S., Renz, M., Rosen, S.D. and Lasky, L.A. (1993) *Science*, 262, 311.

36. Krych, M., Atkinson, J.P. and Holers, V.M. (1992) *Curr. Opin. Immunol.*, 4, 8.

Cell-surface Antigens

37. Greenwalt, D.E., Lipsky, R.H., Ockenhouse, C.F., Ikeda, H., Tandon, N.N. and Jamieson, G.A. (1992) *Blood*, **80**, 1105.

38. Virtaneva, K.I., Angelisova, P., Baumruker, T., Horejsi, V., Nevanlinna, H. and Schroder, J. (1993) *Immunogenetics*, **37**, 461.

39. Drach, J., Zhao, S., Malavasi, F. and Mehta, K. (1993) *Biochem. Biophys. Res. Comm.*, **195**, 545.

40. Gouttefangeas, C., Mansur, I., Schmid, M., Dastot, H., Gelin, C., Mahouy, G., Boumsell, L., Bensussan, A. (1992) *Eur. J. Immunol.*, **22**, 2681.

41. Banchereau, J., Blanchard, D., Briere, F., Galizzi, J.P., van Kooten, C., Liu, Y.J., Rousset, F., Saeland, S. and Bazan, F. (1994) *Ann. Rev. Immunol.*, **12**, in press.

42. Kieffer, N. and Phillips, D.R. (1990) *Ann. Rev. Cell Biol.*, **6**, 334.

43. Roth, G.J. (1991) *Blood*, **77**, 5.

44. Rieu, P., Porteu, F., Bessou, G., Lesavre, P. and Halbwachs-Mecarelli, L. (1992) *Eur. J. Immunol.*, **22**, 3021.

45. Lesley, J., Hyman, R. and Kincade, P.W. (1993) *Adv. Immunol.*, **54**, 271.

46. Trowbridge, I. and Thomas, M.L. (1994) *Ann. Rev. Immunol.*, **12**, in press.

47. Liszewski, M.K., Post, T.W. and Atkinson, J.P. (1991) *Ann. Rev. Immunol.*, **9**, 431.

48. Hadam, M.R. (1989) in *Leucocyte Typing IV: White Cell Differentiation Antigens* (W. Knapp, B. Dörken, W.R. Gilks, E.P. Rieber, R.E. Schmidt, H. Stein and A.E.G.K. von dem Borne, eds). Oxford University Press, Oxford p. 658.

49. Gonzalez Cebrero, J., Freeman, G.J., Lane, W.S. and Reiser, H. (1993) *Proc. Natl. Acad. Sci. USA*, **90**, 3418.

50. Chen, B.M.C. and Hemler, M.E. (1993) *J. Cell Biol.*, **120**, 537.

51. Lozano, F., Alberol-Ila, J., Places, L. and Vives, J. (1992) *Eur. J. Biochem.*, **203**, 321.

52. Hynes, R.O. (1992) *Cell*, **69**, 11.

53. Xia, M.Q., Hale, G. and Waldman, H. (1993) *Mol. Immunol.*, **30**, 1089.

54. Lane, P.J., McConnell, F.M., Clark, E.A. and Mellins, E. (1991) *J. Immunol.*, **147**, 4103.

55. Thomas, D.J. and Lublin, D.M. (1993) *J. Immunol.*, **150**, 151.

56. Edelman, G.M. and Crossin, K.L. (1991) *Ann. Rev. Biochem.*, **60**, 155.

57. Schubert, J., Lanier, L.L. and Schidt, R.E. (1989) in *Leucocyte Typing IV: White Cell Differentiation Antigens* (W. Knapp, B. Dörken, W.R. Gilks, E.P. Rieber, R.E. Schmidt, H. Stein and A.E.G.K. von dem Borne, eds). Oxford University Press, Oxford p. 711.

58. Dustin, M.L. and Springer, T.A. (1991) *Ann. Rev. Immunol.*, **9**, 27.

59. Deckert, M., Kubar, J., Zoccola, D., Bernard-Pomier, G., Angelisova, P., Horejsi, V. and Bernard, A. (1992) *Eur. J. Immunol.*, **22**, 2943.

60. Fox, D.A., Millard, J.A., Kan, L., Zeldes, W.S., Davis, W., Higgs, J., Emmrich, F. and Kinne, R.W. (1990) *J. Clin. Invest.*, **86**, 1124.

61. Lasky, L.A. (1992) *Science*, **258**, 964.

62. Metzelaar, M.J., Wijngaard, P.L., Peters, P.J., Sixma, J.J., Nieuwenhuis, H.K. and Clevers, H.C. (1991) *J. Biol. Chem.*, **266**, 3239.

63. Lund-Johansen, F., Olweus, J., Horjsi, V., Skubitz, K.M., Thompson, J.S., Vilella, R. and Symington, F.W. (1992) *J. Immunol.*, **148**, 3221.

64. Skubitz, K.M., Ducker, T.P. and Goueli, S.A. (1992) *J. Immunol.*, **148**, 852.

65. Holness, C.L. and Simmons, D.L. (1993) *Blood*, **81**, 1607.

66. Yamashita, I., Nagata, T., Tada, T. and Nakayama, T. (1993) *Int. Immunol.*, **5**, 1139.

67. Stein, H., Schwarting, R., Niedobitek, G. and Dallenbach, F. (1989) in *Leucocyte Typing IV: White Cell Differentiation Antigens* (W. Knapp, B. Dörken, W.R. Gilks, E.P. Rieber, R.E. Schmidt, H. Stein and A.E.G.K. von dem Borne, eds). Oxford University Press, Oxford p. 446.

68. Girones, N., Alverez, E., Seth, A., Lin, I.M., Latour, D.A. and Davis, R.J. (1991) *J. Biol. Chem.*, **266**, 19006.

69. Von Hoegen, I., Nakayama, E. and Parnes, J.R. (1990) *J. Immunol.*, **144**, 4870.

70. Lamb, C.A., Yewdell, J.W., Bennink, J.R. and Cresswell, P. (1991) *Proc. Natl. Acad. Sci. USA*, **88**, 5998.

71. Bast, B.J.E.G., Zhou, L.-J., Freeman, G.J., Colley, K.J., Ernst, T.J., Munro, J.M. and Tedder, T.F. (1992) *J. Cell Biol.*, **116**, 423.

72. Kniep, B., Muhlradt, P.F., Dorken, B., Moldenhauer, G., Vilella, R. and Schwartz-Albiez, R. (1990) *FEBS Lett.*, **261**, 347.

73. Mangeney, M., Richard, Y., Coulaud, D., Tursz, T. and Wiels, J. (1991) *Eur. J. Immunol.*, **21**, 1131.

74. Dörken, B., Möller, P., Pezzutto, A., Schwartz-Albiez, R. and Moldenhauer, G. (1989) in *Leucocyte Typing IV: White Cell Differentiation Antigens* (W. Knapp, B. Dörken, W.R. Gilks, E.P. Rieber, R.E. Schmidt, H. Stein and A.E.G.K. von dem Borne, eds). Oxford University Press, Oxford p. 122.

75. Melchers, F., Haasner, D., Grawunder, U., Kalberer, C., Karasuyama, H., Winkler, T. and Rollink, A.G. (1994) *Ann. Rev. Immunol.*, **12,** in press.

76. Wyss-Coray, T., Mauri-Hellweg, D., Baumann, K., Bettens, F., Grunow, R. and Pichler, W.J. (1993) *Eur. J. Immunol.*, **23,** 2175.

77. Schick, M.R., Nguyen, V.Q. and Levy, S. (1993) *J. Immunol.*, **151,** 1918.

78. Freeman, G.J., Gribben, J.G., Boussiotis, V.A., Ng, J.W., Restivo V.A. Jr, Lombard, L.A., Gray, G.S. and Nadler, L.M. (1993) *Science,* **262,** 909.

79. Zhou, L.J., Schwarting, R., Smith, H.M. and Tedder, T.F. (1992) *J. Immunol.*, **149,** 735.

80. Engel, P. *et al.* (1994) in *Leucocyte Typing V: White Cell Differentiation Antigens* (S.F. Schlossman, L. Boumsell, W. Gilks, J.M. Harlan, T. Kishimoto, C. Morimoto, J. Ritz, S. Shaw, R.L. Silverstien, T.A. Springer, T.F. Tedder and R.F. Todd, eds). Oxford University Press, Oxford.

81. Dumler, I., Petri, T. and Schleuning, W.D. (1993) *FEBS Lett.*, **322,** 37.

82. Gerard, C.J. (1994) *Ann. Rev. Immunol.*, **12,** in press.

83. Maliszewski, C.R., March, C.J., Schoenborn, M.A., Gimpel, S. and Shen, L. (1990) *J. Exp. Med.*, **172,** 1665.

84. Draberova, L. and Draber, P. (1993) *Proc. Natl. Acad. Sci. USA*, **90,** 3611.

85. LaMarre, J., Wolf, B.B., Kittler, E.L., Quesenberry, P.J. and Gonias, S.L. (1993) *J. Clin. Invest.*, **91,** 1219.

86. Aramburu, J., Balboa, M.A., Izquierdo, M. and Lopez-Botet, M. (1991) *J. Immunol.*, **147,** 714.

87. King, L.B. and Ashwell J.D. (1993) *Curr. Opin. Immunol.*, **5,** 368.

88. Wang, P.L., O'Farrell, S., Clayberger, C. and Krensky, A.M. (1992) *J. Immunol.*, **148,** 2600.

89. Lumadue, J.A., Glick, A.B. and Ruddle, F.H. (1987) *Proc. Natl. Acad. Sci. USA*, **84,** 9204.

90. Hale, L.P. and Haynes, B.F. (1992) J. *Immunol.*, **149,** 3809.

91. Bougeret, C., Mansur, I.-G., Dastot, H., Schmid, M., Mahouy, G., Bensussan, A. and Boumsell, L. (1991) *J. Immunol.*, **148,** 318.

92. Boumsell, L. *et al.* (1994) in *Leucocyte Typing V: White Cell Differentiation Antigens* (S.F. Schlossman, L. Boumsell, W. Gilks, J.M. Harlan, T. Kishimoto, C. Morimoto, J. Ritz, S. Shaw, R.L. Silverstien, T.A. Springer, T.F. Tedder and R.F. Todd, eds). Oxford University Press, Oxford.

93. Xu, H., Tong, I.L., De Fougerolles, A.R. and Springer, T.A. (1992) *J. Immunol.*, **149,** 2650.

94. Leung, E., Mead, P.E., Yuan, Q., Jiang, W.M., Watson, J.D. and Krissansen, G.W. (1993) *Int. Immunol.*, **5,** 551.

95. Hynes, R.O. (1992) *Cell*, **69,** 11.

96. Bellón, T., Corbí, A., Lastres, P., Catés, C., Cebrián, M., Vera, S., Cheifetz, S., Messague, J., Letarte, M. and Bernabéu, C. (1993) *Eur. J. Immunol.*, **23,** 2340.

97. Ruegg, C., Postigo, A.A., Sikorski, E.E., Butcher, E.C., Pytela, R. and Erle, D.J. (1992) *J. Cell Biol.*, **117,** 179.

98. Sawada, R., Lowe, J.B. and Fukuda, M. (1993) *J. Biol. Chem.*, **268,** 12675.

99. Klickstein, L. *et al.* (1994) in *Leucocyte Typing V: White Cell Differentiation Antigens* (S.F. Schlossman, L. Boumsell, W. Gilks, J.M. Harlan, T. Kishimoto, C. Morimoto, J. Ritz, S. Shaw, R.L. Silverstien, T.A. Springer, T.F. Tedder and R.F. Todd, eds). Oxford University Press, Oxford.

100. Sutherland, D.R., Yeo, E. *et al.* (1994) in *Leucocyte Typing V: White Cell Differentiation Antigens* (S.F. Schlossman, L. Boumsell, W. Gilks, J.M. Harlan, T. Kishimoto, C. Morimoto, J. Ritz, S. Shaw, R.L. Silverstien, T.A. Springer, T.F. Tedder and R.F. Todd, eds). Oxford University Press, Oxford.

101. Vairo, G. and Hamilton, J.A. (1991) *Immunol. Today*, **12,** 362.

102. Nicola, N.A. and Metcalf, D. (1991) *Cell,* **67,** 1.

103. Fleming, W.H., Alpern, E.J., Uchida, N., Ikuta, K. and Weissman, I.L. (1993) *Proc. Natl. Acad. Sci. USA*, **90,** 3760.

104. Farrar, M.A. and Schreiber, R.D. (1993) *Ann. Rev. Immunol.*, **11,** 571.

105. Stout, R.D. (1993) *Curr. Opin. Immunol.*, **5,** 398.

106. Dinarello, C.A. and Thompson, R.C. (1991) *Immunol. Today*, **12,** 404.

107. Harada, N., Yang, G., Miyajima, A. and Howard, M. (1992) *J. Biol. Chem.*, **267,** 22752.

108. Van Snick, J. (1990) *Ann. Rev. Immunol.*, **8,** 253.

109. Dadi, H.K., Ke, S. and Roifman, C.M. (1993) *Biochem. Biophys. Res. Comm.*, **192,** 459.

110. Cerretti, D.P., Kozlosky, C.J., Vanden-Bos, T., Nelson, N., Gearing, D.P. and Beckmann, M.P. (1993) *Mol. Immunol.*, **30,** 359.

111. Garni-Wagner, B., Purohit, A., Mathew, P.A., Bennett, M. and Kumar, V. (1993) *J. Immunol.*, **151,** 60.

Cell-surface Antigens

112. Pollok, K.E., Kim, Y.J., Zhou, Z., Hurtado, J., Kim, K.K., Pickard, R.T. and Kwon, B.S. (1993) *J. Immunol.*, **150**, 771.

113. Dougherty, G.J., Kay, R.J. and Humphries, R.K. (1989) *J. Biol. Chem.*, **264**, 6509.

114. Emi, N., Kitaori, K., Seto, M., Ueda, R., Saito, H. and Takahashi, T. (1993) *Immunogenetics*, **37**, 193.

115. Freeman, G.J., Gribben, J.G., Boussiotis, V.A., Ng, J.W., Restivo, V.A. Jr, Lombard, L.A., Gray, G.S. and Nadler, L.M. (1993) *Science*, **262**, 909.

116. Blikstad, I., Wikstrom, T., Aurivillius, M. and Obrink, B. (1992) *FEBS Lett.*, **302**, 26.

117. Korthauer, U., Graf, D., Mages, H.W., Briere, F., Padayachee, M., Malcolm, S., Ugazio, A.G., Notarangelo, L.D., Levinsky, R.J. and Kroczek, R.A. (1993) *Nature*, **361**, 539.

118. Mackay, C.R. (1993) *Curr. Opin. Immunol.*, **5**, 423.

119. Jackson, D.G., Hart, D.N., Starling, G. and Bell, J.I. (1992) *Eur. J. Immunol.*, **22**, 1157.

120. Schwartz, R.H. (1992) *Cell*, **71**, 1065.

121. Youssoufian, H., Longmore, G., Neumann, D., Yoshimura, A. and Lodish, H.F. (1993) *Blood*, **81**, 2223.

122. Fridman, W.H. (1993) *Curr. Opin. Immunol.*, **5**, 355.

123. Matthews, W., Jordan, C.T., Wiegand, G.W., Pardoll, D. and Lemischka, I.R. (1991) *Cell*, **65**, 1143.

124. Quehenberger, O., Prossnitz, E.R., Cavanagh, S.L., Cochrane, C.G. and Ye, R.D. (1993) *J. Biol. Chem.*, **268**, 18167.

125. Arai, K., Lee, F., Miyajima, A., Miyatake, S., Arai, N. and Yokata, T. (1990) *Ann. Rev. Biochem*, **59**, 783.

126. Seaman, W.E., Niemi, E.C., Stark, M.R., Goldfien, R.D., Pollock, A.S. and Imboden, J.B. (1991) *J. Exp. Med.*, **173**, 251.

127. Arm, J.P., Gurish, M.F., Reynolds, D.S., Scott, H.C., Gartner, C.S., Austen, K.F. and Katz, H.R. (1991) *J. Biol. Chem.*, **266**, 15966.

128. Cambier, J.C., Pleiman, C. and Clark, M.R. (1994) *Ann. Rev. Immunol.*, **12**, in press.

129. Colotta, F., Bussolino, F., Polentarutti, N., Guglielmetti, A., Sironi, M., Bocchietto, E., De Rossi, M. and Mantovani, A. (1993) *Exp. Cell Res.*, **206**, 311.

130. Burke, F., Naylor, M.S., Davies, B. and Balkwill, F. (1993) *Immunol. Today*, **14**, 167.

131. Moore, K.W., O'Garra, A., de Waal Malefyt, R., Vieira, P. and Mosmann, T.R. (1993) *Ann. Rev. Immunol.*, **11**, 165.

132. Perussia, B., Chan, S.H., D'Andrea, A., Tsuji, K., Santoli, D., Pospisil, M., Young, D., Wolf, S.W. and Trinchieri, G. (1992) *J. Immunol.*, 149, 3495.

133. Zurawski, S.M., Vega Jr, F., Huyghe, B. and Zurawski, G. (1993) *EMBO J.*, **12**, 2663.

134. Triebel, F., Jitsukawa, S., Baixeras, E., Roman, S., Genevee, C., Viegas-Pequignot, E. and Hercend, T. (1990) *J. Exp. Med.*, **171**, 1393.

135. Collawn, J.F., Kuhn, L.A., Liu, L.F., Tainer, J.A. and Trowbridge, I.S. (1991) *EMBO J.*, **10**, 3247.

136. Bradbury, L.E., Kansas, G.S., Levy, S., Evans, R.L. and Tedder, T.F. (1992) *J. Immunol.*, **149**, 2841.

137. Gearing, D.P., Comeau, M.R., Friend, D.J., Gimpel, S.D., Thut, C.J., McGourty, J., Brasher, K.K., King, J.A., Gillis, S., Mosley, B. *et al.* (1992) *Science*, **255**, 1434.

138. Kozutsumi, H., Toyoshima, H., Hagiwara, K., Furuya, A., Mioh, H., Hanai, N., Yazaki, Y. and Hirai, H. (1993) *Biochem. Biophys. Res. Comm.*, **190**, 674.

139. Sandrin, M.S., Gumley, T.P., Henning, M.M., Vaughan, H.A., Gonez, L.J., Trapani, J.A. and McKenzie, I.F. (1992) *J. Immunol.*, **149**, 1636.

140. Yokoyama, W.M. and Seaman, W.E. (1993) *Ann. Rev. Immunol.*, **11**, 613.

141. Kasinrerk, W., Fiebiger, E., Stefanova, I., Baumruker, T., Knapp, W. and Stockinger, H. (1992) *J. Immunol.*, **149**, 847.

142. Law, S.K.A., Micklem, K.J., Shaw, J.M., Zhang, X.-P., Dong, Y., Willis, A.C. and Mason, D.Y. (1993) *Eur. J. Immunol.*, **23**, 2320.

143. Truong, M.J., Gruart, V., Kusnierz, J.P., Papin, J.P., Loiseau, S., Capron, A. and Capron, M. (1993) *J. Exp. Med.*, **177**, 243

144. Cohen-Dayag, A., Schneider, H. and Pecht, I. (1992) *Immunobiol.*, **185**, 124.

145. Ezekowitz, R.A.B., Sastry, K., Bailly, P. and Warner, A. (1990) *J. Exp. Med.*, **172**, 1785.

146. Ii, M., Kurata, H., Itoh, N., Yamashina, I. and Kawasaki, T. (1990) *J. Biol. Chem.*, **265**, 11295.

147. Phillips, R.L., Reinhart, A.J. and Van-Zant, G. (1992) *Proc. Natl. Acad. Sci. USA*, **89**, 11607.

148. Lawlor, D.A., Zemmour, J., Ennis, P.D. and Parham, P. (1990) *Ann. Rev. Immunol.*, **8**, 23.

149. Trowsdale, J., Ragoussis, J. and Campbell, R.D. (1991) *Immunol. Today*, **12**, 443.

150. Koehler, M., Goorha, R., Kitchingman, G.R., Ayers, G.D. and Mirro J., Jr (1992) *Blood*, **80**, 367.

151. Vigon, I., Mornon, J.P., Cocault, L., Mitjavila, M.T., Tambourin, P., Gisselbrecht, S. and Souyri, M. (1992) *Proc. Natl. Acad. Sci. USA*, **89,** 5640.

152. McCaughan, G.W., Clark, M.J. and Barclay, A.N. (1987) *Immunogenetics*, **25,** 329.

153. Mallett, S., Fossum, S. and Barclay, A.N. (1990) *EMBO J.*, **9,** 1063.

154. Skubitz, K.M., Stroncek, D.F. and Sun, B. (1991) *J. Leuk. Biol.* **49,** 163.

155. Yabe, T., McSherry, C., Bach, F.H., Fisch, P., Schall, R.P., Sondel, P.M. and Houchins, J.P. (1993) *Immunogenetics*, **37,** 455.

156. Muller, E., Dagenais, P., Alami, N. and Rola-Pleszczynski, M. (1993) *Proc. Natl. Acad. Sci. USA*, **90,** 5818.

157. Rebbe, N.F., Tong, B.D. and Hickman, S. (1993) *Mol. Immunol.*, **30,** 87.

158. Koch, F., Haag, F., Kashan, A. and Thiele, H.G. (1990) *Proc. Natl. Acad. Sci. USA*, **87,** 964.

159. Fleming, T.J., O'Huigin, C. and Malek, T.R. (1993) *J. Immunol.*, **150,** 5379.

160. Kodama, T., Freeman, M., Rohrer, L., Zabrecky, J., Matsudaira, P. and Krieger, M. (1990) *Nature*, **343,** 531.

161. Okazaki, M., Luo, Y., Han, T., Yoshida, M. and Seon, B.K. (1993) *Blood*, **81,** 84.

162. Leppä, S., Mali, M., Miettinen, H.M. and Jalkanen, M. (1992) *Proc. Natl. Acad. Sci. USA*, **89,** 932.

163. Messagué, J. (1992) *Cell*, **69,** 1067.

164. Godfrey, D.I., Masciantonio, M., Tucek, C.L., Malin, M.A., Boyd, R.L. and Hugo, P. (1992) *J. Immunol.*, **148,** 2006.

Cell-surface Antigens

CHAPTER 8
ANTIGEN RECEPTORS
A.R. Venkitaraman

1. STRUCTURE

1.1. General features

Antigen receptors are formed of pairs of polypeptide chains: α/β or γ/δ in the T-cell receptor (TCR) and heavy (H)/light (L) in immunoglobulin (Ig) (*Figure 1*). TCR molecules are heterodimers, and Ig molecules are tetramers of two H and two L chains. The two H chains within any given tetramer are identical to one another, as are the L chains. The individual polypeptide chains of TCR and Ig molecules are subdivided into amino-terminal variable (V) and carboxyl-terminal constant (C) regions.

Figure 1. Structure of Ig and TCR molecules. (a) General features of IgG. One membrane-bound and one secretory H chain are shown for convenience, but such an association does not occur naturally. CDR, complementarity-determining region; TM, transmembrane segment; CYT, cytoplasmic tail. (b) TCR α/β and γ/δ heterodimers. The two allelic forms of the human γ chain are shown; $\gamma2$ does not pair covalently with δ chains. Glycosylation sites are not shown.

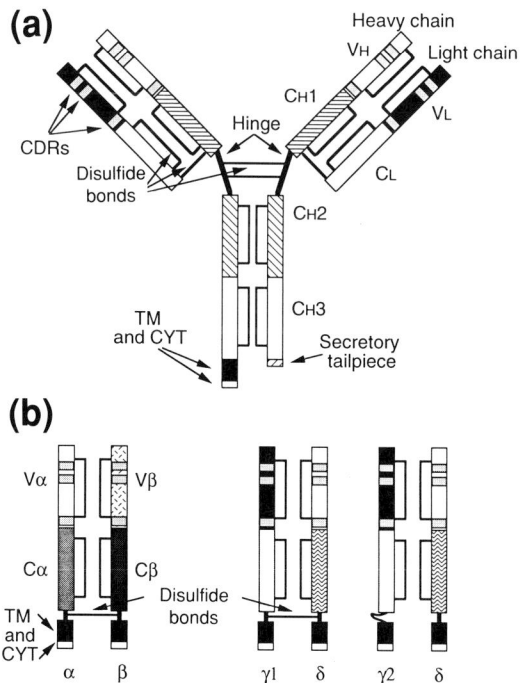

Each chain is modular in its organization, the basic module (termed the immunoglobulin superfamily or IgSF domain) comprising ~110 amino acid residues (1). TCR chains and Ig L chains contain two IgSF domains each, Ig H chains contain either four or five. Individual IgSF domains in a chain are connected by short polypeptide strands which permit them to rotate relative to one another. Each IgSF domain assumes a three-dimensional structure consisting of two β-pleated sheets, approximately parallel to one another and held together by a disulfide bond. In C-region IgSF domains, one of the sheets has four, and the second, three, strands, connected at either end by amino acid loops extending outwards from the sheets (*Figure 2*).

1.2. Immunoglobulin heavy chains

There are five C-region classes or isotypes; μ, δ, γ, ε and α. The γ and α classes are split into subclasses (*Table 1*). Membrane-bound and secretory forms of each class are distinguished by the presence of a transmembrane (TM) anchor sequence and cytoplasmic tail in the membrane-bound form, replaced by a short segment of hydrophilic amino acids (the secretory tailpiece) in the secretory form (*Table 1*).

A typical Ig H chain consists of a V region (V_H) comprising one IgSF domain, which together with the first C region domain (C_H1) forms the antigen-binding 'arm' (*Figure 1a*). The arm is linked to the C_H2 and C_H3 domains through a flexible hinge region (*Table 1*), which has little sequence homology between H chains of different classes, except for the relative abundance of Pro and Cys residues accounting for the structural flexibility. Disulfide bridges between the two H chains are formed through Cys residues in the hinge (*Figure 1a*), and vary in number from 1 (in δ) to 11 (in human γ3). In μ and ε H chains the hinge region is replaced by an entire, additional IgSF domain (numbered C_H2) and so these two H chains have four rather than three C-region domains. All Ig H chains bear N-linked glycosylation sites. They vary in number from one in the C_H2 domain of γ chains to five and six in μ and ε chains, respectively.

1.3. Immunoglobulin light chains

There are two classes, κ and λ. Each may associate with any H-chain isotype. The κ:λ ratio varies from species to species (human, 70:30; mouse, 95:5). Each chain comprises a V_L domain linked to a C_κ or C_λ domain (*Figure 1a*). Subclasses exist (e.g. at least nine C_λ subclasses and a single C_κ class in humans), but there is very little difference between their amino acid sequences.

Figure 2. Structure of a C-region IgSF domain. A *side* view is shown. β-strands are designated by letters, those in the upper plane are thick solid lines and those in the lower (shaded) plane are thick broken lines. The loops connecting the strands are thin lines, as is the disulfide bridge between strands B and F.

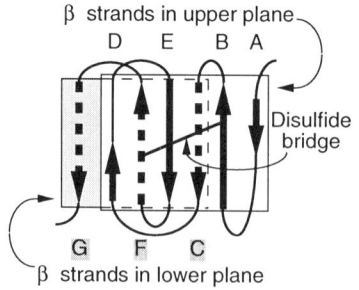

Table 1. Structure of murine (Mo) and human (Hu) heavy chains[a]

	μ	δ	γ	ε	α
C_H domains	C_H1–4	C_H1 and C_H3 (Mo) C_H1–3 (Hu)	C_H1–3	C_H1–4	C_H1–3
Subclasses	–	–	γ1, γ2a, γ2b, γ3 (Mo) γ1–4 (Hu)	–	α1, 2 (Hu)
Size (number of amino acids)					
Hinge region	0	35 (Mo) 70 (Hu)	13–22 (Mo) 12–62 (Hu)	0	14 (Mo) 7–20 (Hu)
Cytoplasmic tail	3	3	28	28	14
Secretory tailpiece	20	22 (Mo) 8 (Hu)	2	8	20
M_r	72	63	40–55	72	57

[a] Data from refs 2, 3.

1.4. Class differences between Ig molecules

Secreted Ig molecules differ in their capacity to: (i) bind to and activate complement components; and (ii) bind to Fc receptors (*Table 2*). These differences in protein binding generally arise in the C_H2 and C_H3 domains. Class-specific differences in the hinge region may give rise to functional differences, e.g. the protease sensitivity of IgD has been attributed to its long, charged hinge. Membrane-bound forms of Ig H chains also differ in their TM and cytoplasmic segments (*Table 1*), but the functional role of these differences is not yet understood.

In secreted IgM and IgA, the basic H_2/L_2 subunits may themselves multimerize. Thus secreted IgM is a pentamer and IgA secreted from mucosal tissues (secretory IgA) is a dimer of H_2/L_2 subunits (*Table 2*). Each multimer is associated with a 137 amino acid J chain encoded by a gene distant from the Ig gene loci. It contains multiple Cys residues which form intrachain disulfide bonds, and linkages to the H chains.

Secretory IgM contains a region in its carboxyl-terminus which binds the endoplasmic reticulum resident protein BiP (binding protein, see also Section 7). Thus, monomeric forms are retained intracellularly; pentamer formation displaces BiP and allows secretion.

Dimeric IgA/J chain is particularly abundant in saliva, breast milk and the secretions of the respiratory, gastrointestinal and genitourinary tracts. In these secretions the IgA/J chain complex is associated with secretory component (sc), which consists of five IgSF domains constituting the extracellular portion of the polymeric Ig receptor on epithelial cells.

1.5. T-cell receptor chains

Each TCR chain is structurally similar to an Ig L chain, except that they are membrane-bound (*Figure 1b*). Each has two extracellular IgSF domains (the amino-terminal V region, and the carboxyl-terminal C region), which are predicted to be very highly homologous in

Table 2. Properties of human secreted Ig molecules[a]

	IgM	IgD	IgG				IgE	IgA	
			G1	G2	G3	G4		A1/2	sec
Composition									
H$_2$/L$_2$ units	5	1	1				1	1	2
Assoc. chains	J	–	–				–	–	J,SC
Approx. M_r	950	175	160 (G3) 150 (other)				190	160	415
Serum conc.[b] (% serum Ig)	1 (5–10)	0.03 (0.5)	15 (70–85)				trace –	2 (5–15)	
Complement activation[c]	+ +	–	+	±	+	–	–	–	
Distribution of Fc receptors	–	–	Macrophages Neutrophils				Mast cells	–	
Placental transfer	–	–	+	–	+	±	–	–	

[a] Data from refs 2, 3.
[b] Average serum concentration in adults (mg ml^{-1}).
[c] Classical pathway activation is shown. IgG4 and IgA also activate the alternative pathway.

structure to the corresponding Ig domains based on amino acid sequence conservation. A short connecting peptide links the C region to the membrane; the single interchain disulfide bond between the α/β or γ/δ chains is made through a Cys residue in this peptide (*Figure 1b*). The TM anchor of 22 amino acids contains one (in β and γ) or two (in α and δ) positively charged amino acids (Section 7.1). Cytoplasmic tails are of just five residues.

C-region subclasses of β and γ exist. The two Cβs differ by just a few, highly conservative, amino acid changes. In contrast, human Cγ2 contains 16–32 additional residues just amino-terminal to the TM segment when compared to Cγ1, and lacks the conserved Cys residue that makes the interchain disulfide bond (*Figure 1b* and *Figure 7*). Therefore human γ-chains range from 40 kDa (γ1) to 55 kDa (γ2); the other TCR chains range between 40 and 45 kDa. Variability in molecular weight, also due to glycosylation differences, varies from chain to chain and exhibits species differences.

1.6. Antigen binding
Each V region domain (*Figure 3a*) assumes a structure similar to the C region IgSF domains depicted in *Figure 2*, except for an extra strand in the lower β-pleated sheet. Four 'framework' regions (FR1-4) surround three highly variable regions ('complementarity determining' regions CDR1-3). The CDRs are loops extending out from the β-sheets of the FRs (*Figure 3*), and form the antigen combining site, which may range from a flat surface to a deep pocket (*Figure 3b*). Analysis of solved structures suggests that only a relatively small number of 'canonical' CDR conformations exists. However, differences in the FR regions (which alter the relative positions of the CDR loops), as well as sequence differences in a few critical residues near the centre of the antigen-combining site, enable specific binding to diverse antigens. FR residues distant from the antigen contact sites can have dramatic effects upon binding.

Figure 3. The antigen-binding site. (a) A V region IgSF domain in *side* view (cf. *Figure 2*). The portions of the polypeptide encoded by the V, D and J segments (see Section 2.1) are marked, as are the CDR loops. (b) The antigen binding site in *top* view. The CDR loops depicted as rectangles surmount the planes of the β-sheets shown in (a) and extend outwards and upwards from the page.

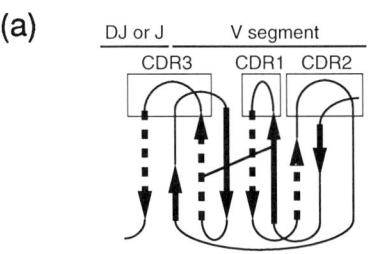

(a)

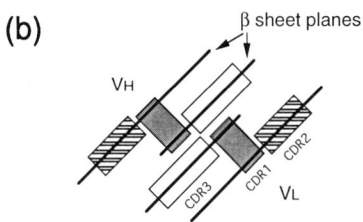

(b)

Although TCR structure has not yet been solved crystallographically, the degree of conservation between the amino acid sequences of Ig and TCR V domains suggests that they will assume a similar configuration. Attempts have been made to dissect the contribution of individual CDR loops to peptide, as distinct from MHC, binding. One proposal has been that CDRs 1 and 2 of the paired V regions form the sides of a shallow cup, the base of which is formed by the two CDR3s; thus the sides of the cup contact the MHC α-helices, while the base (CDR3) binds the antigenic peptide between them (cf. *Figure 3b*).

2. GENES ENCODING ANTIGEN RECEPTORS

2.1. Organization

The antigen-binding V region domains of TCR and Ig are encoded by two or three distinct gene segments (V, variable; D, diversity; J, joining, *Figure 3a* and *Table 3*), separated in the germ line and brought together by a series of DNA rearrangements during lymphocyte development. The number of distinct V segments may run into the hundreds, and when combined with tens of D and J segments, a very large pool of receptor specificities can be generated combinatorially. Not all TCR and Ig chains make use of D segments; the V region domains of Ig L chains and TCR α and γ are encoded by rearranged V–J (and not V–D–J) segments.

Each rearranged V–D–J or V–J segment may be linked to different C region exons, permitting TCR and Ig of a given C region to assume virtually any specificity. For Ig H chains, a given V–D–J segment may be linked to C-region exons of each of the five classes at different stages in the life history of a single B cell (see Section 5.2). The V, D, J gene segments and C-region exons encoding each of the Ig and TCR chains are clustered together, usually 5′

Table 3. Estimated approximate number of V, D and J gene segments (not including pseudogenes). The actual number can vary somewhat between different individuals (e.g. between different inbred strains of mice)

Gene	IgH	Igκ	Igλ	TCRα	TCRβ	TCRγ	TCRδ
V							
Mouse	128	160	2	50	20	7	8
Human	80	100	40	50	55	8	6
D							
Mouse	12	–	–	–	2	–	2
Human	≥ 5	–	–	–	2	–	3
J							
Mouse	4	5	4	> 70	12	3	2
Human	6	4	9	100	13	5	3

to 3′. Varying numbers of these segments (and some C regions) are pseudogenes. The TCR δ genes are embedded within the α locus between the Vα and Jα clusters. None of the other TCR or Ig loci are linked to one another, even if located on the same chromosome (*Table 4*).

2.2. The Ig heavy chain locus

C_H genes
Mouse. Eight C_H genes, span 200 kb (*Figure 4*). Each consists of multiple exons, three or four of which encode the C_H IgSF domains, one the hinge region (where present), and two the TM segment and short cytoplasmic tail of the membrane-bound forms. Two polyadenylation (polyA) sites are present.

Human. The 350 kb C_H locus contains nine genes and three pseudogenes (ψ) in the following 5′ to 3′ order: J_H–9 kb–μ–8 kb–δ–60 kb–γ3–26 kb–γ1–19 kb–ψε1–13 kb–α1–35 kb–ψγ–25 kb–γ2–19 kb–γ4–23 kb–ε–10 kb–α2. Thus a gene duplication event has given rise to two copies of the γ–γ–ε–α set at the 3′ end. In humans additional C_H pseudogenes are found outside the IgH locus (e.g. ψε2 on chromosome 9, and at least one more ψγ).

D and J segment clusters
The D_H and J_H segments together encode 17–38 amino acids at the carboxyl-terminal end of the V-region domain corresponding to CDR3 (*Figure 3a*). D_H segments are 3–40 nucleotides in length; J_H segments, 45–60.

Table 4. Chromosomal location of Ig and TCR loci

	IgH	IgL		TCR			
		κ	λ	α	β	γ	δ
Mouse	12	6	16	14	6	13	14[a]
Human	14	2	22	14	7	7	14[a]

[a]The δ locus is embedded within the α locus.

Figure 4. The Ig H-chain gene locus. The murine Ig H gene locus (4,5) is shown in the top half of the figure, with intergenic distances between the J$_H$ cluster and C region genes in kb (not to scale). The exon/intron structure of the murine Cμ gene (4) is shown below. Exons are shown as boxes, and introns as lines. M1 and M2 denote the two membrane exons.

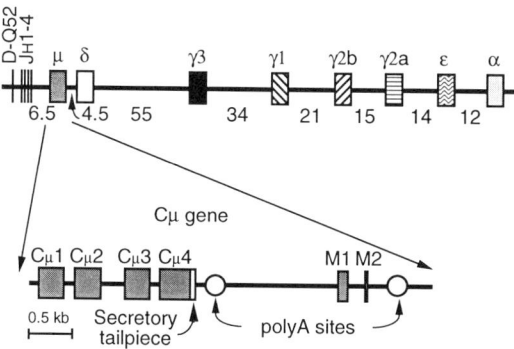

Mouse. Eleven D$_H$ segments (D-SP2 family with nine members, D-FL16 with two) map within 60 kb, located 20 kb 5′ to J$_H$ cluster (J$_H$1–4). An additional D$_H$ segment (D-Q52) lies 700 bp 5′ of J$_H$1 (*Figure 4*).

Human. The J$_H$ cluster (∼3 kb) contains J$_H$1-6 and three pseudogenes (ψJ$_H$1–3), and lies 9 kb 5′ to Cμ. A single D segment homologous to the murine D-Q52 is contained within the J$_H$ cluster between the 5′-most pseudogene, ψJ$_H$1, and J$_H$1. The number and organization of the remaining D$_H$ segments has not yet been established. At least four D$_H$ segments lie in a 33 kb region 5′ of the J$_H$ cluster.

V$_H$ genes
Each V$_H$ segment consists of a 5′ exon of ∼80 bp encoding the leader sequence, and a V$_H$ exon of 300 bp encoding the first 100 amino acids of the V region, including CDRs 1 and 2 (*Figure 3a*).

Mouse. Approximately 128 V$_H$ segments are grouped into 14 families (1–60 members each). Within each family, nucleotide sequence homologies are of the order of 80%. Members of each V$_H$ family cluster together on the chromosome.

Human. There are more than 120 V$_H$ segments, of which ∼40 are pseudogenes. Some of the functional V$_H$S represent different alleles of the same segment, which vary from individual to individual, thus the number of unique segments is likely to be smaller. Six families are interspersed on the chromosome (in contrast to their murine counterparts). Most human V$_H$ segments map within a region extending some 1000 kb 5′ of the J$_H$ cluster. Additional human V$_H$ segments are located distant from the Ig loci on chromosomes 15 and 16; however, not all have associated D$_H$ and J$_H$ segments.

2.3. The Ig light-chain gene loci

One V$_L$ and one J$_L$ rearrange to encode the V region. The C$_L$ domain is encoded by a single exon. The murine κ locus includes five Jκ segments located 3 kb 5′ to a single Cκ exon. An estimated 160 murine Vκ segments (similar in structure to V$_H$ segments) are grouped into 19 families. The human κ locus is similar in organization, but is likely to contain no more than

Antigen Receptors

100 Vκ and four Jκ segments. The murine λ locus contains two repeats of a Vλ–Jλ–Cλ-Jλ-Cλ unit (i.e. two Vλ segments, and four Cλs – each with a Jλ just 5'). The restricted repertoire of V and J segments may be related to the fact that only 5% of murine antibodies use a λ L chain. In contrast, nearly one-third of human antibodies use a λ L chain and the human λ gene locus is correspondingly more diverse. At least 40 Vλ segments exist, grouped into seven families. Nine Cλ genes, each with an adjacent Jλ, have been identified, and six are linked in a 50 kb fragment.

2.4. The TCR gene loci
The human and murine loci have a conserved organization and, except in a few specific instances, will be considered together.

Organization of TCR gene loci
Just 5' to each of two Cβ genes (*Figure 5*) are seven Jβ segments (Jβ1.1–1.7, Jβ2.1–2.7). One Dβ lies 5' of each Jβ cluster (Dβ1 and Dβ2); Dβ1 may rearrange to any Jβ, Dβ2 only to Jβ2.1–2.7. One Jβ in each cluster is a pseudogene.

The α gene locus contains >70 Jα segments over a 60 kb region 5' of Cα. They are separated from the Vα segments by the interposed TCR δ gene locus, spanning 40 kb (*Figure 6*). The δ locus contains three functional Dδ and three functional Jδ segments located 5' to the Cδ exons.

Each of the two human Cγ genes is located 3' from a cluster of Jγ segments. The Jγ1 cluster contains Jγ1.1–1.3, of which the first and third are highly homologous to the two Jγ segments (Jγ2.1 and 2.3) in the second Jγ cluster. There are three separate Cγ genes in the mouse, each with a small cluster of Vγ and Jγ segments just 5' to it; and a fourth Cγ pseudogene.

Figure 5. The TCR β-chain gene locus. The murine TCR β gene locus (not to scale) is shown above, with the exon/intron structure of the Cβ1 gene below (6). The exon encoding the C region IgSF domain is marked C.

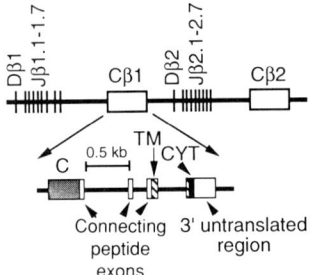

Figure 6. Organization of the TCR α and δ gene loci. The position of the Vα and Vδ segments 5', and the Jα segments 3', to the δ gene locus is shown (6). The Vδ3 segment is in reverse transcriptional orientation to Cδ.

TCR V segments

Although there are somewhat fewer Vβs (∼20 in mice, 55 human) and Vαs (∼50 in both species) than V$_H$s, there is greater sequence diversity in TCR V regions. The Vβs are grouped into 22 families, the Vαs into > 30; each contains 1–20 members which are typically little more than 75% homologous. In contrast, the γ and δ loci have very limited V segment diversity. Fourteen Vγ segments (four families) are found in humans, six are pseudogenes. Only six Vδ segments have been identified; one is 3' to Cδ in an inverted transcriptional orientation (*Figure 6*). There is no reason *a priori* why Vα segments (which lie 5' to the Dδ cluster) could not be used in Vα–Dδ–Jδ rearrangements and vice versa. Although some overlap between the α and δ loci in V-segment usage has been detected, in general such rearrangements do not occur.

Structure of the TCR C-region genes

Each of the two Cβ genes contains four exons (*Figure 5*). The Cα and Cδ genes have a similar exon–intron structure, except that the first exon contains only the IgSF domain and the third exon includes the cytoplasmic tail, leaving only the 3' untranslated (UT) region to the fourth exon.

The human Cγ genes comprise 3–5 exons (*Figure 7*). Cγ1 has a single copy of exon 2; Cγ2 occurs in two allelic forms which contain two or three copies, respectively. Only the Cγ1 connecting peptide contains a Cys residue which enables disulfide linkage to the δ chain. The three functional murine Cγs all contain the disulfide-linked Cys residue, although the connecting peptide varies from 10 to 33 amino acids.

Figure 7. Human Cγ genes. The Cγ1 gene and the two allelic forms of Cγ2 are shown.

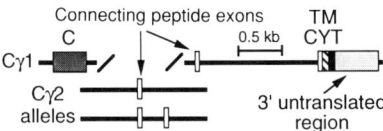

3. REARRANGEMENT OF ANTIGEN-RECEPTOR GENES (7)

3.1. Joining signals and the 12/23 rule

Conserved heptamer and nonamer sequences are separated by a degenerate spacer element of either 12 or 23 base pairs. The joining signals, which may take the form 5'–heptamer–spacer–nonamer–3' (or its complement), are found immediately 3' of each V segment either side of each D segment, and immediately 5' of each J segment. The consensus heptamer sequence is 5'–CACAGTG–3' (with the first four bases essentially invariant); and that of the nonamer is 5'–ACAAAAACC–3' (with positions 2, 5, 6 and 7 highly conserved). Two gene segments may only rearrange if they are flanked by joining signals of different spacer length (the so-called 12/23 rule) which point in opposite directions (*Figure 8a*). The signal organization is such that V–D–J rearrangement will only occur in that order, and V–V, D–D, J–J or V–J joins cannot occur (except for D–D joins in the TCR δ-chain, see Section 4.1).

3.2. Mechanisms of joining *(Figure 8)*

(i) Deletion joining: occurs between segments in the same 5'–3' transcriptional orientation and results in the production of rearranged segments (the 'coding joint') and a DNA circle containing the fused joining signals (the 'signal joint').

(ii) Inversion joining: occurs between segments that are in opposite transcriptional orientations. The signal joint is left linked to the rearranged segments on the same DNA strand.

Figure 8. Mechanisms of joining. Joining signals are depicted by triangles, with filled triangles denoting signals containing a 23 bp spacer, and hatched triangles those with a 12 bp spacer. Triangles point 'towards' the nonamer sequence. Transcriptional orientation of V, D and J segments is shown by arrows. Note that inversion joining (b) occurs between segments in the opposite orientation.

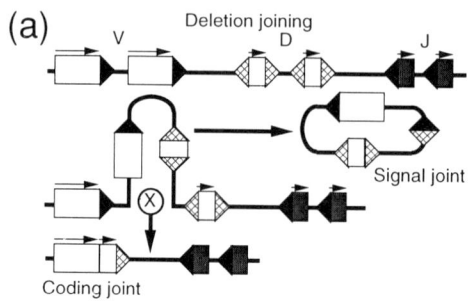

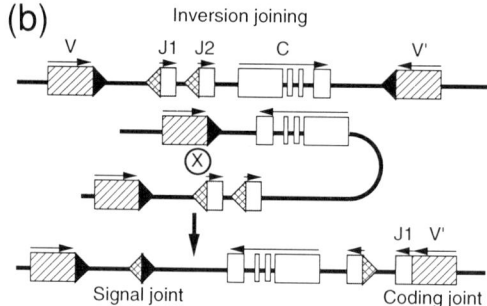

(iii) Rarely, rearrangements occur between gene segments on different chromosomes by sister chromatid exchange.

3.3. Imprecision and nucleotide addition at coding joints

The formation of coding joints (but not signal joints) is very imprecise, with nucleotide deletions and additions of 1–12 bases. This contributes greatly to diversity, since the junctional regions encode part of CDR3 (*Figure 3a*). 'N regions': nucleotides added by terminal deoxyribonucleotidyl transferase (TdT) to the ends of segments; G and C additions are preferred. 'P regions': 1–3 base additions, exactly complementary to the last few nucleotides of the segments to which they are added (the enzyme(s) responsible have not yet been identified).

3.4. Components of the rearrangement machinery

The enzymatic machinery that mediates rearrangement steps is not yet clear, although it has been shown to be the same both in B and T cells. In addition to TdT, the recombination activating genes RAG-1 and RAG-2 are essential for rearrangement to occur (7).

3.5. Order of rearrangement

D_H to J_H, followed by V_H to D_H–J_H. Successful IgH rearrangement allows L-chain rearrangement to proceed. The κ locus is the first to rearrange, with Vκ to Jκ joining. In some murine B cells, a Vκ segment joins to a heptamer–nonamer joining signal located about 20 kb distal to the Cκ gene (the κ deleting element) resulting in deletion of the κ locus. Cells which have undergone κ deletion proceed to initiate Vλ to Jλ rearrangement.

In general, TCR rearrangements are similar to those of Ig, but the order of gene rearrangement is less well understood. In populations, rather than individual cells, δ rearrangements are the earliest to be detected in fetal thymi, soon followed by γ and β locus rearrangements. The α locus is the last to rearrange.

This picture is complicated by increasing evidence that γ/δ and α/β T cells represent independent lineages (8). Although some α/β lymphocytes may contain γ locus rearrangements, and γ/δ cells may contain β rearrangements, it is believed that commitment to the lineages is determined before, or regardless of, these rearrangements. Because of the location of the δ locus between the Vα and Jα clusters, rearrangements at the α locus must be prevented in γ/δ cells; lineage commitment may therefore be regulated at this level.

3.6. Allelic exclusion

During IgH rearrangement, B-lymphocyte progenitors express two nonrearranging genes, V pre-B and $\lambda 5$ (located together on murine chromosome 16, human chromosome 22), which are important for allelic exclusion to occur (9). The V pre-B gene encodes a single IgSF domain similar in structure to V_L, and the $\lambda 5$ protein is somewhat larger than, but similar to, a C_L domain. The V pre-B and $\lambda 5$ proteins together form a 'surrogate' light chain, which can associate with the newly synthesized membrane-bound μ heavy chain and as yet unidentified proteins to form a μ/surrogate light-chain complex. It is believed that complex formation transmits a signal to induce IgH allelic exclusion, whereupon the cell proceeds to IgL rearrangement and synthesis of a κ or λ light chain. Association of membrane-bound μ with κ or λ could then signal for IgL allelic exclusion.

Allelic exclusion at the TCR loci has been less well studied. Introduction of a fully rearranged β-chain transgene into the mouse germ line induces allelic exclusion at the TCR β locus. However, allelic exclusion is not absolute at the TCR α and γ loci, and mature, peripheral T cells expressing more than one functional α- or γ-chain protein have been detected. There is some recent evidence that β chains in immature thymocytes may form a complex with other, non-TCR proteins but homologs of the surrogate light chains have not yet been described in T cells.

4. GENERATION OF DIVERSITY

4.1. Combinatorial and junctional diversity

It has been estimated that antigen receptors must recognize in the order of 10^9–10^{11} antigenic specificities. Much of the diversity (*Table 5*) arises as a result of gene rearrangement:

(i) Combinatorial diversity, generated by rearrangement of hundreds of distinct V segments with tens of J (and sometimes D) segments.
(ii) Junctional diversity, generated by the addition or deletion of nucleotides at the ends of the rearranging segments with resultant changes in the encoded protein (see Section 3.3).
(iii) Pairing of Ig H/L chains or TCR α/β and γ/δ, each of which contributes to antigen specificity, multiplies the number of specificities that can potentially be recognized.

Although the Ig H chain locus has more V_H segments than the α or β loci, the overall sequence diversity of the TCR V segments is greater, most notably in the FR regions which are fairly well-conserved in V_H segments. The complexity of the TCR loci is multiplied by the presence of larger numbers of Jα and Jβ segments, which are also longer and more divergent in their sequences. Combinatorial diversity in TCR γ/δ receptors is limited by the small numbers of V segments. However, junctional diversity is found to be particularly extensive and is augmented in the δ locus by the occasional use of two Dδ segments in tandem (i.e. V–D–D–J).

Table 5. Potential repertoires of Ig and TCR specificities[a]

	IgH/κ	TCR α/β	TCR γ/δ
Combinations of V-segment pairing between chains	2×10^4 (Hu)– 250×10^4 (Mo)	2.5×10^3 (Mo)– 8×10^3 (Hu)	50 (Mo)– 70 (Hu)
Diversity in D–J junctional regions[b]	10^{11}	10^{15}	10^{18}

[a]Data from ref. 10.
[b]Including the contributions of: D/J segment usage, nucleotide additions at junctions, differences in translational reading frame at junctions.
Abbreviations: Hu, human; Mo, mouse.

Allelic variation is another feature of Ig and TCR loci. Many human and murine V segments occur in different allelic forms which vary from individual to individual, or strain to strain. Inbred mouse strains may also carry genomic deletions which encompass the V-segment clusters; similar deletions have been reported in humans.

4.2. Somatic hypermutation of Ig genes

Two processes account for the improvement in antibody affinity generally seen in secondary immune responses. First, the V segments used in secondary antibodies are often completely different to those used in a primary response. Secondly, amino acid sequence analysis reveals mutations clustered in and around the CDRs; at the nucleotide sequence level these mutations predominantly take the form of base substitutions ('somatic hypermutation') in the rearranged V–D–J or V–J segments (11).

The observed frequency of hypermutation is in the order of 10^{-3}/nucleotide/cell division, which is $\sim 10^4$–10^5 times greater than the frequency of random mutation in chromosomal DNA. Mutations are specifically targeted to rearranged V–D–J or V–J segments; with the caveat that the frequency of mutation in germ line V segments is difficult to measure. Mutations cluster at so-called 'hotspots' in and around the regions encoding CDRs 1–3, and exhibit a distinct predilection for certain types of base substitution. The mechanism of somatic hypermutation is poorly understood. Ig transcriptional enhancers (Section 5.1) have been implicated. That such enhancers are highly B-lineage specific (and therefore nonfunctional in T cells) is in keeping with the fact that somatic hypermutation does not occur in the TCR gene loci.

5. EXPRESSION OF Ig AND TCR GENES

The expression of rearranged Ig and TCR genes is tightly regulated at many levels (12,13).

5.1. Transcriptional regulation of Ig and TCR genes

Each V_H and V_L segment in the germ line carries its own promoter element just 5′ to the leader exon (Section 2.2), containing a well-conserved octanucleotide sequence which is essential for its function. The long intron between the J_H cluster and $C\mu$ (*Figure 4*) contains a strong, B-lineage-specific, transcriptional enhancer element (the IgH intron enhancer) of ~ 400 bp. The intron enhancer contains at least four short-sequence motifs, which serve as binding sites for transcription factors, as well as a copy of the octanucleotide sequence.

A similar intron enhancer is found between the Jκ cluster and Cκ. Additionally, all three Ig loci contain a second strong enhancer element at their far 3' end. In the IgH locus the 3' enhancer is distal to the Cα gene, and in the κ locus, to Cκ. The λ locus contains several 3' enhancer elements downstream of the multiple Cλ genes.

The V segments in the TCR loci also carry promoter elements just upstream. Strong, T-cell-specific enhancers are found 3' to the Cα and Cβ genes, and an enhancer has also been detected distal to the murine Cγ1 gene. The δ enhancer is located in the Jδ3–Cδ intron (*Figure 6*). The location of the promoter and enhancer elements within the Ig and TCR gene loci ensures that they remain intact after gene rearrangement.

The α/β versus γ/δ lineage specificity of TCR gene transcription remains puzzling because the enhancer elements are active in all T cells. There is evidence that dominant transcriptional 'silencers' within each locus may prevent inappropriate transcription and thereby maintain lineage specificity.

5.2. Isotype switching in IgH genes
Mature B cells initially coexpress IgM and IgD, and upon secondary antigenic challenge may switch to the expression of IgG, IgE or IgA. Isotype switching does not occur in T lymphocytes.

Coexpression of IgM and IgD
While definitive proof is lacking, there is much evidence that B cells coexpressing IgM and IgD produce separate transcripts by differential splicing of a single long precursor mRNA (*Figure 9*). When B cells coexpressing IgM and IgD first encounter antigen, many of them differentiate into antibody-secreting plasma cells. In these cells, IgM secretion predominates (although some IgD is also synthesized) and the choice between IgM and IgD synthesis may be exerted at an additional level, with most IgH gene transcription terminating at the distal end of the Cμ gene without proceeding into Cδ.

Figure 9. Ig H-chain isotype switching. The central panel shows the murine IgH locus (not to scale), with a rearranged leader (L)-VDJ segment 5' to the C-region genes. Switch regions are marked by filled circles.

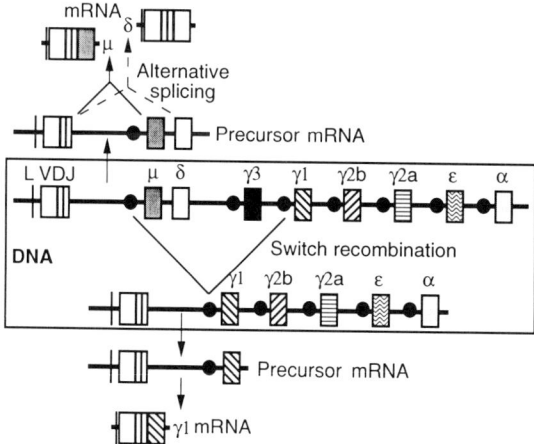

Switch recombination
Primary antigenic challenge of IgM/IgD coexpressing B lymphocytes induces differentiation not only into plasma cells but also into long-lived memory cells, in which the process of somatic hypermutation (Section 4.2) occurs. Secondary antigenic challenge of memory B cells induces class switching to other isotypes (14), involving a process of DNA rearrangement in the C_H genes.

Upstream to the $C\mu$ gene and each of the other C_H genes, apart from $C\delta$ are regions of 2–10 kb, consisting of multiple repeats of short conserved sequence motifs called switch regions (termed $S\mu$, $S\gamma$, etc.). Switch recombination occurs at these regions; a rearranged V–D–J segment is moved from its location 5' of $C\mu$ to a new location 5' of another C_H gene (*Figure 9*). Intervening sequences between $S\mu$ and the switch region to which recombination takes place are deleted. For example, once switch recombination from $S\mu$ to $S\alpha$ occurs, expression of intervening isotypes such as IgG and IgE is no longer possible.

Cytokines secreted by T cells play an important role in regulating switch recombination (14). T_H2-derived IL-4 induces switching to IgG1 and IgE, whereas IFN-γ, a product of T_H1 cells, induces switching to IgG2a and IgG3.

Synthesis of membrane-bound and secreted forms of IgH chains
Production of membrane-bound versus secreted Ig molecules is regulated primarily at the level of mRNA processing. Regulation has been studied most extensively at the murine $C\mu$ gene (12). Use of the first polyadenylation site produces a secretory transcript; whereas use of the second site, together with splicing out of the secretory tailpiece codons, produces a membrane-bound transcript (*Figure 10*). A similar mechanism has been invoked for γ, ϵ and α isotypes, in which the C_H genes have a similar organization to $C\mu$. In $C\delta$ the secretory tailpiece is encoded in a separate exon located between $C\delta3$ and M1 and so, unlike the other isotypes, regulation occurs exclusively by differential mRNA splicing.

B lymphocytes synthesize slightly more of the membrane-bound than the secretory transcript, whereas plasma cells contain a vast excess of the secretory transcript. Yet both cell types produce both forms of transcript, implying that further layers of regulation (at the translational and post-translational levels) must exist to explain why plasma cells lack membrane-bound Ig at the cell surface (Section 7.4).

6. THE EXPRESSED V-REGION REPERTOIRE OF TCR AND Ig

The expressed (as opposed to potential) V-region repertoire of TCR and Ig genes is modified by several factors.

Figure 10. Origin of secretory and membrane-bound Ig H-chain transcripts. Splice junctions between the exons are shown.

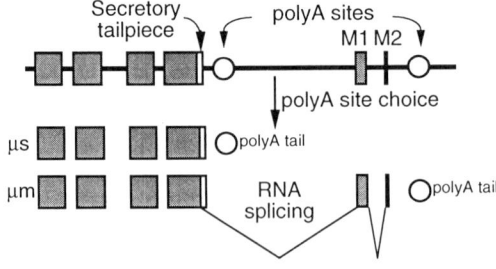

(i) Allelic polymorphism.
(ii) Not all segments are used with equal frequency.
(iii) Expressed specificities are subject to negative and positive selection.

6.1. The Ig V-region repertoire

See also Section 2.2. The largest murine V_H family is V_H1 with an estimated 60 members, followed by V_H2, V_H5, V_H6 and V_H8, with between 10 and 15 members each. Eight different haplotypes exist in inbred strains. Human V_H segments, too, display a great deal of allelic variation, and a number of different deletions and insertions have been found. By current estimates, human V_H1 and V_H3 families contain 25–30 members each; V_H4, 14; and V_H6, just one. Family assignment of murine and human Vκs is difficult because of the large numbers of segments and their limited homology. Little is known about the extent of variation in human Vλ segments.

There is evidence in mice that early in development V_H usage is biased towards those families ($V_H2,5$) that are most proximal to the J_H cluster, and a similar bias has been reported in humans (15). In mice these families may encode many autoreactive specificities, and are particularly prominent in the $CD5^+$ B-cell lineage (Chapter 2). These views are not universally accepted. Similarly, biased usage of V_H segments in human autoimmune diseases and their murine models is a subject of much debate (15).

6.2. The TCR V-region repertoire

The germ line diversity of the human Vβ and Vα segments has not been completely characterized, but estimates are presented in Section 2.4 and *Table 3*. Allelic variations in both humans and mice contribute to biased V-segment usage: for example, certain mouse strains have large deletions encompassing the TCR β locus. TCR molecules must bind to polymorphic self-MHC molecules (Chapter 9); allelic variations may, in part, be due to the selective pressure thus exerted.

Selection of the Vβ repertoire by superantigens

Nonrandom Vβ segment usage is observed in human and murine mature peripheral T lymphocytes (16). In mice it has been found that these variations are largely due to deletion during thymic development of T cells bearing particular Vβ segments by superantigens. Superantigens are molecules that bind to MHC class II on antigen-presenting cells (Chapter 9) and stimulate T-cell proliferation by specific interaction with Vβ segments. Many mouse strains carry genomic retroviral insertions which encode superantigens, thus T cells bearing Vβs reactive with these endogenous superantigens are deleted during development. It is not known whether similar endogenous superantigens account for skewed Vβ usage in humans. However, exogenous superantigens such as toxins encoded by *Staphylococcus* and *Streptococcus* species certainly have the ability to bind to specific human Vβ segments and activate the cells that carry them.

α/β V-region repertoire and autoimmune disease

Analysis of infiltrating T-cell clones derived from individuals with certain autoimmune diseases has revealed very marked preferences for the use of specific V segments. In some cases preferential V usage is so marked that administration of anti-V antibodies ameliorates the disease. The occurrence of conserved sequences in the CDR3 regions of rearranged V–D–J segments from autoreactive T cells has also been reported.

Repertoires of γ/δ T cells

Junctional diversity makes a major contribution to the γ/δ repertoire because of the small number of V, D or J segments in the γ and δ TCR loci. However, γ/δ T cells which arise early in thymic ontogeny have very limited junctional diversity, and use just a few of the available V segments (8). These cells are produced in 'waves' intrathymically, and appear to home to epithelial layers in the skin, gastrointestinal and genitourinary tracts. γ/δ T cells which arise later in ontogeny have extensive junctional diversity and are the predominant type found in the spleen, lymph nodes and blood.

7. ASSEMBLY AND CELL-SURFACE TRANSPORT OF ANTIGEN RECEPTORS

7.1. Components of the TCR complex

The mature TCR complex at the cell surface includes at least five subunits other than the α/β and γ/δ chains (*Figure 11*). These are the CD3 chains γ, δ, ε, ζ and η (η is an alternatively spliced product of the ζ chain gene) (*Table 6* and ref. 17). The CD3 chains contain negatively charged amino acids in their TM segments which may facilitate their interaction with the TCR chains (Section 1.5).

The stoichiometry of subunit association with the α/β and γ/δ chains has not yet been established with certainty. Available evidence suggests that there are two noncovalent CD3 dimers (εγ, εδ) and one covalent ζζ homodimer associated with each TCR heterodimer (*Figure 11*). It is not yet clear whether εγ and εδ dimers coexist in the same complex. Rarely, a ζη heterodimer may replace the ζζ homodimer.

7.2. Assembly and surface transport of CD3/TCR complexes

Incomplete CD3/TCR complexes do not reach the cell surface in cell lines which fail to express any one of the subunits. The fate of partial complexes is determined at several stages in assembly (18). Each of the TCR and CD3 chains contains amino acid sequence motifs which specify their intracellular fate unless these motifs are 'masked' by subunit assembly (*Figure 11*); indeed, these motifs often mediate subunit associations. The extracellular domains of all TCR and CD3 chains contain a region which causes intracellular

Figure 11. Assembly of the surface CD3/TCR complex. Note that only the ζζ homodimer and the αβ heterodimer are covalently bonded. The intracellular fate of single chains and partially assembled complexes is shown.

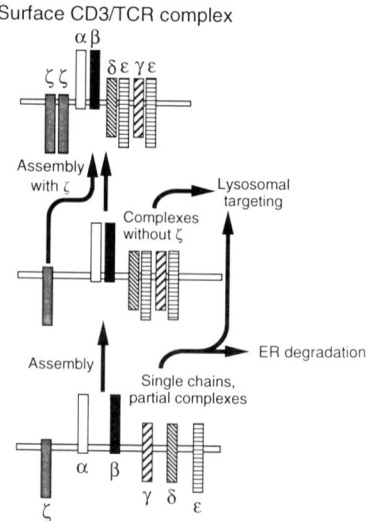

Table 6. Properties of the CD3 chains[a]

	CD3-γ	CD3-δ	CD3-ε	CD3-ζ	CD3-η
M_r (kDa)	25–28	20	20	16	22
Extracellular IgSF domains	1	1	1	– (8 aa)	– (8 aa)
N-glycosylation sites	2	2	–	–	–
Cytoplasmic residues	44 aa	44 aa	55 aa	112 aa	154 aa
Chromosomal location of gene	9 (Mo) 11 (Hu)	9 (Mo) 11 (Hu)	9 (Mo) 11 (Hu)	1 (Mo) 1 (Hu)	1 (Mo) 1 (Hu)

[a]Data from ref. 1.

retention, perhaps by interaction with the resident endoplasmic reticulum (ER) protein BiP (binding protein). Assembly of CD3-γ[δ]/ε complexes allows escape from BiP retention; these then associate with TCR α- or β-chains to yield CD3–γ[δ]/ε, α or CD3–γ[δ]/ε, β complexes. The TCR α- and β-chains, and the CD3 γ- and δ-chains contain motifs surrounding charged residues in their TM segments which cause rapid ER degradation if these partial complexes are not formed.

Disulfide linkage of TCR α- and β-chains associated with the partial CD3 complexes is next to occur. If ζ-chain is absent, the incomplete CD3–γ/δ/ε, α/β complexes are targeted for lysosomal destruction (*Figure 11*) by sequence motifs in the cytoplasmic domains of the CD3 chains. Formation of a mature CD3–TCR complex by association with a ζζ or ζη dimer directs surface expression.

7.3. Components of the membrane-bound Ig complex

All five isotypes are expressed at the cell surface in complex with at least two TM proteins which bear sequence homology to the CD3 chains (*Figure 12*). These proteins, CD79a and CD79b, have been termed the Ig-α and Ig-β subunits, are the products of the *mb-1* and *B29* genes, respectively (19), and contain a single extracellular IgSF domain with multiple

Figure 12. The membrane-bound Ig complex. The composition of the canonical complex of all five Ig isotypes is shown, as are the additional forms of membrane-bound IgD. The stoichiometry of subunit associations has not been determined and is arbitrarily depicted.

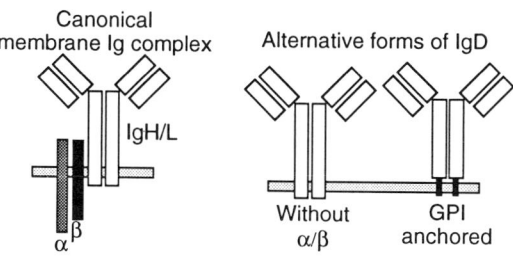

N-glycosylation sites. Variable glycosylation of Ig-α is observed, depending on the isotype of Ig with which it is associated. Thus membrane-bound IgM receptor complexes in murine cells contain a 32 kDa Ig-α subunit, and IgD complexes, a 34 kDa Ig-α subunit. The human Ig-α protein is extensively glycosylated and larger in molecular weight (40–45 kDa) than murine Ig-α, whereas both human and murine Ig-βs are 36–38 kDa.

7.4. Assembly and surface transport of membrane-bound Ig

Membrane-bound Ig heavy chains do not exit the ER without L-chain association. ER retention is mediated by the binding of BiP to a region of the C_H1 domain; L-chain association displaces BiP. In pre-B cells (which have not rearranged their L-chain genes) a similar function may be served by the surrogate light-chain proteins V pre-B and λ5 (Section 3.6).

Membrane-bound IgM, IgE and IgA molecules fail to reach the cell surface in the absence of the Ig-α and Ig-β subunits, and are retained and degraded in the ER. Membrane-bound IgD and IgG2b molecules may not require Ig-α/β association for surface transport, although they associate with Ig-α/β when present. Membrane IgD in the absence of Ig-α/β may also be inserted into the cell membrane by a GPI anchor (*Figure 12*). Whereas Ig-β is expressed at all stages of B-cell differentiation, Ig-α is found only in pre-B and B cells. Thus plasma cells, which lack Ig-α, do not transport membrane-bound Ig to the cell surface. Information related to signal transduction by antigen receptors and associated molecules can be found in Chapter 12.

8. REFERENCES

1. Barclay, A.N., Birkeland, M.L., Brown, M.H., Beyers A.D., Davis, S.J., Somoza, C. and Williams, A.F. (1993) *The Leucocyte Antigen FactsBook*. Academic Press, London.

2. Burton, D.R. (1987) in *Molecular Genetics of Immunoglobulin* (F. Calabi and M.S. Neuberger, eds). Elsevier Science Publishers, Amsterdam, p. 1.

3. Hasemann, C.A. and Capra, J.D. (1989) in *Fundamental Immunology* (W.E. Paul, ed.). Raven Press, New York, p. 209.

4. Brüggeman, M. (1987) in *Molecular Genetics of Immunoglobulin* (F. Calabi and M.S. Neuberger, eds). Elsevier Science Publishers, Amsterdam, p. 51.

5. Max, E.E. (1989) in *Fundamental Immunology* (W.E. Paul, ed.). Raven Press, New York, p. 234.

6. Davis, M.M. (1990) *Ann. Rev. Biochem.*, **59**, 475.

7. Schatz, D., Oettinger, M. and Schlissel, M. (1992) *Ann. Rev. Immunol.*, **10**, 359.

8. Haas, W., Pereira, P. and Tonegawa, S. (1993) *Ann. Rev. Immunol.*, **11**, 637.

9. Melchers, F., Karasuyama, H., Haasner, D., Bauer, S., Kudo, A., Sakaguchi, N., Jameson, B. and Rolink, A. (1993) *Immunol. Today*, **14**, 60.

10. Davis, M.M. and Bjorkman, P.J. (1988) *Nature*, **334**, 395.

11. Berek, C. and Milstein, C.M. (1987) *Immunol. Rev.*, **96**, 23.

12. Cook, G.P., Mason, J.O. and Neuberger, M.S. (1987) in *Molecular Genetics of Immunoglobulin* (F. Calabi and M.S. Neuberger, eds). Elsevier Science Publishers, Amsterdam, p. 153.

13. Leiden, J.M. (1993) *Ann. Rev. Immunol.*, **11**, 539.

14. Harriman, W., Völk, H., Defranoux, N. and Wabl, M. (1993) *Ann. Rev. Immunol.*, **11**, 361.

15. Möller, G. (1992) *Immunol. Rev.*, **128**, 1.

16. Moss, P.A.H., Rosenberg, W.M.C. and Bell, J.I. (1992) *Ann. Rev. Immunol.*, **10**, 71.

17. Fraser, J.D., Strauss, D. and Weiss, A. (1993) *Immunol. Today*, **14**, 357.

18. Klausner, R.D., Lippincott-Schwartz, J. and Bonifacino, J. (1990) *Ann. Rev. Cell Biol.*, **6**, 403.

19. Reth, M. (1992) *Ann. Rev. Immunol.*, **10**, 97.

CELLULAR IMMUNOLOGY LABFAX

CHAPTER 9
ANTIGEN PROCESSING AND PRESENTATION

B.M. Chain, L. Sealy, D.R. Katz and M. Binks

The study of antigen processing and presentation, by which we mean the complete ensemble of events leading up to the activation of an antigen-specific T lymphocyte, has been a focus of analysis by immunologists since the discovery of the T cell itself, and indeed even before. Three characteristics have maintained the study of these particular interactions in the forefront of research over more than 30 years. First, the increasing realization that the antigen-presenting event is a key point in the regulation of the immune system; secondly, the high degree of complexity of the antigen-presenting event; and thirdly, the continuing interdisciplinary nature of antigen-presentation research. Thus the field was dominated initially by classical genetics, then by the development of sophisticated *in vitro* cell culture techniques, by the rapid expansion of molecular genetics, and, most recently, by newer technological advances in cell biology, crystallography and peptide analysis.

An overview illustrating the tremendous breadth of the field of antigen processing and presentation is shown in *Table 1*, together with a few of the more recent reviews covering each topic.

Table 1. Specialties and subspecialties in antigen processing and presentation

	Refs
Genetics	
MHC locus	1
MHC and disease	2, 3
Intracellular biology	
Antigen uptake	4
Antigen degradation	5, 6
Intracellular cycling	6–8
MHC/peptide interaction	9–11
Intercellular biology	
Antigen-presenting cell heterogeneity	12–14
Intercellular interaction	15–17
Integrating the system	
Antigen presentation and the repertoire	18
Therapeutic intervention	
MHC/peptide/TCR interaction	19
Accessory molecules	20, 21

The aim of the present compilation is neither to review the subject, nor to provide an exhaustive bibliography. Still less is it possible to collect together all the information now available about antigen-processing systems. Rather, we have sought to draw together collections of information regarding some of the best-studied aspects of antigen processing/presentation, which may be of value to those trying to establish experimental systems of their own, or those seeking illustrative examples of our knowledge of this field.

1. GENETICS

Genetics, both classical and molecular, has played a key role in understanding antigen processing and presentation. The most recent impetus in this area has come from the application of transgenic technology, and dozens of novel transgenic strains, either overexpressing antigens or intrinsic molecules of the immune system, or using homologous recombination to 'knock-out' specific gene expression, are now available. The essential observation that T-cell recognition can be quantitatively and qualitatively altered by a multitude of host genes still seeks final explanation. Nevertheless, attention for more than three decades has focused on the genes of the major histocompatibility complex (MHC, known as human leukocyte antigen (HLA) in humans, or the H-2 region in mouse), which regulate the magnitude and nature of all T-cell-dependent responses. This region of the chromosome has been studied intensively in both animals and man, the most recent impetus coming from the powerful techniques of chromosome walking. This analysis has revealed that the MHC, as a locus, contains many genes with as yet no known function and others with no obvious role in antigen processing and presentation. The MHC is covered in Chapter 10, and therefore a genetic map of the human MHC showing *only those genes* believed to have an antigen processing/presentation function is shown in *Figure 1*. The fine details of the maps continue to change, and novel genes within the MHC region still remain to be discovered, but the general outline of the map is unlikely to alter substantially.

A unique feature of the MHC locus is its high degree of polymorphism. This makes analysis of immune responses in human populations very difficult. Hundreds of alleles have already been described, and more are continually reported as analysis moves from serology to sequence; the sequences of many human and mice alleles are already available in sequence databanks such as that provided by the EMBL database. For this reason, the study of antigen processing and presentation has been helped enormously by the availability of inbred mouse strains homozygous at the MHC. Many of the commonly used strains, as well as a number of MHC recombinants , are illustrated in *Table 2*. Of particular value are the sets of congenic strains, which carry different sets of MHC alleles on identical genetic backgrounds. Comparison of immune responses in such mice allows the identification of the MHC contribution to immune response regulation to be determined directly. Despite the importance of the MHC, it is important to note that in the case of most responses to complex antigens, regulation by non-MHC polymorphic loci, many of which remain to be identified, are of considerable importance.

2. THE ANTIGENS

A second approach to the study of antigen processing and presentation has been the detailed dissection of T-cell epitopes within well-characterized model protein antigens. A key element of these studies has been the use of short synthetic peptides to identify the site recognized by *in vitro* cultured clonal populations of antigen-specific T cells. Indeed, the demonstration that such short linear epitopes can substitute for the intact protein antigen (in sharp contrast to the predominantly configuration-dependent epitopes recognized by antibody) was itself a major

Figure 1. Map of the human MHC, showing the genes involved in antigen processing and presentation (adapted from ref. 1).

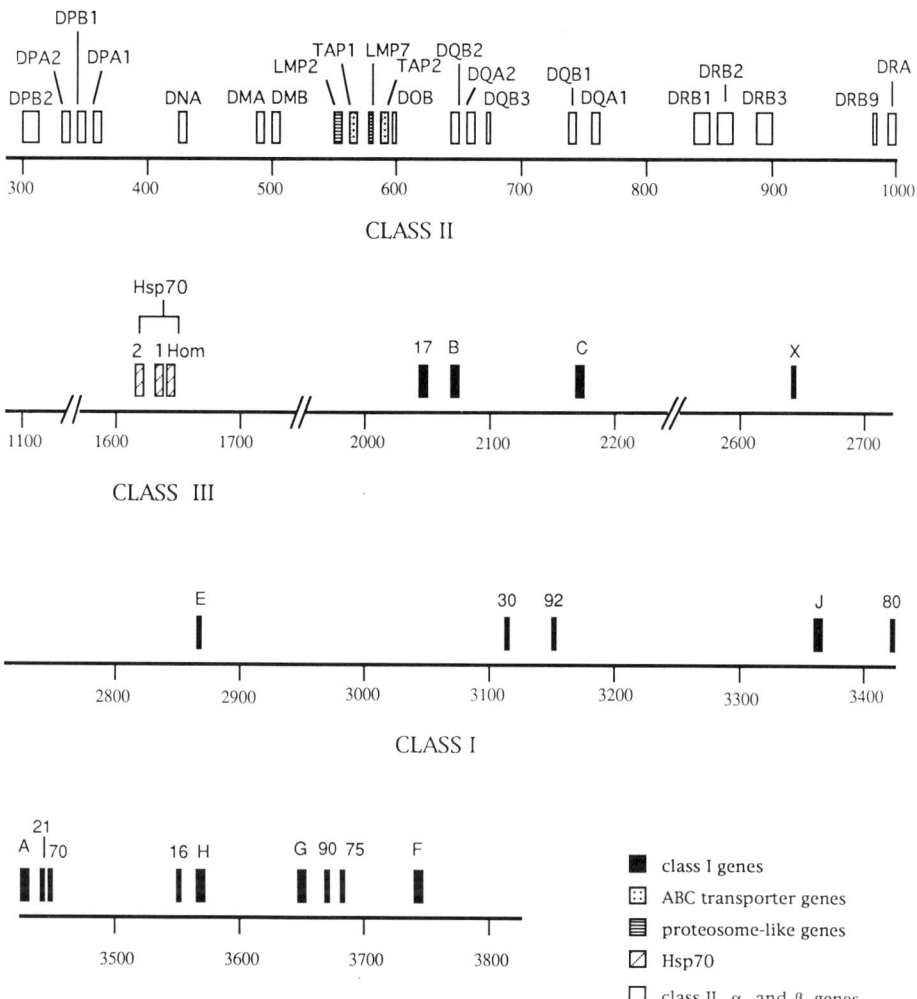

landmark in identifying the nature of the antigen-processing event. Several dozen antigens have been analyzed in this way. An analysis of some of the best-characterized T-cell epitopes identified by T lymphocytes within a very small set of intensively studied proteins is shown in *Table 3*. In the case of class II MHC restricted epitopes, the epitope can be mimicked by peptides of a variety of different lengths, and the region shown is only an approximation to the 'natural' peptide(s) produced by the antigen-presenting cell. In the case of class I peptides, the length restrictions are much tighter, and a very sharp fall in potency is often observed when using longer than optimal peptides. In both cases, presentation of specific epitopes is largely, although not absolutely, restricted to particular alleles of specific MHC molecules. A second key feature is that the range of epitopes actually recognized by T cells immunized to

Table 2. Inbred mouse strains commonly used in immunological studies (data taken from ref. 22)

The mouse strains representing the major MHC haplotypes in inbred laboratory strains[1]

C57Bl/10, C57Bl/6	$H-2^b$
DBA/2, BALB/c	$H-2^d$
C3H, CBA, AKR	$H-2^k$
B10.P	$H-2^p$
DBA/1, B10.Q, B10.G	$H-2^q$
B10.RIII, RIII	$H-2^r$
A.SW, B10.S	$H-2^s$
NZW	$H-2^z$

Congenic sets of MHC matched mouse strains[2]

Mouse strains	H-2 haplotype	Mouse strains	H-2 haplotype
BALB/c	d	C3H	k
BALB.B10	b	C3H.B10	b
BALB/AKR	k	C3H.NB	p
BALB.K	k	C3H.SW	s
AKR	k	C57Bl/6	b
AKR.B6/1 or 2	b	C57Bl/6bm set	b mutants[3]
C57Bl/10	b	C57Bl/10.CBA	k
C57Bl/10.AKR	k	C57Bl/10.CNB	p
C57Bl/10.ASW	s	C57Bl/10.D1	q
C57Bl/10.Br	k	C57Bl/10.D2	d
C57Bl/10.NB	p	C57Bl/10.F	p
C57Bl/10.Q	q	C57Bl/10.G	q
C57Bl/10.NZW	z	C57Bl/10.S	s
C57Bl/10.RIII	r	C57Bl/10.M	f

[1]Each designated strain represents an independently derived line, expressing homozygous MHC genes of the appropriate haplotype at all MHC class I and II loci.
[2]Each set of strains expresses the appropriate different sets of MHC genes in the context of an identical background provided by the designated parental strain. Thus all C57Bl/10 congenics express the same complex of non-MHC genes, and differ only at the MHC locus. Each strain has been obtained by crossing two homozygous MHC disparate strains and then repeatedly backcrossing against the background parental strain.
[3]This set of mouse strains expresses the same MHC genes, but with a variety of well-characterized small mutations in the parental $H-2^b$ genes.

an intact protein antigen is usually very small, and that many factors, including the affinity of processed peptides for MHC, selection by the processing machinery itself (e.g. the specificity of enzyme cleavage sites) and interantigenic competition, all play a part in shaping the population of responding T cells.

More recently, a complementary approach to the identification of T-cell epitopes has become possible, involving the elution and sequencing of MHC-bound peptides. A major advantage of this system is clearly the ability to identify directly the 'naturally' processed epitope. However, the technology required for the micro-isolation and sequencing is still very complex and costly, and this has limited the number of laboratories able to carry out such analyses. The

Table 3. Some T-cell epitopes identified by peptide analysis of T-cell responses

Antigen	Sequences	Restriction element	Comments
Lysozyme	35–43	I-A^k	
	46–61	I-A^k	
	116–129	I-A^k	Dominant
	74–86	I-A^k	Cryptic
	94–100	I-A^k	Cryptic
	20–28	I-A^b	
	38–45	I-A^b	
	52–69	I-A^b	
	81–96	I-A^b	
	1–18	I-E^k	
	25–43	I-E^k	
	107–116	I-E^d	
	31–50	I-E^q	
	93–113	I-E^q	
Ovalbumin	323–339	I-A^d	Dominant
	257–264	K^b	
Myoglobin	108–117	I-A^d	
	132–146	I-E^d	
	112–118	I-E^d	
	37–53	SJL class II	
	61–77	SJL class II	
	73–101	SJL class II	
	109–125	SJL class II	
	133–149	SJL class II	
Tetanus toxoid	830–843	DR5	Promiscuous
	953–967	DR5	
	947–960	DR7 and DR9	
	949–960	DP2 and DP4	
	1273–1284	DR52a/52c	
Influenza hemagglutinin HA1			
Strain H3N2/AX31	56–76	I-A^d	
	71–91	I-A^d	
	81–97	I-A^d	
	177–199	I-A^d	
	186–205	I-A^d	
	206–227	I-A^d	
	257–271	I-A^d	
	54–62	I-A^k	
	68–83	I-A^k	
	120–139	I-A^k	
	226–245	I-E^k	
	246–265	I-E^k	

Table 3. Continued

Antigen	Sequences	Restriction element	Comments
A/JAP/57	202–212	H-2^d class I	
	211–221	H-2^d class I	
A/PR/8/34	523–545	K^k	
	259–266	K^k	

For further details see refs 23–26.
Dominant epitope: the major epitope recognized in the particular antigen/strain combination.
Cryptic epitope: an epitope recognized in isolation, but not recognized when presented to the immune system in the context of the whole antigen molecule.
Promiscuous epitope: an epitope recognized in the context of several different HLA types.

published data on peptides identified so far, which identify only a tiny percentage of the peptides actually bound to any individual MHC haplotype at one time, are listed in *Tables 4–7*. In parallel with this development has been the ability to use sequencing to identify 'conserved' residues, or motifs, which can be used to predict likely T-cell epitopes within a complex protein. It is important to realize, however, that these predictive motifs should only be used as a guide. There are examples of peptides with motifs not being recognized, since MHC–peptide binding is necessary but not sufficient for a response. Conversely, peptides with low affinity for MHC (whose MHC binding cannot be readily detected) may function as strong T-cell epitopes. Prediction cannot, at present, replace the more laborious task of identifying epitopes experimentally.

3. ANTIGEN-PRESENTING CELLS

The stimulation of CD4$^+$ T cells, which are the predominant regulatory element of the immune system, occurs during interaction with a class II bearing antigen-presenting cell. Class II MHC expression under normal conditions is restricted to rather few cell types, including the lymphoid (interdigitating) dendritic cell and its related family, the B cell, and the thymic epithelium. This latter is believed to play an important role in thymic eduction, probably by regulating 'positive selection'; rather little is known about the physiology or cell biology of this process. Certain macrophage subpopulations appear to express low levels of class II MHC constitutively, and the expression of class II on macrophages is readily up-regulated by a number of inflammatory mediators. Other cell types, including cells of epithelial and endothelial origin, have been shown to express class II MHC molecules either *in vivo*, during strong T-cell-dependent inflammatory responses, or *in vitro* under the influence of cytokine regulation. The physiological function of this 'aberrant' class II expression is still debated, and this class of cells are collectively termed 'nonprofessional' antigen-presenting cells. The characteristics of the three major antigen-presenting cell types are described below.

3.1. The interdigitating (lymphoid) dendritic cell

The lymphoid dendritic cell forms part of an antigen-presenting cell family, comprising the Langerhans cell of the skin, the veiled cell of the lymph, and the interdigitating cell of lymphoid tissue. In addition, related cells are present within surface epithelia and most solid organs. Together this bone marrow-derived family comprises the 'professional' antigen-presenting cell population of the body.

Table 4. Sequences of peptides eluted from MHC molecules: human class I

Allele		Peptide sequence 1 2 3 4 5 6 7 8 9[a]	Protein source	Ref.
HLA- A2.1	Anchor residues	L　　　　　　　　V		9
	Frequent residues	M E V K 　K		
	Ligands	S X P S G G X G V	–	
		L L D V P T A A V	Human pp30 signal peptide	
		LLL D V P T AAVQA	Human pp30 signal peptide	
		G X V P F X V S V	–	
		S L L P A I V E L	p61 (regulatory subunit of PP2A)	
		S X X V R A X E V		
		K X N E P V X X X	–	
		Y L L P A I V H I	–	
		T L W V D P Y E V	–	
HLA- A2.5	Anchor residues	L		9
	Frequent residues	V Y G V I　　K L P E Y I F D L Q I　I		
HLA- B27	Anchor residues	R		9
	Ligands	R R Y Q K S T E L	Histone H3 H3.3	
		R R I K E I V K K	Hsp 89α	
		R R V K E V V K k	Hsp 89β	
		R R W L P A G d a	Elongation factor 2	
		R R S K E I T V R	RNA helicase	
		G R I D K P I L K	Ribosomal protein	
		f R Y N G L i H r	Rat 60S ribosomal protein	
		K R F E G L T Q R	–	
		R R F T R P E H –	–	
		R R I S G V D R Y	–	
		A R L F G I R A K	–	

[a]Lower-case letters indicate residues of lower confidence.

Antigen-presenting-cell function

All members of the family are strong stimulators of CD4 and CD8 T-cell responses, including alloresponses, responses to soluble and particulate foreign antigens, and haptens. These cells are the only ones capable of initiating a primary T-cell immune response. Dendritic cells in the thymus are believed to regulate 'negative' selection, in the acquisition of self-tolerance. Langerhans cells, but not dendritic cells of lymphoid tissue, are phagocytic and express receptors for the Fc region of Ig. However, dendritic cells may be capable of processing particulate antigens by cell-membrane proteinases.

ANTIGEN PROCESSING AND PRESENTATION

Table 5. Sequences of peptides eluted from MHC molecules: human class II

Allele	Protein source	Peptide sequence	Ref.
HLA-DR1	Consensus	1XXXX2XXX3[a]	27
	HLA-A2 (105–117)	SDWRFLRGYHQYA	
	Invariant chain (105–118)	KMRMATPLLMQALP	
	(Na$^+$K$^+$) ATPase (199-216)	IPADLRIISANGCKVDNS	
	Transferrin receptor (680–696)	RVEYHFLSPYVSPKESP	
	Bovine fetuin (56-73)	YKHTLNQIDSVKVWPRRP	
HLA-DR2	HLA-DQα chain (97–119)	NIVIKRSNSTAATNEV(PEVTVFS)	28
	HLA-DQβ chain (42–59)	(S)DVGVYRAVTPQGRPD(AE)	
	HLA-DR2bβ chain (94–111)	RVQPKVTVYPSKTQP(LQH)	
	FnRα chain (586–616)	LSPIHIALNFSLDPQAPVDSHGLR PALHYQ	
	K$^+$-channel protein (173–190)	DGILYYYQSGGRLRRPV(N)	
	Mannose binding protein (174–193)	IQNLIKEEAFLGITDEKTEG	
	MET (59–81)	EHHIFLGATNYIYVLNEEDLQKV	
	GPB-2 (434–450)	QELKNKYYQVPRKGIQA	
	Apo B-100 (1200–1220)	FPKSLHTYANILLDRRVPQ(TD)	
	Factor VIII (1775-1790)	LWDYGMSSSPHVLRNR	
HLA-DR3	HLA-A30 (28–?)	QDDTQFVRFDSDAASQ...[b]	28
	HLA-DRα chain (111–129)	PPEVTVLTNSPVELREPN(V)	
	Invariant chain (131–149)	ATKYGNMTEDHVMHLLQNA	
	Acetylcholine receptor (289–304)	VFLLLLADKVPETSLS	
	Glucose transporter (459–474)	TFDEIASGFRQGGASQ	
	Na$^+$ channel protein (384–397)	YGYTSYDTFSWAFL	
	CD45 (1071–1084)	GQVKKNNHQEDKIE	
	IFN receptor (128–148)	GPPKLDIRKEEKQIMIDIFH(P)	
	EBV gp220 (592–606)	TGHGARTSTEPTTDY	
	IP-30 (38–59)	SPQALDFFGNGPPVNYKTG(NL)	
	Cyt-*b5* (155–172)	GFAIRPDKKSNPIIRTV	
	Apo B-100		
	(1273–1295)	(IPD)NLFLKSDGRIKYTL(NKNSLK)	
	(1207–1224)	YANILLDRRVPQTDMTF	
	(1794–1810)	VTTLNSDLKYNALDLTN	

Table 5. Continued

Allele	Protein source	Peptide sequence	Ref.
HLA-DR4	HLA-A2 (28–50)	(VDD)TQFVRFDSDAAS(QRMEPRAP)	28
	HLA-Cw9 (28–50)	(VDD)TQFVRFDSDAASPR (GEPRAPWV)	
	(130–150)	(D)LRSWTAADTAAQIT(QRKWEAA)	
	HLA-Bw62 (129–150)	DLSSWTAADTAAQIT(QRKWEAA)	
	VLA-4 (229–248)	GSLFVYNITTNKYKAF(LDKQ)	
	HLA-DQ3.2β chain (24–38)	SPEDFVYQFKGMCYF	
	PAI-1 (261–281)	AAPYEKEPVLSALTNILS(AQL)	
	Cathepsin C (151–167)	YDHNFVKAINADQKSW(T)	
	IgG heavy chain (121–?)	GVYFYLQWGRSTLVSVS...[b]	
	Bovine hemoglobin (26–41)	AEALERMFLSFPTTKT	
HLA-DR7	HLA-A29 (234–261)	(RPAGD)GTFQKWASVVV (PSGQEQRYTCHV)	28
	HLA-B44 (83–99)	RETQISKTNTQTYRE(NL)	
	HLA-DRα chain		
	(101–126)	RSNYTPITNPPEVTVLTNSPVELREP	
	(58–78)	GALANIAVDKANLEIMTKRSN	
	HLA-DQα chain	SLQSPIVTEWRAQSESAQSKWLS GIGGFVL...[b]	
	(179–?)		
	4F2 (318–338)	VTQYLNATGNRWCSWSL(SQAR)	
	LIF receptor (854–866)	TSILCYRKREWIK	
	Thromboxane-A synthetase (406–420)	PAFRFTREAAQDCEV	
	K$^+$-channel protein (492–516)	GDMYPKTWSGMLVGALCALAGVLTI	
	Hsp70 (38–54)	TPSYVAFTDTERLIG(DA)	
	EBV MCP (1264–1282)	VPGLYSPCRAFFNK(EELL)	
	Apo B-100		
	(1586–1608)	KVDLTFSKQHALLCS(DYQADYES)	
	(1942–1954)	FSHDYRGSTSHRL	
	(2077–2089)	LPKYFEKKRNTII	
	Complement C9 (465–483)	APVLISQKLSPIYNLVPVK	
HLA-DR8	HLA-DRα chain (158–180)	SETVFLPREDHLFRKFHYLPFLP	28
	HLA-DPβ chain (80–92)	RHNYELDEAVTLQ	
	LAM Blast-1		
	(88–108)	(DPS)GALYISKVQKEDNSTYI	
	(129–146)	DPVPKPVIKIEKIED(MDD)	

Processing, Presentation

Table 5. Continued

Allele	Protein source	Peptide sequence	Ref.
	Ig κ chain (63–80)	FTFTISRLEPEDFAV(YYC)	
	LAR (1302–1316)	DPVEMRRLNYQTPG	
	LIF receptor (709–726)	YQLLRSMIGYIEELAPIV	
	IFN-α receptor (271–287)	GNHLYKWKQIPDCENVK	
	IL-8 receptor (169–188)	LPFFLFRQAYHPNNSSPVCY	
	Ca^{2+} release channel (2614–2623)	RPSMLQHLLR	
	CD35 (359–380)	DDFMGQLLNGRVLFPVNLQLGA	
	CD75 (106–122)	IPRLQKIWKNYLSMNKY	
	Calcitonin receptor (38–53)	EPFLYILGKSRVLEAQ	
	TIMP-1 (101–118)	(NR)SEEFLIAGKLQDGLL(H)	
	TIMP-2 (187–214)	QAKFFACIKRSDGSCAWYR (GAAPPKQEF)	
	PAI-1 (378–396)	DRPFLFVVRHNPTGTVLFM	
	(133–148)	MPHFFRLFRSTVKQVD	
	Cathepsin E (89–112)	QNFTVIFDTGSSNLWV(PSVYCTSP)	
	Cathepsin S (189–205)	TAFQYIIDNKGIDSDAS	
	Cystatin SN (41–58)	DEYYRRLLRVLRAREQIV	
	Tubulin α-1 chain (207–223)	EAIYDICRRNLDI(ERPT)	
	Myosin β heavy chain (1027–1047)	HELEKIKKQVEQEKCEIQAAL	
	α-enolase (23–?)	AEVYHDVAASEFF...[b]	
	c-myc (371–385)	KRSFFALRDQIPDL	
	K-ras (164–180)	RQYRLKKISKEEKTPGC	
	Apo B-100		
	(1724–1743)	KNIFHFKVNQEGLKLS(NDMM)	
	(1780–1799)	YKQTVSLDIQPYSLVTTLNS	
	(2646–2664)	(S)TPEFTILNTLHIRSFT(ID)	
	(2885–2900)	SNTKYFHKLNIPQLDF	
	(2072–2088)	LPFFKFLPKYFEKKR(NT)	
	(4022–4036)	WNFYYSPQSSPDKKL	
	Bovine transferrin (261–281)	DVIWELLNHAQEH(FGKDKSKE)	
	von Willebrand factor (617–636)	IALLLMASQEPQRM(SRNFVR)	

[a]The putative HLA-DR1 peptide binding motif includes three key amino acids; 1, positively charged; 2, hydrogen-bond donor; 3, hydrophobic residue.
[b]Partial sequence not verified by mass spectrometry.
LAM = L-selectin (CD62L); LCMV, lympho-choriomeningitis virus; LIF, leukemia inhibitory factor; PAI, plasminogen activator inhibitor; TIMP, tissue inhibitor of metalloproteinase; VSV, vesicular stomatitis virus.
Brackets indicate the presence of a nested set of peptides in addition to the basic core sequence.

Table 6. Sequences of peptides eluted from MHC molecules: mouse class I

Allele		1	2	3	4	5	6	7	8	9	Protein source	Ref.
H-2K^d	Anchor residues		Y							L		9
										I		
	Frequent residues			N	P	M	K	T				
				I			F	N				
				L								
	Ligands	T	Y	Q	R	T	R	A	L	V	Influenza nucleoprotein (147–155)	
		S	Y	F	P	E	I	T	H	I	Protein kinase JAK1	
		K	Y	Q	A	V	T	T	T	L	Tumor antigen of P815	
		G	Y	K	D	G	N	E	Y	I	Listeriolysin (*Listeria monocytogenes*) (91–99)	
H-2D^b	Anchor residues					N				M		9
	Frequent residues	M	I	K	L			I				
		L	E				F					
		P	Q									
		V	V									
	Ligands	A	S	N	E	N	M	E	T	M	Influenza nucleoprotein (366–374)	
H-2K^b	Anchor residues					F			L			9
						Y						
	Frequent residues			Y			M					
	Ligands	R	G	Y	V	Y	Q	G	L		VSV (52–59)	
		S	I	I	N	F	E	K	L		Chicken ovalbumin (257–264)	
		H	I	Y	E	F	P	Q	L		Self protein of P815	
H-2K^k	Anchor residues		E			I						9
	Frequent residues		K									
			N									
			Y									
			M									
H-2K^{kml}	Anchor residues					I						9
	Frequent residues	E	K									
H-2L^d	Ligands	P	Q	A	S	G	V	Y	M	G	LCMV nucleoprotein (119–126)	9
		Y	P	H	F	M	P	T	N	L	pp89 (168–176)	
		I	S	T	Q	N	H	R	A	L	tum-antigen P91A (190–198)	
		L	S	P	F	P	F	D	L		Mouse spleen protein	

JAK, Janus kinase; LCMV, lympho-choriomeningitis virus.

Processing, Presentation

Table 7. Sequences of peptides eluted from MHC molecules: mouse class II

Allele	Protein source	Peptide sequence	Ref.
I-A^s	Consensus	XXXXITXXXXHXXX	29
	MuLV env protein	IRLKITDSGPRVPIGpn	
	IgG2a	WPSQSITCNVAHPASST	
	IgG2a	NVEVHTAQTQTHREDY	
	TfR	KPTEVSGKLVHANFGT	
	–	XPYMFADKVVHLPGSQ	
I-A^b	Consensus	XXNX XXXXPXXXX	29
	MuLV env	HNEGFYVCPGPHRP	
	I-E α chain	ASFEAQG ALANIAVDKA	
	Invariant chain	KPVSQM RMATPLLMR	
	I-Aβ chain	RPDA EYWNSQPE	
	IgG VH	XNA DFKTPATLTVDkp	
	–	NYNA YNATPATLAVD	
I-E^b	Consensus	XXYLYXXXXRRXXYX	29
	MuLV env	PSYVYHQFERRAKYK	
	Bovine serum albumin	GKYLYEIARRHPYFyap	
	–	QSYLIHEXXXIS	
	DR consensus	AAYAAAAAAKAAA	
I-A^d	Apo E	WANLMEKIQASVATNPI	30
	Cys C	DAYHSRAIQVVRARKQ	
	I-E^d α chain	ASFEAQGALANIAVDKA	
	Apo E	EEQTQQIRLQAEIFQAR	
	Invariant chain	KPVSQMRMATPLLMRPM	
	Transferrin receptor	VPQLNQMVRTAAEVAGQX	
	Ovalbumin	ISQAVHAAHAEINE	
	λ repressor	LEDARRLKAIYEKKK	
I-A^k	Hen egg lysozyme	DGSTDYGILQINSRWW	31
	Hsp70	IIANDQGNRTTPSY	
	I-A^k β chain	TPRRGEVYTCHVEHP	
	s30 ribosomal protein	KVHGSLARAGKVRGQTPKVAKQ	
	s30 ribosomal protein	AGKVRGQTPKVAKQEKKKKKT	
	Ryudocan	EPLVPLDNHIPENAQPG	
I-E^d	Hen egg lysozyme	AWVAWRNRCK	32

MuLV, murine leukemia virus.

Cell-surface markers

Lymphoid dendritic cells express high levels of both class I and class II antigens, and in addition constitutively express high levels of a variety of co-stimulatory and adhesion molecules, including B7, CD54, CD11a/18 and CD58. They do not synthesize IL-1, or indeed any other known cytokine. They express low levels of CD4 and hence are susceptible to infection by HIV-1.

Differentiation

The cells of the lymphoid dendritic cell family are believed to be related in a differentiation pathway: Langerhans cells migrate out of the skin, and migrate through the lymphatics (in the form of veiled cells) to lymphoid tissue, where they differentiate into interdigitating dendritic cells. The spleen and lymph nodes may also contain a second resident population which functions to trap and process antigens passing through the tissue. The signals which regulate this differentiation pathway are still not completely elucidated, but the cytokines GM-CSF, TNFα, IL-1 and IL-4 have all been shown to modulate dendritic cell maturation *in vitro*.

Methods of isolation

A wide variety of isolation methods have been reported in the literature, but all exploit the relatively low density (high cytoplasmic to nuclear ratio) of dendritic cells during density fractionation, and negative selection using antibodies to B, T and macrophage markers. Detailed methods can be found in refs 12 and 33.

3.2. The macrophage

The macrophage/monocyte family comprises a group of bone marrow-derived phagocytic cells, which are found populating all major tissues and body cavities, and whose major role is scavenging and degrading senescent or damaged 'self', any foreign particles which penetrate the internal environment of the host, and antigen–antibody complexes. The differentiation pathways of the macrophage lineage are extremely complex, and macrophage phenotypes specific for different anatomical locations are found. The extent to which different macrophage phenotypes represent true developmental subpopulations, or alternatively simply represent the influence of the local microenvironment, remains largely unresolved. In addition, the phenotype of the macrophage is under complex regulatory control, via a host of soluble mediators produced by the specific and the innate immune system. The macrophage is not, primarily, an antigen-presenting cell. Thus expression of class I and class II MHC under resting conditions is low, and macrophages frequently suppress, rather than enhance, a T-cell-dependent immune response, via the release of inhibitory inflammatory mediators including arachidonic acid metabolites.

The best-studied macrophage types are the peritoneal macrophages, particularly in rodents, and the circulating macrophage precursors, the blood monocytes in humans. In both these cell types the normal low levels of class I and class II MHC expression can be dramatically up-regulated under the influence of T-cell-derived cytokines, particularly interferon-γ. TNF and GM-CSF also up-regulate MHC expression in these cell types. Under these conditions, activated macrophages can be demonstrated to show strong antigen-presenting function, a function which may be particularly important in processing and then presenting micro-organisms relatively resistant to cellular degradation, such as many intracellular parasites. The regulatory relationship between T cells and macrophages is therefore complex and reciprocal – this is likely to be a key interaction underlying the immunopathology associated with chronic immunological responsiveness (13).

3.3. The B cell

Antigen processing and presentation by B lymphocytes forms the molecular basis for T/B co-operation, and gives rise to the fundamental phenomenon of 'linked' help, whereby T-cell help for B-cell epitopes is restricted to those epitopes which physically form part of one molecular unit.

Processing and presentation by B cells follows the same general rules as those of other cell types, with the important exception that antigen uptake and entry into the processing pathway

ANTIGEN PROCESSING AND PRESENTATION

is specifically promoted by interaction of antigen with antibody at the cell surface. Each B cell therefore takes up antigen specifically via its surface Ig at concentrations several orders of magnitude lower than uptake of nonspecific antigen. Once bound via surface Ig, antigen is internalized and processed normally, thus releasing internal antigen structures to interact with MHC class II molecules, and stimulate appropriate T-cell responses. This mechanism ensures that B cells can recognize antigen predominantly via interactions with conformational epitopes on the surface of protein antigens, but can then stimulate T cells specific for linear processed epitopes, which often lie within the globular protein structure. The uptake of antigen as a complex with surface Ig imposes additional constraints on the degradation and processing of antigen, since epitopes within or close to the antibody-binding sites may be partially protected from the action of proteinases (determinant protection) – thus the B-cell (antibody) repertoire may play a major role in shaping the repertoire of the T-cell response, and conversely, the nature of the dominant epitopes recognized by the T-cell pool will play a major role in shaping the B-cell repertoire by selecting which B cells will receive maximum 'help'.

B-cell differentiation and antigen presentation
B cells constitutively express low levels of MHC class II molecules, but resting B cells fail to express effective co-stimulator signals for T-cell activation and hence can induce nonresponsiveness or tolerance. In contrast, activation of B cells (via T-cell-derived cytokines, and accessory molecule interaction including the CD40/gp39 interaction) induces up-regulation of MHC class II, and simultaneous expression of strong co-stimulatory activity, including expression of the CD28 ligand B7. Thus activated B cells are potent antigen-presenting cells, at least for secondary immune responses. Further differentiation of B cells to plasma cells is accompanied by the shutting off of MHC class II synthesis (14).

4. INHIBITORS OF ANTIGEN PROCESSING

A key to the dissection of the antigen-processing pathway has been the availability of specific inhibitors of individual steps of the pathway. The properties of the more commonly used inhibitors is shown in *Table 8*. A number of other inhibitors have been used occasionally, but the data on their action are too limited to include, and it is advisable to collaborate with a laboratory experienced in the analysis of such inhibitors, as new molecules are being developed constantly. It is important to appreciate that, although the inhibitors are usually used with the purpose of blocking one specific reaction in the antigen-processing pathway, this specificity is often not achieved. Thus chloroquine, which blocks acidification of intracellular organelles, not only inhibits protein degradation in these organelles but invariant chain degradation and thus MHC-peptide assembly. Similar caveats operate for all the molecules listed, and appropriate controls for such secondary effects must be included. There has obviously been very considerable interest in the possibilities of using inhibitors of processing/presentation as potential novel therapeutic approaches to autoimmunity or transplant rejection. So far, with the exception of chloroquine, which is partially effective in the treatment of rheumatoid arthritis, but whose mode of action in this disease is not known, no such strategy has proved feasible, perhaps because of the relatively nonspecific nature of many of the processes being targeted.

In the absence of suitable small molecular weight inhibitors (still the molecules of choice for pharmaceutical development) much attention has focused on the use of biological response modifiers to block T-cell activation at the level of antigen presentation. One approach has used antibodies to cell-surface structures (e.g. CD4, CTLA-4, CD58), with some limited success in preliminary clinical trials. An alternative strategy has been to identify specific peptide analogs which can block the tripartite antigen–MHC–T-cell receptor interaction, either at the level of

Table 8. The pharmacological tools used in the study of antigen processing and presentation

Molecule	MW (Da)	Source	Solubility	Working concn (µM)	Function	Ref.
Cycloheximide	281	Fungal antibiotic	Methanol, ethanol	3–30	Inhibits protein synthesis, and hence especially *de novo* MHC class I synthesis	34,35
Brefeldin A	280	Fungal antibiotic	Methanol	3–30	Inhibits egress from ER, and hence especially processing for class I MHC antigen presentation	36,37
Chloroquine	320	Synthetic organic hetero-cyclic	Water	100–500	Inhibits endosomal acidification, and hence processing for class II MHC	38,39
Primaquine	259	Synthetic organic hetero-cyclic	Water, ethanol	100–500	As for chloroquine	40
Ammonium chloride	53	Inorganic	Water	10 000	As for chloroquine	41
Monensin	671	Fungal antibiotic	Ethanol	30 000	As for chloroquine	42
Leupeptin	457	Fungal peptide	Water	10–100	Inhibits cysteine and some serine proteinases, and hence processing for class II MHC	43
Pepstatin	686	Fungal peptide	DMSO	10–100	Inhibits aspartic proteinases, and hence processing for class II MHC	43
E-64	357	Fungal peptide	DMSO	10–100	Inhibits cysteine proteinases and hence processing for class II MHC	43

Abbreviations: DMSO, dimethyl sulfoxide; E-64, *N*-*N*-L-3-transcarboxyoxizan-2-carbonyl-L-leucyl-agmatine.

ANTIGEN PROCESSING AND PRESENTATION

the binding of processed antigen to MHC, or by anergizing specific T-cell clones, or by blocking the subsequent interaction with the specific T-cell receptor. Some promising results have been obtained *in vitro* and in animal model systems, but the real potential of such strategies in terms of clinical value remains to be evaluated.

5. THE ANTIGEN-PROCESSING PATHWAY

The dissection of the detailed cell biology of antigen processing has proved to be one of the most difficult aspects of the field. Significant progress in this field has occurred mainly within the past 5 years, and many details remain to be established. Nevertheless, the major elements of the intracellular pathways followed by antigen during processing are shown in *Figure 2*. The use of antigen-processing mutants, improved immunoelectron microscopy, pulse-chase metabolic labeling of proteins followed by immunoprecipitation, and transfection of genes coding individual components of the pathway into nonexpressing cell lines, have all proved particularly valuable techniques.

Perhaps the most fundamental, and unexpected, finding has been the realization that class I and class II processing pathways are generally distinct (although division between the two pathways may not be quite as rigorous as was at first thought). In addition to this unexpected dichotomy, it is gradually becoming appreciated that antigen processing is a distinct cellular function, with its own components, and its own intracellular compartments, although making use of more general elements of cell physiology. Examples of such specializations are the TAP (transporter associated with antigen processing) genes of the class I pathway, and the 'MHC loading' compartment of the class II pathway. The outdated idea that antigen processing is simply an unregulated by-product of protein catabolism is slowly being abandoned.

Figure 2. The MHC processing pathways: (a) class I, (b) class II.

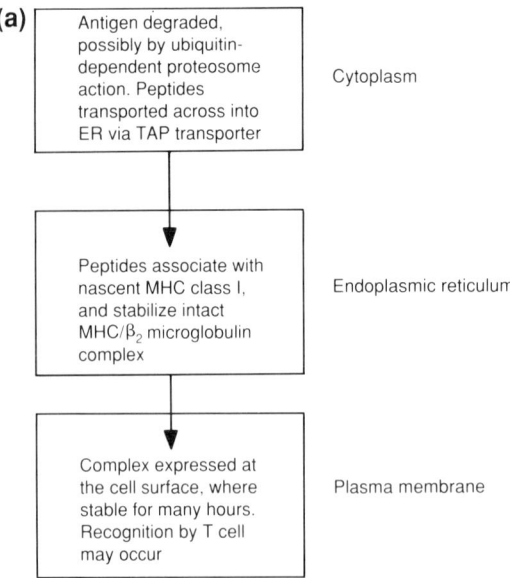

6. THE MOLECULES

The triumph of cellular immunology over the past decade has been its ability to dissect complex cellular phenomena into collections of well-characterized molecules. Nowhere has this progress been more apparent than in the fields of antigen processing and presentation. As a tribute to this effort, and perhaps to allow those who have become fixated on single molecules to adopt a more balanced approach, *Table 9* lists all the molecules believed to play an important role in antigen processing or presentation. All the molecules listed have been cloned and sequenced, and in most cases also identified by specific antibody reactions. Analysis of the function of such molecules has remained much more difficult: approaches commonly used include trying to modulate antigen processing/presenting function by addition of blocking antibody to functional assays, or in the case of soluble molecules (cytokines) by addition of the molecule itself to the assays; by transfection of the relevant gene into appropriate negative cell lines; or, most recently, by the production of transgenic mice in which the relevant gene is either overexpressed, or expression is 'knocked-out' by homologous recombination. Once again, we have omitted those molecules whose involvement in antigen

Figure 2. (b)

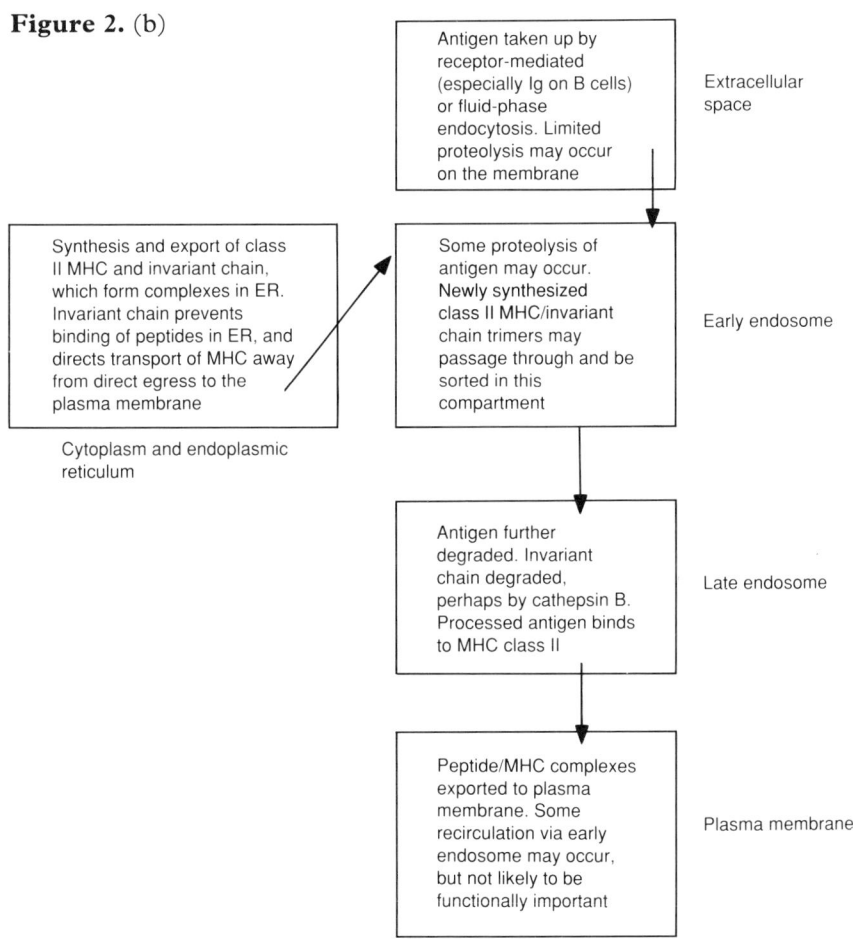

▶ p. 193

ANTIGEN PROCESSING AND PRESENTATION

Table 9. The molecular components of the antigen-processing and presentation pathways

Molecule	Other names	MW (kDa)	Main function in antigen presentation	Ref.
Intracellular				
Cathepsin B		27	Cysteine proteinase involved in antigen and/or invariant chain degradation	43
Cathepsin D		gp 38–40	Aspartic lysosomal proteinase involved in antigen degradation	44
Cathepsin E	Slow-moving aspartic protease	gp 42/84	Aspartic nonlysosomal proteinase, esp. in gastrointestinal epithelium; role in antigen processing	45
Invariant chain	CD74	gp 43/41/35/33 isoforms	Binds to, and regulates the folding and transport of MHC class II molecules; also prevents binding of processed peptides to class II MHC in the endoplasmic reticulum	46
LMP1			Proteosome subunit, perhaps involved in antigen degradation for class I MHC processing	47
LMP2			As for LMP1	47
TAP1			Peptide transporter subunit, carrying processed peptides from cytoplasm into the ER, for presentation by class I MHC	48
TAP2			As for TAP1	48
Cell-surface molecules				
MHC class I		α gp 44 β_2m p12 dimer	Peptide-binding molecules presenting antigen to CD8 T cells	49
MHC class II		α gp 35 β gp 28 dimer	Peptide-binding molecules presenting antigen to CD4 T cells	10
DMA			Nonpolymorphic class II homolog. Function unknown	50
DMB			As for DMA	50
Ig		gp 150–185	Antigen-specific receptor mediating uptake and processing of antigen by B lymphocytes	14
B7	BB1	60 gp	Expressed on activated B cells, and professional antigen-presenting cells, and providing co-stimulatory activity to T cells via interaction with its ligands CD28 or CTLA-4 on T cells	51

CELLULAR IMMUNOLOGY LABFAX

Table 9. Continued

Molecule	Other names	MW (kDa)	Main function in antigen presentation	Ref.
CTLA-4		2 × gp 26 homodimer	See B7	21
HSP-70		p 70	Peptide-binding protein facilitating antigen–MHC interaction	52
CD1		gp 43–49	Putative peptide-presenting molecule for γ/δ T cells	53
CD2		gp 50	Adhesion molecule on T cells, binding CD58 (LFA-3) on antigen-presenting cells, and enhancing antigen presentation	51
CD3–TCR complex		CD3: γ gp 26, δ gp 20, ε p 20, ζ p16, η p 22 TCR: α gp 40–45, β gp 38–45	T-cell antigen-specific receptor complex, interacting with MHC–peptide complex and providing specificity element to T cell–antigen-presenting cell interaction	
CD4		gp 55	MHC class II binding receptor on T cells, enhancing antigen presentation	15
CD5		gp 67	Marker of a subset of B cells, that is particulary effective in antigen-presenting function	51
CD8		α gp 32 β gp 32 dimer	MHC class I binding receptor on T cells, enhancing antigen presentation	15
CD11a	LFA-1, α chain to CD18	gp 180	One chain of the LFA-1 complex, which binds ICAMs, and forms an adhesive interaction between T cell and antigen-presenting cell, which enhances antigen presentation	51
CD13	Amino-peptidase N	gp 150	Cell-surface aminopeptidase, which may have a role in trimming MHC-bound peptides	
CD14		gp 55	Macrophage marker; receptor for the LPS–LPS binding protein complex	54
CD18	β chain to CD11a (LFA-1)	gp 95	See CD11a	51
CD23	FcεRII	gp 45–50	IgE Fc receptor; putative role in modulating antigen processing and presentation of antigens for IgE production	55
CD28		2 × gp 44 homodimer	See B7	17
CD40		gp 50	Receptor for gp39 on T cells, regulating B-cell activation and survival	56

Processing, Presentation

ANTIGEN PROCESSING AND PRESENTATION **191**

Table 9. Continued

Molecule	Other names	MW (kDa)	Main function in antigen presentation	Ref.
CD44	Pgp-1	gp 85–250	Down-regulates the adhesive interaction between T cells and dendritic cells	57
CD45 (RA,B and O)		gp 180–220	Binds to an unspecified ligand, and down-regulates antigen presentation via its cytoplasmic tyrosine phosphatase activity	57
CD54	ICAM-1	gp 85–110	See CD11a	51
CD55	DAF	gp 64–73		57
CD58	LFA-3	gp 56–70	See CD2	51
CD59		gp 18–20		58
CD64	FCγRI	gp 75	Modulates antigen processing via its effect on antigen uptake	4
Cytokines				
IFN-α and -β		gp 17–40	Up-regulate class I MHC (occasionally also class II)	59
IFN-γ	2 × gp 20–25 homodimer		Converts macrophages to efficient antigen-presenting cells by up-regulating expression of MHC and a variety of adhesion molecules; also induces MHC on a variety of other cell types	59
TNF-α and β	α: 3 × p 17 homotrimer β: 3 × gp 25 homotrimer		Synergizes with IFN-γ, in regulating macrophage function; regulates MHC expression; regulates Langerhans cell/dendritic cell differentiation	60, 61
IL-1α and β	gp 17		Co-stimulatory molecules produced by macrophages, and enhancing antigen presentation; regulates differentiation along the Langerhans cell/dendritic cell pathway	62
IL-4	gp 20		Up-regulates class II MHC on B cells and macrophages	63
IL-6	gp 26		Macrophage derived co-stimulator (with IL-1) of T-cell activation	64
IL-10	2 × p 18 homodimer		Inhibits the ability of antigen-presenting cells to stimulate a TH-1 type cytokine response	65
GM-CSF	gp 14–35		Regulates differentiation along the Langerhans cell/dendritic cell pathway, and also along the monocyte/macrophage pathway; in both cases enhances class I MHC expression	66

LMP, low molecular mass polypeptide complex.

processing is either very indirect, or inadequately documented at the time of writing. Data on the function of many of the molecules continue to be collected apace, and inaccuracies are bound to be revealed. Finally, most of the molecules of the immune system are pleiotropic, and the functions briefly listed in *Table 9* are meant as a guide to their involvement in antigen presentation only.

7. CONCLUSIONS

The past 10 years have seen an explosive growth in the study of antigen processing and presentation. In common with many other branches of biology, the impact of molecular biology has had a disproportionate effect, leading to the identification of several dozen molecules directly involved. For this reason, the focus of this collection has been molecular – molecular genetics, molecular analysis of antigens and molecular dissection of antigen-presenting systems. There is clearly considerably more of this type of analysis to come, and the number of molecules likely to be uncovered may well double or triple in the next 10 years.

A different type of analysis of antigen presentation/processing has hardly begun, however; by this is meant the synthesis of the disparate components of the system into an integrated model of the whole, and more particularly, the ability to identify rules which govern the behavior of the system in respect to its parts. An example of the type of question to be addressed would be the impact of antigen-processing enzyme specificity on the repertoire of the T-cell response, or at least on the repertoire of the set of MHC-bound peptides. Another would be an analysis of the interactive hierarchy of co-stimulator signals which govern the antigen-processing event – what is the contribution of each, relative to the contribution to each other, and what governs the relative change in contribution of each, in relation to changes in activity of each other? Many other such integrative questions remain to be answered.

8. REFERENCES

1. Campbell, R.D. and Trowsdale, J. (1993) *Immunol. Today*, **14**, 349.

2. Nepom, G.T. and Erlich, H. (1991) *Ann. Rev. Immunol.*, **9**, 494,

3. Svejgaard, A., Platz, P. and Ryder, L.P. (1983) *Immunol. Rev.*, **70**, 193.

4. Lanzavecchia, A. (1990) *Ann. Rev. Immunol.*, **8**, 773.

5. Chain, B.M. and Levine, T.P. (1993) in *Blood Cell Biochemistry, Volume 5: Macrophages and Related Cells* (M.A. Horton, ed.). Plenum Press, New York, p. 161.

6. Yewdell, J.W. and Bennink, J.R. (1992) *Adv. Immunol.*, **52**, 1.

7. Germain, R.N. and Margulies, D.H. (1993) *Ann. Rev. Immunol.*, **11**, 403.

8. Brodsky, F.M. (1992) *Trends in Cell Biol.*, **2**, 109.

9. Rammensee, H.G., Falk, K. and Rotzscke, O. (1993) *Ann. Rev. Immunol.*, **11**, 213.

10. Brown, J.H., Jardetzky, T.S., Gorga, J.C., *et al.* (1993) *Nature*, **364**, 33.

11. Zhang, W., Young, A.C., Imarai, M.,

Nathenson, S.G. and Sacchettini, J.C. (1992) *Proc. Natl. Acad. Sci. USA*, **89**, 8403.

12. Steinman, R. (1992) *Ann. Rev. Immunol.*, **10**, 271.

13. Weaver, C.T. and Unanue, E.R. (1990) *Immunol. Today*, **11**, 49.

14. Lanzavecchia, A. (1987) *Immunol. Rev.*, **99**, 39.

15. Miceli, M.C. and Parnes, J.R. (1993) *Adv. Immunol.*, **53**, 59.

16. Larson, R.S. and Springer, T.A. (1990) *Immunol. Rev.*, **114**, 181.

17. Linsley, P.S. and Ledbetter, J.A. (1993) *Ann. Rev. Immunol.*, **11**, 191.

18. Sercarz, E., Lehman, P.V., Ametoni, A. and Benichou, G. (1993) *Ann. Rev. Immunol.*, **11**, 729.

19. Guery, J.C., Neagu, M., Rodriguez-Tarduchy, G. and Adorini, L. (1993) *J. Exp. Med.*, **177**, 1461.

20. Haug, C.E., Colvin, R.B., Delmonico, F.L., *et al.* (1993) *Transplantation*, **55**, 766.

21. Lenschow, D.J., Zengh, Y., Thistlethwaite, J.R., *et al.* (1992) *Science*, **257**, 789.

22. Klein, J., Figueroa, F. and David, C.S. (1983) *Immunogenetics*, **17**, 553.

23. Milich, D.R. (1989) *Adv. Immunol.*, **45**, 195.

24. Panina-Bordignon, P., Tan, A., Termijtelen, A., Demotz, S., Corradin, G. and Lanzavecchia, A. (1989) *Eur. J. Immunol.*, **19**, 2237.

25. Sweetser, M.T., Braciale, V.L. and Braciale, T.J. (1989) *J. Exp. Med.*, **170**, 1357.

26. Gould, K.G., Scotney, H. and Brownlee, G.G. (1991) *J. Virol.*, **65**, 5401.

27. Chicz, R.M., Urban, R.G., Lane, W.S. *et al.* (1992) *Nature*, **358**, 764.

28. Chicz, R.M., Urban, R.G., Gorga, J.C., Vignali, A.A., Lane, W.S. and Strominger, J.L. (1993) *J. Exp. Med.*, **178**, 27.

29. Rudensky, A.Y., Preston-Hurlburt, P., Hong, S.-C., Barlow, A. and Janeaway, C.A. (1992) *Nature*, **359**, 429.

30. Hunt, D.F., Michel, H., Dickinson, T.A., Shabanowitz, J., Cox, A.L., Sakaguchi, E.A., Appella, E., Grey, H.M. and Sette, A. (1992) *Science*, **256**, 1817.

31. Vignali, D.A.A., Urban, R.G., Chic, R.M. and Strominger, J.L. (1993) *Eur. J. Immunol.*, **23**, 1602.

32. Overington, J., Donnelly, D., Johnson, M., Sali, A. and Blundell, T. (1993) *Protein Sci.*, **1**, 216.

33. King, P.D. and Katz, D.R. (1990) *Immunol. Today*, **11**, 206.

34. Jensen, P.E. (1988) *J. Immunol.*, **141**, 2545.

35. Morrison, L.A., Lukacher, A.E., Braciale, V.L., Fan, D.P. and Braciale, T.J. (1986) *J. Exp. Med.*, **163**, 903.

36. Yewdell, J.W. and Bennink, J.R. (1989) *Science*, **244**, 1072.

37. Nuchtern, J.G., Bonifacino, J.S., Biddison, W.E. and Klausner, R.D. (1989) *Nature*, **339**, 223.

38. Chesnut, R.W., Colon, S.M. and Grey, H.M. (1982) *J. Immunol.*, **128**, 1764.

39. Lotteau, V., Teyton, L., Peleraux, A., *et al.* (1990) *Nature*, **348**, 600.

40. Reid, P.A. and Watts, C. (1990) *Nature*, **346**, 655.

41. Ziegler, H.K. and Unanue, E.R. (1982) *Proc. Natl. Acad. Sci. USA*, **79**, 175.

42. Machamer, C.E. and Cresswell, P. (1984) *Proc. Natl. Acad. Sci. USA*, **81**, 1287.

43. Diment, S. (1990) *J. Immunol.*, **145**, 417.

44. Rodriguez, G.M. and Diment, S. (1992) *J. Immunol.*, **149**, 2894.

45. Bennett, K., Levine, T., Ellis, J.S. *et al.* (1992) *Eur. J. Immunol.*, **22**, 1519.

46. Bikoff, E.K., Huang, L.Y., Episkopou, V., van-Meerwijk, J., Germain, R.N. and Robertson, E.J. (1993) *J. Exp. Med.*, **177**, 1699.

47. Glynne, R., Powis, S.H., Beck, S., Kelly, A., Kerr, L.A. and Trowsdale, J. (1991) *Nature*, **353**, 357.

48. Trowsdale, J., Hanson, I., Mockridge, I., Beck, S., Townsend, A. and Kelly, A. (1990) *Nature*, **348**, 741.

49. Silver, M.L., Guo, H.-C., Strominger, J.L. and Wiley, D.C. (1992) *Nature*, **360**, 367.

50. Kelly, A., Monaco, J.J., Cho, S.G. and Trowsdale, J. (1991) *Nature*, **353**, 571.

51. Young, J.W., Koulova, L., Soergl, S.A., Clark, E.A. and Steinman, R.M. (1992) *J. Clin. Invest.*, **90**, 229.

52. Naget, D.C. and Pierce, S.K. (1991) *Sem. Immunol.*, **3**, 65.

53. Porcelli, S., Morita, C.T. and Brenner, M.B. (1992) *Nature*, **360**, 593.

54. Wright, S.D., Ramos, R.A., Tobias, P.S., Ulevitch, R.J. and Mathison, J.C. (1991) *Science*, **249**, 1431.

55. Grenin-Brossette, N., Bourget, I., Akomdi, C. and Bonnefsy, J.Y. (1992) *Eur. J. Immunol.*, **22**, 1573.

56. Marshall, L.S., Aruffo, A., Ledbetter, J.A. and Noelle, R.J. (1993) *J. Clin. Immunol.*, **13**, 165.

57, King, P.D., Batchelor, A.H., Lawlor, P. and Katz, D.R. (1990) *Eur. J. Immunol.*, **20**, 363.

58. Venneker, G.T. and Asghar, S.S. (1992) *J. Exp. Clin. Immunogenetics*, **9**, 33.

59. Rosa, F., Hatat, D., Abadie, A. and Fellous, M. (1985) *Ann. Inst. Pasteur Immunol.*, **136C**, 103.

60. Caux, C., Dezutter-Dambuyant, C., Schmitt, D. and Banchereau, J. (1992) *Nature*, **360**, 258.

61. Arenzana-Seisdedos, F., Mogensen, S., Vuillier, F., Fiers, W. and Virelizier, J.L. (1988) *Proc. Natl. Acad. Sci. USA*, **85**, 6087.

62. Dinarello, C.A. (1989) *Adv. Immunol.*, **44**, 153.

63. Cao, H., Wolff, R.G., Meltzer, M.S. and Crawford, R.J. (1989) *J. Immunol.*, **143**, 3524.

64. Van Snick, J. (1990) *Ann. Rev. Immunol.*, **8**, 253.

65. Moore, K.W., O'Garra, A., dr Waal Malefyt, R., Vieira, P. and Mosmann, T.R. (1993) *Ann. Rev. Immunol.*, **11**, 165.

66. Inaba, K., Steinman, R.M., Pack, M.W. *et al.* (1992) *J. Exp. Med.*, **175**, 1157.

CHAPTER 10
THE MAJOR HISTOCOMPATIBILITY COMPLEX
Y. Satta and N. Takahata

The major histocompatibility complex (*Mhc* in standardized genetic nomenclatures, but more commonly referred to as *MHC*) is a generic name of a set of clustered loci. *MHC* genes proper are found in all vertebrates except jawless fish (*Agnatha*), but not in invertebrates. Functional MHC molecules bind self and/or foreign peptides and present them to T-cell receptors. Some of the *MHC* loci are incomparably polymorphic; with a large number of alleles, a large number of nucleotide differences between alleles, and a long persistence time of allelic lineages.

1. MOLECULES ENCODED WITHIN THE *MHC*

The complex comprises two classes of *MHC* loci, class I and II. Each MHC molecule has two chains, α and β, which form heterodimers with each other. The β-chain of the class I *MHC*, *β₂-microglobulin (β₂m)*, is set apart from the other *MHC* loci in chromosomal location. Class I molecules are expressed on all somatic cells of an adult individual, whereas the expression of class II molecules is limited to only certain selected cells, such as B lymphocytes and, in man, activated T lymphocytes. There are many poorly expressed or nonfunctional *MHC* loci. The *MHC* also contains unrelated (sometimes called class III) loci. Tables 1 and 2 list the names of genes so far identified within the *H-2* and *HLA* complex.

2. STRUCTURE OF CLASS I AND CLASS II MOLECULES

The three-dimensional structure of class I molecules (HLA-A2, HLA-Aw68, HLA-B27 and H-2K^b) and class II molecules, (HLA-DR1) have been determined (7–11). The molecules contain the peptide-binding region (PBR) of β-strands topped by α-helices, one or two immunoglobulin-like domains, a trans-membrane domain and a cytoplasmic tail. *Figure 1* shows the three-dimensional structure of the PBR and immunoglobulin-like domains of HLA-A2 and HLA-DR1. The latter crystallized as a dimer of heterodimers.

The large groove between the α-helices contains six pockets into which amino acid residues of processed peptides are noncovalently embedded. *Figure 2* shows the groove of HLA-A2 and HLA-DR1. *Table 3* shows pocket-forming residues of HLA-A2 and class II molecules.

The binding of peptides is specific and degenerate. Class I molecules bind peptides of 8–11 amino acid residues, one or two of which are allele-specific (anchor residues). Naturally processed peptides bound to class II molecules are 13-17 amino acids long. No simple sequence motifs in class II-associated molecules have yet been found. Sequences of peptides eluted from MHC molecules are given in Chapter 9, *Tables 4–7*.

3. MAPS OF *H-2* AND *HLA* GENES (AND OTHER SPECIES)

The *MHC* occupies a chromosomal region 2–4 Mb long. Both the length and the number of genes in the region vary, not only from one species to another but also from individual to individual within species. The *H-2* complex of about 2 cM length is at a distance of about

▶ p. 198

MHC

Table 1. Genes in the *H-2* complex on chromosome 17

K region	I region	S region	D region	Q region	T region	M region
H-2Ke1	Col11a2	Int-3	**H-2D**	**H-2Q1**	**H-2T24**	D17Leh171★
H-2Ke2	**H-2Pb**	D17MIT13★	Hh-1	**H-2Q2**	**H-2T23**	D17Leh108A★
Rps18	**H-2Na**	Cyp21	**H-2D2**	**H-2Q3**	**H-2T22**	D17Leh108B★
H-2K2	Ring3	C4	**H-2D3**	**H-2Q4**	**H-2T21**	D17Tu24★
H-2K	4.24A	Cyp21-ps	**H-2D4**	**H-2Q5**	**H-2T20**	D17Tu32★
Ring1	4.24F	C4Slp	**H-2L**	**H-2Q6**	**H-2T19**	D17Tu34★
Ring2	**H-2Ma**	Rd		**H-2Q7**	**H-2T18**	tcl-Lub1
H-2Ke4	**H-2Mb2**	Bf		**H-2Q8**	**H-2T17**	**H-2M1**
H-2Ke5	**H-2Mb1**	C2		**H-2Q9**	**H-2T16**	**H-2M7**
	Lmp-2	Bat-8		**H-2Q10**	grr	**H-2M8**
	Tap-1	Bat-9		Otf3-rs7	**H-2Tlev**	**H-2M6**
	Tap-2	Bat-7			tcl-12	**H-2M4**
	Lmp-7	D17Tu2★			Tse	**H-2M5**
	H-2Ob	D17Tu3★			Tctex-4	D17Tu19★
	H-2Ab	D17Tu4★			Tctex-5	**H-2M2**
	H-2Aa	Neu-1			Tctex-6	**H-2M3**
	H-2Eb1	Hsp70			**H-2T15**	
	H-2Eb2	Hsc70t			**H-2T14**	
	H-2Ea	Rnu14			**H-2T13**	
		Csp-1			**H-2T12**	
		Orch-1			**H-2T11**	
		Bat-6			**H-2T10**	
		Bat-5			**H-2T9**	
		Bat-4			**H-2T8**	
		Bat-2			**H-2T7**	
		Bat-3			**H-2T6**	
		D17Slk1			**H-2T5**	
		Tnfa			**H-2T4**	
		Tnfb			**H-2T3**	
		Bat-1			**H-2T2**	
					H-2T1	

Modified from ref. 1 and references therein. See also refs 2–5.
H-2 genes proper are in bold, and DNA markers are marked by asterisks.
The *H-2* gene organization is haplotype-dependent.

Table 2. Genes in the *HLA* complex on chromosome 6

Class II region	Unrelated region	Class I region
KE3(D6S219)	G18(D6S214E)(X6)	17(1.7p)*
RING1(D6S111E)	G17(D6S213E)	**B(HLA-B)**
RING2(D6S112E)	G16(D6S212E)(X5)	**C(HLA-C)**
KE4	G15(D6S211E)(X4)	OCT-3
KE5(D6S218)	G14(D6S210E)(X3)	TUBB
COL11A2	G13(D6S209E)(X1,X2)	**X(HLA-X)***
HLA-DPB2*	G12(D6S208E)	P5-1
HLA-DPA2*	OSG(D6S103E)(XB)	HSR1
HLA-DPB1	CYP21(CYP21B)	**E(HLA-E)**
HLA-DPA1	C4B	**30(HLA-30)***
HLA-DNA(DZA/DOA)	XA	**92(HLA-92)***
RING3(D6S113E)(Y5/Y4)	CYP21P(CYP21A)	VI(HCG-VI)
DMA(RING6)	C4A	I(HCG-I)
DMB(RING7)	G11(D6S60E)	III(HCG-III)
LMPZ(RING12)	RD(D6S45)	V(HCG-V)
TAP1(D6S114E)(RING4/Y3/PSF1)	Bf(Factor B)(BF)	VII-3(HCG-VII-3)
RING9(D6S215)	C2	**J(D6S203)(HLA-J)**
LMP7(D6S16E)(RING10/Y2)	G10(D6S59E)(BAT9)	**(HLA-59)(cda12)***
TAP2(D6S217E)(RING11/Y1/PSF2)	G9a(BAT8)	P5-2
HLA-DOB	G9(D6S58E)(BAT7)	V-2(HCG-IV-2)
HLA-DQB2*	G8(D6S57E)	III-I(HCG-II-10)
HLA-DQA2*	Hsp70-2(HSPA1L)	80(HLA-80)*
HLA-DQB3(D6S205)(DVβ)*	Hsp70-1(HSP70)(HSPA1)	P5-3
HLA-DQB1	Hsp70-Hom	A(HLA-A)
HLA-DQA1	G7a(BAT6)(VARS2)	21(HLA-21)*
HLA-DRB1-9	G7b	70(HLA-70)*
(2*, 6*, 7*, 8*, 9*)	G7(D6S56E)	P5-4
HLA-DRA	G6(D6S55E)	IV-3(HCG-IV-3)
	BAT5(D6S82E)	**16(HLA-16)***
	G5(D6S54E)(BAT4)	**H(HLA-H)(AR/12.4/**
	G3(G6S53E)	**HLA-54)***
	G2(G6S52E)(BAT3)	P5-5
	G1(G6S51E)(BAT2)	**G(HLA-G)**
	B144(D6S49)	**90(HLA-90)***
	LTB	**75(HLA-75)***
	TNFA	**F(HLA-F)**
	TNFB	
	BAT1(D6S81E)	

Class I and II genes are in bold.
*Gene fragments, pseudogenes, truncated genes of *HLA*.
The number of *DRB1* and *CYP21* loci varies among haplotypes.
Modified from ref. 6 and see references therein.

MHC

Figure 1. Schematic representation of (a) the four domains of HLA-A2 (7) and (b) a dimer of HLA-DR1 (DRA/DRB1*0101) heterodimers (8).

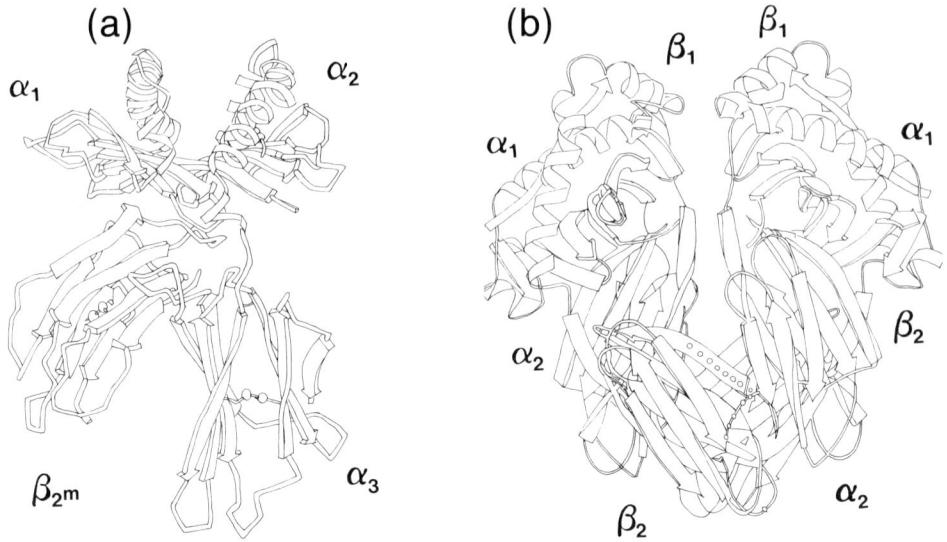

30 cM from the centromere of chromosome 17. All nonmouse mammals except the rat have the human type of *MHC* organization. The *HLA* complex of 4 Mb length is located on chromosome 6. *Figures 3–5* show the *MHC* maps of human, mouse and other mammals.

4. NOMENCLATURE OF *H-2* AND *HLA* ALLELES AND HAPLOTYPES

Table 4 lists the names of *MHC* in various species and the years of their discovery.

The *H-2* complex is divided into seven regions, *K, I, S, D, Q, T* and *M* (4, 37, 38). The *I* region is divided further into two subregions, *I-A* and *I-E*. Alleles are designated by small superscript letters (e.g. $H\text{-}2K^k$, $H\text{-}2D^k$, etc.). Mutant alleles are designated by the superscript *m1*, *m2*, etc., in the order of discovery, and the superscript is added to the allele symbol (e.g. K^{bm1}, K^{bm2}, etc.). The H-2 haplotypes, alternative forms of the whole *H-2* complex, are designated by *H-2* with a superscript consisting of one or two small letters and, in some instances, an Arabic numeral (e.g. $H\text{-}2^b$, $H\text{-}2^{ap1}$, etc.). The haplotypes of wild mice are designated by the superscript letter '*w*', followed by an Arabic numeral assigned in the order of discovery (e.g. $H\text{-}2^{w1}$, $H\text{-}2^{w2}$, etc.). *Tables 5–8* show *HLA* nomenclatures.

The designations of nonhuman primate MHC are standardized according to the rules proposed by refs 40 and 41.

5. TYPING METHODS

Cellular and humoral immune response to MHC alloantigens can be measured in a variety of ways, referred to collectively as histogenetic (those based on T-lymphocyte activation) and serological (those based on the production of antibodies in the serum and other body fluids). The two most commonly used histogenetic methods of MHC detection are mixed lymphocyte reaction (MLR) and cell-mediated lymphocytotoxicity (CML). To detect an antigen serologically, a specific antibody is required. Standard immunization procedures produce polyclonal antibodies, whereas B-cell hybridomas, a single clone of B cells, can produce

▶ p. 204

Figure 2. Diagrams of (a) HLA-A2 and (b) HLA-DR1 binding clefts (7, 8). Domain α1 of A2 and DR1 is in solid lines.

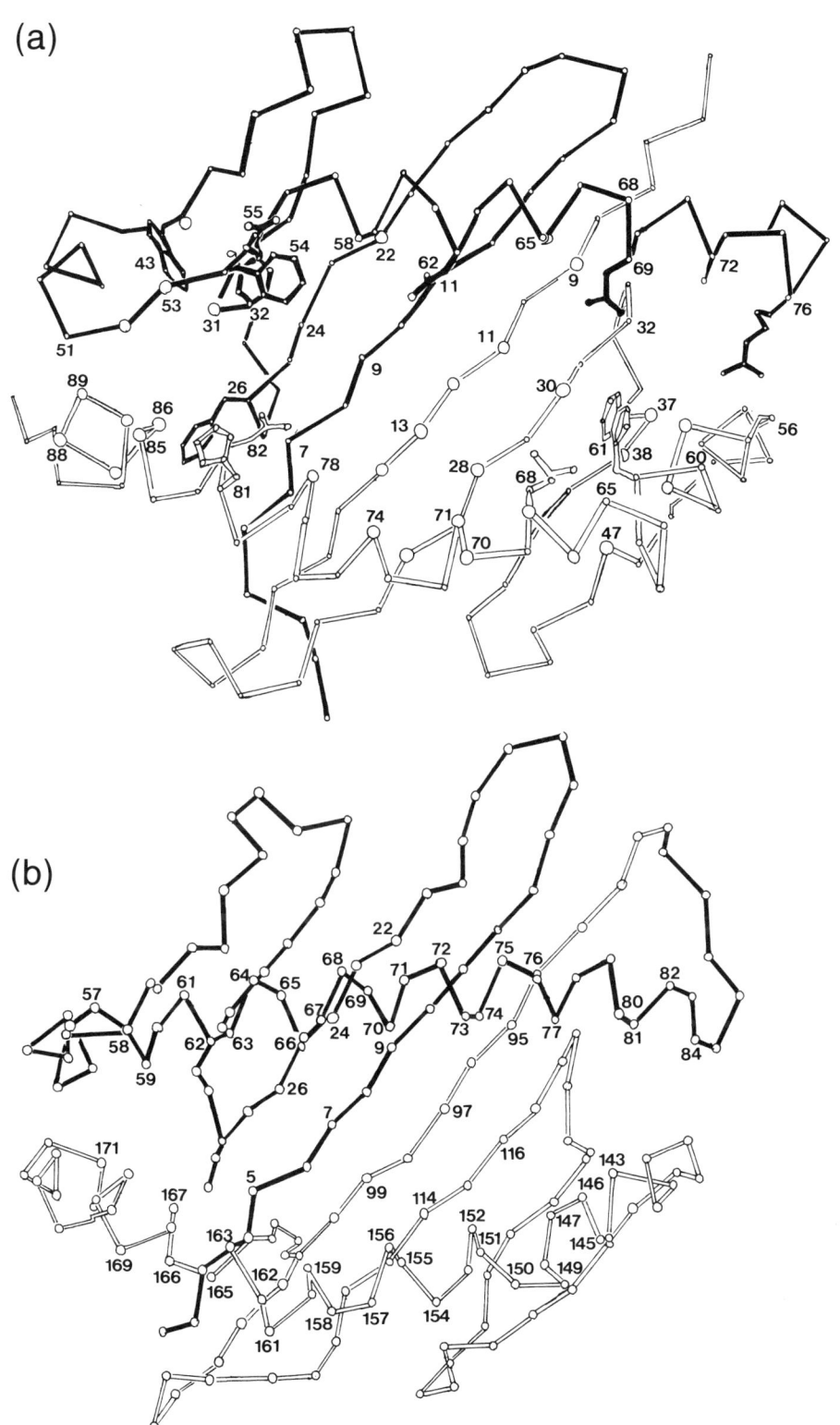

Table 3. Pocket-forming residues of HLA-A2 and class II molecules

Pocket A

A2	Y59	E63	K66	Y7	Y99	Y159	M5	T163	W167	Y171
DR1	F51	E55	G58	Q9	F13	A74	I7	Y78	N82	G86
DQw58	F54	D58	F61	G1	G13	E74	S10	V78	N82	E86
DP2	A51	E55	G58	Y9	Q13	V72	S7	M76	N80	G84

Pocket B

A2	K66	Y7	G26	E63	V34	M45	V67	A24	F9	H70	Y99
DR1	G58	Q9	F24	E55	H33	A37	A59	F22	E11	N62	F13
DQw58	F61	G12	H27	D58	Y36	E40	A62	Y25	N14	N65	G13
DP2	G58	Y9	F24	E55	Y36	D37	G59	F22	A11	N62	Q13

Pocket C

A2	H70	H74	T73	R97	Y99	Y116
DR1	N62	D66	V65	L11	F13	C30
DQw58	N65	L69	V68	F11	G13	Y30
DP2	N62	L66	I65	G11	Q13	Y30

Pocket D

A2	Y99	H114	L156	Q155	L160	Y159
DR1	F13	E28	R71	Q70	V75	A74
DQw58	G13	T28	T71	R70	L75	E74
DP2	Q13	E26	E69	E68	P73	V72

Pocket E

A2	R97	W147	V152	L156	H114
DR1	L11	W61	L67	R71	E28
DQw58	F11	W61	V67	T71	T28
DP2	G11	W59	I65	E69	E26

Pocket F

A2	Y116	D77	L81	T80	Y84	K146	T143	Y123	W147
DR1	C30	N69	M73	I72	R76	Y60	D57	S37	W61
DQw58	Y30	N72	V76	I75	R79	Y60	A57	Y37	W61
DP2	Y28	N69	L73	T72	R76	Y58	D55	F35	W59

Modified from ref. 12. β-chain residues of class II molecules are underlined. The single letter code for amino acids is used.

Figure 3. *HLA* map (6) of 4 Mb on chromosome 6. The *HLA* and complement component-coding genes are shown on the left-hand side of each vertical bar corresponding to 1 Mb, and other unrelated genes are on the right.* The number of *DRB*, *CYP21* and *C4* loci is haplotype-dependent. Alternative names of genes, if available, are given in parentheses.

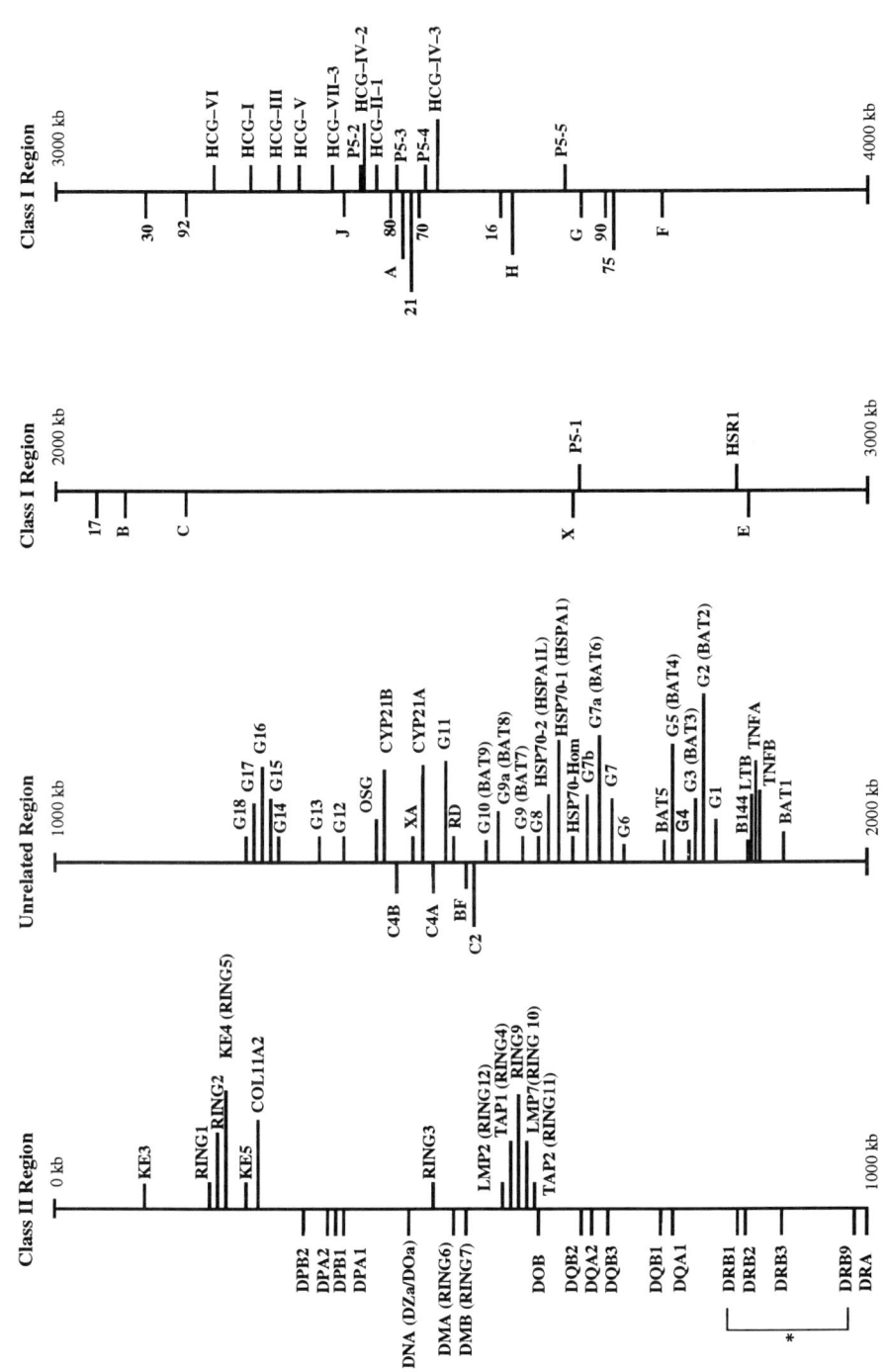

MHC

Figure 4. *H-2* map (1) of c. 2 cM (~ 4 Mb) on chromosome 17. The *H-2* and complement component-coding genes are drawn on the left-hand side of each region, whereas other unrelated genes are on the right. The number of *H-2* class I genes differs from haplotype to haplotype, as in *HLA*.

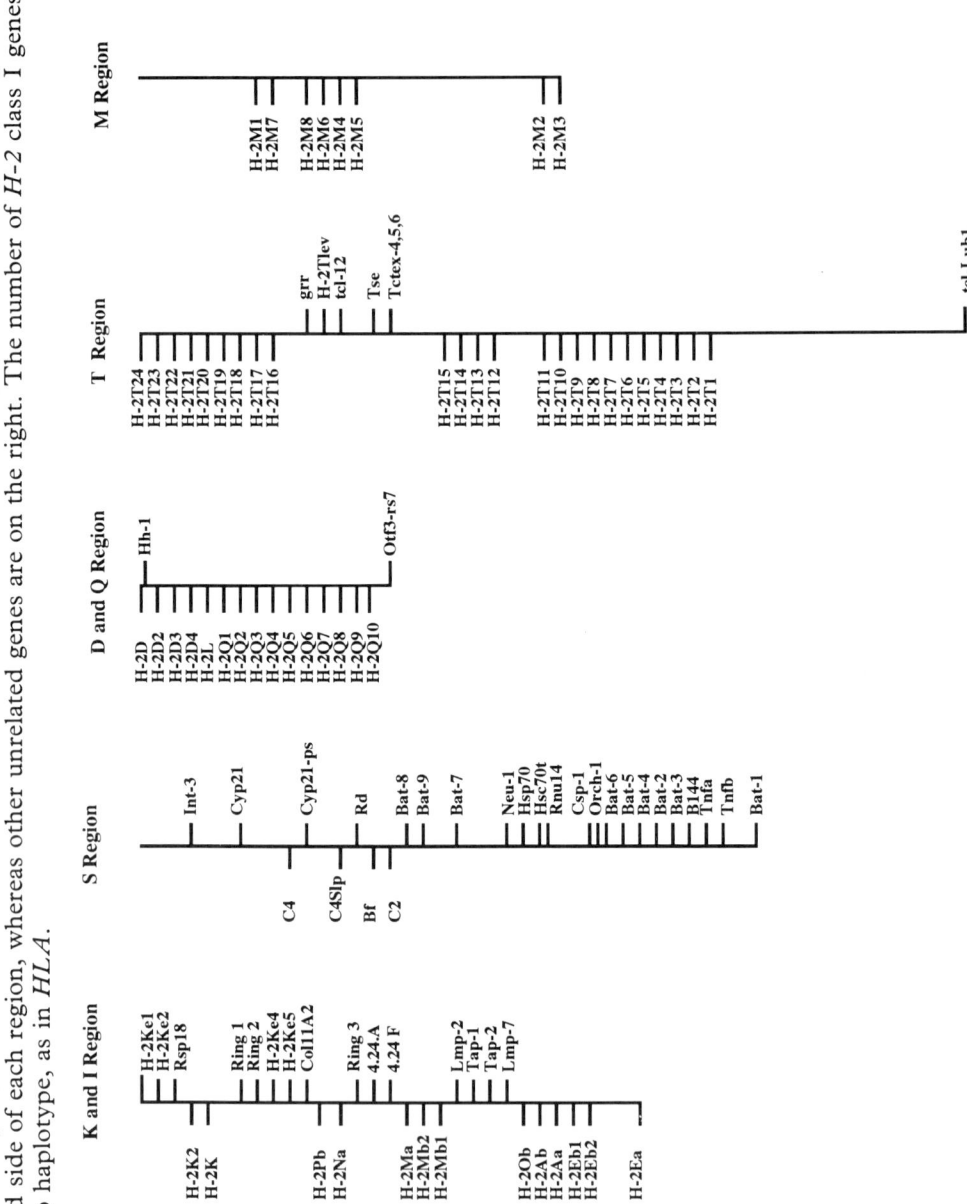

CELLULAR IMMUNOLOGY LABFAX

Figure 5. *MHC* maps of other mammals (13–33). Dotted and hatched rectangles are class I and II genes, respectively. Speculated genes are in open rectangles. The class II gene order of nonhuman vertebrates is assumed to be the same as in *HLA*. No homolog of the domestic fowl *G* gene has been found in other vertebrates (32). There are two α and five β coding genes in *Xenopus* class II, but their relationship to *DP*, *DQ* and *DR* is unknown (33). In general, the number of class II genes in these species is haplotype-dependent.

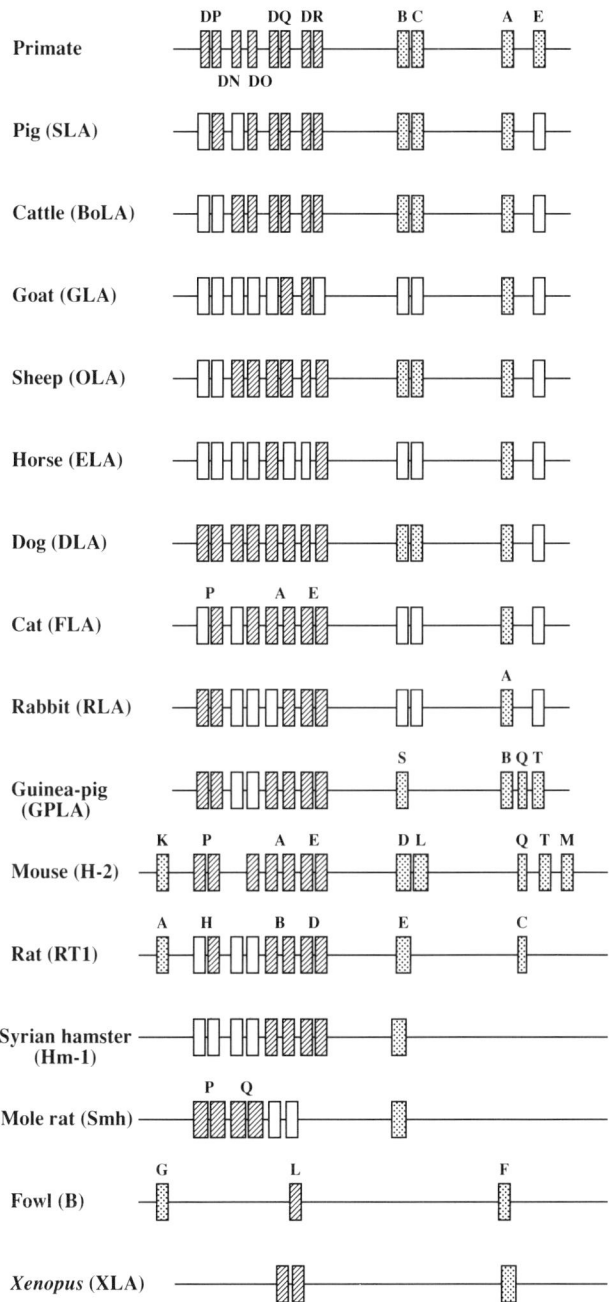

monoclonal antibodies. However, a principal difficulty of serological typing, even with monoclonal antibodies, arises from the fact that MHC molecules are particularly rich in antigenic sites (14, 42).

DNA typing is more precise than serological typing. A target region of an *MHC* gene is amplified using the polymerase chain reaction (PCR) (see *Table 9* for specific primers) and the gene fragment identified by: (i) hybridization with SSO (sequence-specific oligonucleotides), (ii) RFLP (restriction fragment length polymorphism), or (iii) SSCP (single-strand conformation polymorphism). Care should be taken about elimination of cross-hybridization in the SSO, availability of restriction enzyme sites for RFLP analysis, and reliable relationships between sequence differences and conformational changes in SSCP. SSCP may be used as an auxiliary tool to either SSO or RFLP (43–48).

Table 4. Discovery of *MHC* in different species

Year	Species	MHC	Discovered by
1936	Mouse	H-2	P.A. Gorer
1950	Fowl	B	W.E. Briles *et al.*
1958	Human	HLA	J. Dausset
1960	Rat	RT1	A.E. Bogden and P.M. Aptekman
1970	Pig	SLA	M. Vaiman *et al.*
1971	Rhesus monkey	RhLA	H. Balner *et al.*
1971	Dog	DLA	J.W. Templeton and E.D. Thomas
1972	Rabbit	RLA	R.H. Tissot and C. Cohen
1972	Guinea-pig	GPLA	W. Sato and A.L. de Weck
1974	Sheep	OLA	P. Millot, H.C.H.J. Ford
1974	Chimpanzee	ChLA	H. Balner *et al.*
1975	Clawed frog	XLA	L. du Pasquier *et al.*
1976	Goat	GLA	R.H. van Dam *et al.*
1978	Cattle	BoLA	B. Amorena and W.H. Stone
1978	Syrian hamster	Hm-1	W.R. Duncan and J.W. Streilein
1979	Horse	ELA	E. Bailey *et al.*
1982	Cat	FLA	M.S. Pollack *et al.*
1983	Cynomolgus monkey	CyLA	C.A. Keever and E.R. Heise
1984	Mole-rat	Smh	D. Nižetić *et al.*

See ref. 34 for *RT1*, ref. 35 for *RhLA* and ref. 36 for *DLA*. For more about the history of *MHC* studies, see ref. 14 and references therein.

Table 5. Designations of *HLA-A*, *-B* and *-C* alleles (39). See ref. 39 for the individual or cell line from which the sequence was derived, and references or submitting author(s)

HLA allele	HLA specificity	Previous equivalents	Accession number
A*0101	A1	–	X55710
A*0201	A2	A2.1	
A*0202	A2	A2.2F	
A*0203	A203	A2.3	
A*0204	A2	–	X57954, M86404
A*0205	A2	A2.2Y	
A*0206	A2	A2.4a	
A*0207	A2	A2.4b	
A*0208	A2	A2.4c	
A*0209	A2	A2-OZB	
A*0210	A210	A2-LEE	
A*0211	A2	A2.5	X60764
A*0212	A2	–	M84378
A*0301	A3	A3.1	
A*0302	A3	A3.2	
A*1101	A11	A11E	
A*1102	A11	A11K	
A*2301	A23(9)	–	M64742
A*2401	A24(9)	–	
A*2402	A24(9)	–	M64740
A*2403	A2403	A9.3	M64741
A*2501	A25(10)	–	
A*2601	A26(10)	–	
A*2901	A29(19)	–	
A*2902	A29(19)	A29.2	X60108
A*3001	A30(19)	A30.3	
A*3002	A30(19)	A30.2	X61702
A*31011	A31(19)	–	
A*31012	A31(19)	–	M84375, M86405
A*3201	A32(19)	–	
A*3301	A33(19)	Aw33.1	
A*3401	A34(10)	–	X61704
A*3402	A34(10)	–	X61705
A*3601	A36	–	X61700
A*4301	A43	–	X61703
A*6601	A66(10)	–	X61711
A*6602	A66(10)	–	X61712
A*6801	A68(28)	Aw68.1	
A*6802	A68(28)	Aw68.2	
A*6901	A69(28)	–	
A*7401	A74(19)	–	X61701
B*0701	B7	B7 1	
B*0702	B7	B7.2	

MHC

Table 5. Continued

HLA allele	HLA specificity	Previous equivalents	Accession number
B*0703	B703	BPOT	X64454
B*0801	B8	–	M59841
B*1301	B13	B13.1	
B*1302	B13	B13.2	
B*1401	B14	–	
B*1402	B65(14)	–	M59840
B*1501	B62(15)	–	
B*1502	B75(15)	–	M75138
B*1503	B72(70)	–	X61709
B*1504	B62(15)	Bw62-G	M84382, M86402
B*1801	B18	–	
B*2701	B27	27f	
B*2702	B27	27e, 27K, B27.2	
B*2703	B27	27d, 27J	
B*2704	B27	27b, 27C, B27.3	
B*2705	B27	27a, 27W, B27.1	
B*2706	B27	27D, B27.4	
B*2707	B27	B27-HS	M62852
B*3501	B35	–	
B*3502	B35	–	
B*3503	B35	–	M81798
B*3504	B35	–	M86403
B*3505	B35	B35-G	M84385
B*3506	B35	B35-K	M84381
B*3701	B37	–	
B*3801	B38(16)	B16.1	
B*3901	B3901	B16.2	
B*3902	B3902	B39.2	
B*4001	B60(40)	–	
B*4002	B40	B40*	
B*4003	B40	B40-G1	M84383
B*4004	B40	B40-G2	M84384
B*4005	B4005	BN21	M84694
B*4101	B41	–	
B*4201	B42	–	
B*4401	B44(12)	B44.1	
B*4402	B44(12)	B44.2	
B*4403	B44(12)	B44.1:New	X64366
B*4501	B45(12)	–	X61710
B*4601	B46	–	
B*4701	B47	–	
B*4801	B48	–	M84380
B*4901	B49(21)	–	
B*5001	B50(21)	–	X61706
B*5101	B51(5)	–	
B*5102	B5102	B5.35	M68964

Table 5. Continued

HLA allele	HLA specificity	Previous equivalents	Accession number
B*5103	B5103	BTA	M80670
B*5201	B52(5)	–	
B*5301	B53	–	
B*5401	B54(22)	–	M77774
B*5501	B55(22)	–	M77778
B*5502	B55(22)	–	M77777
B*5601	B56(22)	–	M77776
B*5602	B56(22)	–	M77775
B*5701	B57(17)	–	X55711
B*5702	B57(17)	Bw57.2	X61707
B*5801	B58(17)	–	
B*7801	B7801	B'SNA', Bx1	X61708
B*7901	–	B'X'-HS	M62852
Cw*0101	Cw1	Cw1.1	
Cw*0102	Cw1	Cw1.2	M84171
Cw*0201	Cw2	Cw2.1	
Cw*02021	Cw2	Cw2.2	
Cw*02022	Cw2	Cw2.2	
Cw*0301	Cw3	–	
Cw*0302	Cw3	–	M84172
Cw*0401	Cw4	–	M84386
Cw*0501	Cw5	–	
Cw*0601	Cw6	–	
Cw*0701	Cw7	–	
Cw*0702	Cw7	JY328	
Cw*0801	Cw8	–	M84174
Cw*0802	Cw8	–	M59865, M84173
Cw*1201	–	Cx52	
Cw*1202	–	Cb-2	
Cw*1301	–	CwBL18	
Cw*1401	–	Cb-1	
E*0101	–	JTW15	
E*0102	–	HLA-6.2	
E*0103	–	–	M32507
E*0104	–	–	M32508

MAJOR HISTOCOMPATIBILITY COMPLEX

MHC

Table 6. Designations of *HLA-DP* alleles (39)

HLA allele	Serological specificity	Previous equivalents	Accession number
DPA1*0101	–	LB14/LB24, DPA1	
DPA1*0102	–	pSBα-318	
DPA1*0103	–	DPw4α1	
DPA1*0201	–	DPA2, pDAα13B	
DPA1*02021	–	2.21	M83906
DPA1*02022	–	2.22	M83907
DPA1*0301	–	3.1	M83908
DPA1*0401	–	4.1	M83909
DPB1*0101	DPw1	DPB1, DPw1a	M83129
DPB1*0201	DPw2	DPB2.1	
DPB1*02011	DPw2	DPB2.1	
DPB1*02012	DPw2	DPB2.1	
DPB1*0202	DPw2	DPB2.2	
DPB1*0301	DPw3	DPB3	
DPB1*0401	DPw4	DPB4.1, DPw4a	
DPB1*0402	DPw4	DPB4.2, DPw4b	
DPB1*0501	DPw5	DPB5	
DPB1*0601	DPw6	DPB6	
DPB1*0801	–	DPB8	
DPB1*0901	–	DPB9, DP'Cp63'	
DPB1*1001	–	DPB10	M85223
DPB1*1101	–	DPB11	
DPB1*1301	–	DPB13	
DPB1*1401	–	DPB14	
DPB1*1501	–	DPB15	
DPB1*1601	–	DPB16	
DPB1*1701	–	DPB17	
DPB1*1801	–	DPB18	
DPB1*1901	–	DPB19	
DPB1*2001	–	Oos, DPB-JA	M58608
DPB1*2101	–	DPB-GM, DPB30, NewD	M77659, M83915, M84621
DPB1*2201	–	DPB1*AB1, NewH	M77674, M83919
DPB1*2301	–	DPB32, NewB	M83913, M84014
DPB1*2401	–	DPB33, NewC	M83914
DPB1*2501	–	DPB34, NewE	M83916
DPB1*2601	–	DPB31, WA2	M86229
DPB1*2701	–	DPB23, WA3	M84619, M86230
DPB1*2801	–	DPB21	M84617
DPB1*2901	–	DPB27, NewG	M84625, M83918
DPB1*3001	–	DPB28	M84620
DPB1*3101	–	DPB22, NewF	M84618, M83917
DPB1*3201	–	DPB24, NewI	M84622, M85222
DPB1*3301	–	DPB25	M84623
DPB1*3401	–	DPB26	M84624
DPB1*3501	–	DPB29	M84626
DPB1*3601	–	NewA, SSK2	M83912, D10479

CELLULAR IMMUNOLOGY LABFAX

Table 7. Designations of *HLA-DQ* alleles

HLA allele	Serological specificities	T-cell-defined specificities	Previous equivalents	Accession number
DQA1*0101	–	Dw1, w9	DQA1.1, 1.9	
DQA1*0102	–	Dw2, w21, w19	DQA1.2, 1.19, 1.AZH	
DQA1*0103	–	Dw18, w12, w8, Dw'FS'	DQA1.3, 1.18, DRw8-DQw1	
DQA1*0104	–			M34314
DQA1*0201	–	Dw7, w11	DQA2, 3.7	
DQA1*03011	–	Dw4, w10, w13-15	DQA3, 3.1, 3.2	
DQA1*03012	–	Dw23	DQA3, 3.1, 3.2, DR9-DQw3	
DQA1*0302	–	Dw23	DQA3, 3.1	
DQA1*0401	–	Dw8, Dw'RSH'	DQA4.2, 3.8	
DQA1*0501	–	Dw3, w5, w22	DQA4.l, 2	
DQA1*05011	–	Dw3	DQA4.1, 2	
DQA1*05012	–	Dw5	DQA4.1, 2	
DQA1*05013	–	Dw22	DQA4.1, 2	
DQA1*0601	–	Dw8	DQA4.3	
DQB1*0501	DQ5(1)	Dw1	DQB1.1, DRw10-DQw1.1	
DQB1*0502	DQ5(1)	Dw21	DQB1.2, 1.21	
DQB1*05031	DQ5(1)	Dw9	DQB1.3, 1.9, 1.3.1	
DQB1*05032	DQ5(1)	Dw9	DQB1.3, 1.9, 1.3.2	
DQB1*0504	–	–	DQB1.9	
DQB1*0601	DQ6(1)	Dw12, w8	DQB1.4, 1.12	
DQB1*0602	DQ6(1)	Dw2	DQB1.5, 1.2	
DQB1*0603	DQ6(1)	Dw18, Dw'FS'	DQB1.6, 1.18	
DQB1*0604	DQ6(1)	Dw19	DQB1.7, 1.19	
DQB1*0605	DQ6(1)	Dw19	DQB1.8, DQBSLE, 1.19b, 2013-24	
DQB1*0606	–	–	DQB1*WA1	M86226
DQB1*0201	DQ2	Dw3, w7	DQB2	
DQB1*0301	DQ7(3)	Dw4, w5, w8, w13	DQB3.1	
DQB1*0302	DQ8(3)	Dw4, w10, w13-14	DQB3.2	
DQB1*03031	DQ9(3)	Dw23	DQB3.3	
DQB1*03032	DQ9(3)	Dw23, w11	DQB3.3	
DQB1*0304	DQ7(3)	–	DQB1*03HP,*03new	M74842, M83770
DQB1*0401	DQ4	Dw15	DQB4.1, Wa	
DQB1*0402	DQ4	Dw8, Dw'RSH'	DQB4.2, WA	

Table 8. Designations of *HLA-DR* alleles (39)

HLA allele	Serological specificities	T-cell-defined specificities	Previous equivalents	Accession number
DRA*0101	–	–	DRα, PDR-α-2	J00194, J00196, J00203
DRA*0102	–	–	DR-H	J00201
DRB1*0101	DR1	Dw1	–	
DRB1*0102	DR1	Dw20	DR1-NASC	
DRB1*0103	DR103	Dw'BON'	DR1-CETUS, DRB1*BON	
DRB1*1501	DR15(2)	Dw2	DR2B Dw2	
DRB1*1502	DR15(2)	Dw12	DR2B Dw12	
DRB1*1503	DR15(2)	–	–	M35159
DRB1*1601	DR16(2)	Dw21	DR2B Dw21	
DRB1*1602	DR16(2)	Dw22	DR2B Dw22	
DRB1*0301	DR17(3)	Dw3	–	
DRB1*0302	DR18(3)	Dw'RSH'	–	
DRB1*0303	DR18(3)	–	–	M81743
DRB1*0401	DR4	Dw4	–	
DRB1*0402	DR4	Dw10	–	
DRB1*0403	DR4	Dw13	DR4 Dw13A, 13.1	
DRB1*0404	DR4	Dw14	DR4 Dw14A, 14.1	
DRB1*0405	DR4	Dw15	–	
DRB1*0406	DR4	Dw'KT2'	–	
DRB1*0407	DR4	Dw13	DR4 Dw13B, 13.2	
DRB1*0408	DR4	Dw14	DR4-CETUS, Dw14B, 14.2	
DRB1*0409	DR4	–	–	
DRB1*0410	DR4	–	DR4.CB	M81670
DRB1*0411	DR4	–	DR4.EC	M81700
DRB1*0412	DR4	–	AB2	M77672
DRB1*11011	DR11(5)	Dw5	DRw11.1	
DRB1*11012	DR11(5)	Dw5	–	M34316
DRB1*1102	DR11(5)	Dw'JVM'	DRw11.2	
DRB1*1103	DR11(5)	–	DRw11.3	
DRB1*11041	DR11(5)	Dw'FS'	–	
DRB1*11042	DR11(5)	–	–	M34317
DRB1*1105	DR11(5)	–	–	M84188
DRB1*1201	DR12(5)	Dw'DB6'	–	
DRB1*1202	DR12(5)	–	DRw12b	
DRB1*1301	DR13(6)	Dw18	DRw6a I	
DRB1*1302	DR13(6)	Dw19	DRw6c I	
DRB1*1303	DR13(6)	Dw'HAG'	–	
DRB1*1304	DR13(6)	–	RB1125-14	
DRB1*1305	DR13(6)	–	DRw6'PEV'	
DRB1*1306	DR13(6)	–	DRB1*13.MW	M61899
DRB1*1401	DR14(6)	Dw9	DRw6b I	

Table 8. Continued

HLA allele	Serological specificities	T-cell-defined specificities	Previous equivalents	Accession number
DRB1*1402	DR14(6)	Dw16	–	
DRB1*1403	DR1403	–	JX6	
DRB1*1404	DR1404	–	DRB1*LY10, DRw6b.2	
DRB1*1405	DR14(6)	–	DRB1*14c	
DRB1*1406	DR14(6)	–	DRB1*14.GB, 14.6	M63927, M74032
DRB1*1407	DR14(6)	–	14.7	M74030
DRB1*1408	DR14(6)	–	A01, 14.8	M77673, M74031
DRB1*1409	DR14(6)	–	AB4	M77671
DRB1*1410	–	–	AB3	M77670
DRB1*0701	DR7	Dw17	–	
DRB1*0702	DR7	Dw'DB1'	–	
DRB1*0801	DR8	Dw8.1	–	
DRB1*08021	DR8	Dw8.2	DRw8-SPL	
DRB1*08022	DR8	Dw8.2	DRw8b	
DRB1*08031	DR8	Dw8.3	DRw8-TAB	
DRB1*08032	DR8	Dw8.3	–	
DRB1*0804	DR8	–	RB1066–1, DR8–V86	M84446
DRB1*0805	DR8	–	DR8–A74	M84357
DRB1*09011	DR9	Dw23	–	
DRB1*09012	DR9	Dw23	–	
DRB1*1001	DR10	–	–	
DRB3*0101	DR52	Dw24	DR3 III, DRw6a III	
DRB3*0201	DR52	Dw25	DRw6b III	
DRB3*0202	DR52	Dw25	pDR5b.3	
DRB3*0301	DR52	Dw26	–	
DRB4*0101	DR53	Dw4, Dw10, Dw13-15, Dw17, Dw23		
DRB5*0101	DR51	Dw2	DR2A Dw2	
DRB5*0102	DR51	Dw12	DR2A Dw12	
DRB5*0201	DR51	Dw21	DR2A Dw21	
DRB5*0202	DR51	Dw22	DR2A Dw22	
DRB6*0101	–	–	DRBσ*0101, DRBX11	X53357, M83892
DRB6*0201	–	–	DRBX21, DRBVI	M77284, X53358
DRB6*0202	–	–	DRBσ*0201, DRBX22, DRB6III	M83204 M83894

MHC

Table 9. List of generic and group-specific primers

Locus or group	5′ primer	3′ primer
DRB	CCCCA CAGCA CGTTT CTTG	CCGCT GCACT GTGAA GCTCT
	TGTCA TTTCT TCAAT GGGAC G	TCGCC GCTGC ACTGT GAAG
	CCGGA TCCTT CGTGT CCCCA CAGCA CG	CTCCC CAACC CCGTA GTTGT GTCTG CA
DR1	TTCTT GTGGC AGCTT AAGTT	CCGCT GCACT GTGAA GCTCT
DR2	GGTTG CTGGA AAGAT GCATC T	CCGCT GCACT GTGAA GCTCT
DR4	TTCCT GTGGC AGCCT AAGAG G	CCGCT GCACT GTGAA GCTCT
DR7	GTTTC TTGGA GCAGG TTAAA C	CCGCT GCACT GTGAA GCTCT
DR9	AGTTC CTGGA AAGAC TCTTC T	CCGCT GCACT GTGAA GCTCT
DR10	GAAGC AGGAT AAGTT TGAGT G	CCGCT GCACT GTGAA GCTCT
DR52[a]	GGTTG CTGGA AAGAC GCGTC C	CCGCT GCACT GTGAA GCTCT
	CACGT TTCTT GGAGT ACTCT AC	CCGCT GCACT GTGAA GCTCT
DR52[b]	ACGTT TCTTG GAGTA CTCTA CG	CCGCT GCACT GTGAA GCTCT
	CCCAG CACGT TTCTT GGAGC T	CCGCT GCACT GTGAA GCTCT
DQA	ATGGT GTAAA CTTGT ACCAG T	TTGGT AGCAG CGGTA GAGTT G
	GTGCT GCAGG TGTAA ACTTG TACCA G	CACGG ATCCG GTAGC AGCGG TAGAG TTG
DQB	CATGT GCTAC TTCAC CAACG G	CTGGT AGTTG TGTCT GCACA C
	AGGGA TCCCC GCAGA GGATT TCGTG T (A/T) AC	GAGCT GCAGG TAGTT GTGTC TGCA (C/T) AC
	CTCGG ATCCG CATGT GCTAC TTCAC CAACG	GAGCT GCAGG TAGTT GTGTC TGCAC AC
DQw1	GCATG TGCTA CTTCA C CAAC G	CACCT GCAGA TCCCG CGGTA CGCCA CCTC
DQw2,3,4	GCATG TGCTA CTTCA C CAAC G	CACCT GCAGT GCGGA GCTCC AACTG GTA
DPA	GCGGA CCATG TGTCA ACTTA T	GCCTG AGTGT GGTTG GAACG
DPB	GAGAG TGGCG CCTCC GCTCA T	GCCGG CCCAA AGCCC TCACT C
	GTGAA GCTTT CCCCG CAGAG ATTAC	CACCT GCAGT CACTC ACCTC GGCGC CGCTG

[a] *DRB1* alleles of *DR52*. [b] *DRB3* alleles of *DR52*. See refs 43–47 and references therein.

6. REFERENCES

1. Silver, L.M., Artzt, K., Barlow, D., Fischer-Lindahl, K., Lyon, M.F., Klein, J. and Snyder, L. (1992) *Mamm. Genet.*, **3**, S241.

2. Steinmetz, M., Moore, K.W., Frelinger, J.G., Sher, B.T., Shen, F.-W., Boyse, E.A. and Hood, L. (1981) *Cell*, **25**, 683.

3. Steinmetz, M., Winoto, A., Minard, K. and Hood, L. (1982) *Cell*, **28**, 489.

4. Steinmetz, M., Stephan, D. and Fischer-Lindahl, K. (1986) *Cell*, **44**, 895.

5. O'Brien, S.J., Womack, J.E., Lyons, L.A., Moore, K.J., Jenkins, N.A. and Copeland, N.G. (1993) *Nature Genetics*, **3**, 103.

6. Cambell, R.D. and Trowsdale, J. (1993) *Immunol. Today*, **14**, 349.

7. Bjorkman, P.J., Saper, M.A., Samraoui, B., Bennett, W.S., Strominger, J.L. and Wiley, D.C. (1987) *Nature*, **329**, 512.

8. Brown, J.H., Jardetzky, T.S., Gorga, J.C., Stern, L.J., Urban, R.G., Strominger, J.L. and Wiley, D.C. (1993) *Nature*, **364**, 33.

9. Garrett, T.P.J., Saper, M.A., Bjorkman, P.J., Strominger, J.L. and Wiley, D.C. (1989) *Nature*, **342**, 692.

10. Madden, D.R., Gorga, J.C., Strominger, J.L. and Wiley, D.C. (1991) *Nature*, **353**, 321.

11. Fremont, D.H., Matsumura, M., Stura, E.A., Peterson, P.A. and Wilson, I.A. (1992) *Science*, **257**, 919.

12. Gorga, J.C. (1992) *Crit. Rev. Immunol.*, **11**, 305.

13. Klein, J., Ohlligin, C., Figueroa, F., Mayer, W.E. and Klein, D. (1993) *Mol. Biol. Evol.*, **10**, 48.

14. Klein, J. (1986) *Natural History of the Major Histocompatibility Complex*. John Wiley & Sons, New York.

15. Satz, M.L., Wang, L.-C., Singer, D.S. and Rudikoff, S. (1985) *J. Immunol.*, **135**, 2167.

16. Hirsch, F., Sachs, D.H., Gustafsson, K., Pratt, K., Germana, S. and LeGuern, C. (1990) *Immunogenetics*, **31**, 52.

17. Garber, T.L., Hughes, A.L., Letvin, N. L., Templeton, J.W. and Watkins, D.I. (1993) *Immunogenetics*, **38**, 11.

18. Sigurdardóttir, S., Borsch, C., Gustafsson, K. and Andersson, L. (1992) *Immunogenetics*, **35**, 205.

19. Alexander, A. J., Bailey, E. and Woodward, J.G. (1987) *Immunogenetics*, **25**, 47.

20. Albright, D., Bailey, E. and Woodward, J.G. (1991) *Immunogenetics*, **34**, 136.

21. Grossberger, D., Hein, W. and Marcuz, A. (1990) *Immunogenetics*, **32**, 77.

22. Scott, P.C., Gogolin-Ewens, K.J., Adams, T.E. and Brandon, M.R. (1991) *Immunogenetics*, **34**, 69.

23. Cameron, P.U., Tabarias, H.A., Pulendran, B., Robinson, W. and Dawkins, R.L. (1990) *Immunogenetics*, **31**, 253.

24. Mann, A. J., Abraham, L.J., Cameron, P.U., Robinson, W., Giphart, M.J. and Dawkins, R.L. (1993) *Immunogenetics*, **37**, 292.

25. Sarmiento, U.M. and Storb, R. (1990) *Immunogenetics*, **31**, 400.

26. Sarmiento, U.M., DeRose, S., Sarmiento, J.I. and Storb, R. (1992) *Immunogenetics*, **35**, 416.

27. Marche, P.N., Tykocinski, M.L. Max, E.E. and Kindt, T.J. (1985) *Immunogenetics*, **21**, 71.

28. Laverriere, A., Kulaga, H., Kindt, T.J., LeGuern, C. and Marche, P.N. (1989) *Immunogenetics*, **30**, 137.

29. Yuhki, N., Heidecker, G.F. and O'Brien, S.J. (1989) *J. Immunol.*, **142**, 3676.

30. Yuhki, N. and O'Brien, S.J. (1988) *Immunogenetics*, **27**, 414.

31. Kindt, T.J. and Singer, D.S. (1987) *Immunol. Res.*, **6**, 57.

32. Pharr, G.T., Bacon, L.D. and Dodgson J.B. (1993) *Immunogenetics*, **37**, 381.

33. Flajnik, M.F. and Pasquier, L.D. (1990) *Immunol. Rev.*, **113**, 47.

34. Křen, V., Frenzl, B. and Štark, O. (1960) *Folia Biologia (Praha)*, **6**, 121.

35. Rogentine Jr, G.N., Vaal, L., Ellis, E.B. and Darrow II, C.C. (1971) *Transplantation*, **12**, 267.

36. Vriesendorp, H.M., Rothengatter, C., Bos, E.,Westbroek, D.L. and van Rood, J.J. (1971) *Transplantation*, **11**, 440.

37. Klein, J., Bach, F.H., Festenstein, H., McDevitt, H.O., Shreffler, D.C., Snell, G.D. and Stimfling, J.H. (1974) *Immunogenetics*, **1**, 184.

38. Klein, J. *et al.* (1990) *Immunogenetics*, **32**, 147.

39. The WHO Nomenclature Committee for factors of the *HLA* system. (1992) *Immunogenetics*, **36**, 135.

40. Watkins, D.I., Zemmour, J. and Parham, P. (1993) *Immunogenetics*, **37**, 317.

41. O'hUigin, C., Bontrop, R. and Klein, J. (1993) *Immunogenetics*, **38**, 165.

MHC

42. Klein, J. (1990) *Immunology*. Blackwell, Boston.

43. Kimura, A. and Sasazuki, T. (1992) in *HLA 1991* (K. Tsuji, M. Aizawa and T. Sasazuki, eds). Oxford University Press, Oxford, Vol. 1, p. 397.

44. Vaughan, R.W., Lanchbury, J.S.S., Marsh, S.G.E., Hall, M.A., Bodmer, J.G. and Welsh, K.I. (1990) *Tissue Antigens*, **36**, 149.

45. Salazar, M., Yunis, J.J., Delgado, M.B., Bing, D. and Yunis, E.J. (1992) *Tissue Antigens*, **40**, 116.

46. Ota, M., Seki, T., Fukushima, H., Tsuji, K. and Inoko, H. (1992) *Tissue Antigens*, **39**, 187.

47. Hviid, T.V.F., Madsen, H.O. and Morling, N. (1992) *Tissue Antigens*, **40**, 140.

48. Hoshino, S., Kimura, A., Fukuda, Y., Dohi K. and Sasazuki, T. (1992) in *HLA 1991* (K. Tsuji, M. Aizawa and T. Sasazuki, eds). Oxford University Press, Oxford, Vol. 1, p. 335.

CHAPTER 11
SOLUBLE IMMUNOREGULATORY MOLECULES

D.L. Gibbons and F.M. Brennan

1. INTRODUCTION

An immune response to an invading pathogen is an essential mechanism to provide specific defence against any particular antigen. Furthermore, the immune response retains memory, enabling a faster response to be made to a secondary infection by the same antigen. Regulation of the magnitude and duration of this response is essential to prevent hypersensitivity and/or autoimmune reactions. This immunoregulation is controlled by soluble factors which perform the functions of the immune response and mediate signaling between the various cells involved. Such families of regulatory molecules include the cytokines and their antagonists, immunoglobulins, serum proteins, complement and other inflammatory mediators. The contribution of these factors, in particular cytokines, to immune regulation is outlined in the following sections.

2. CYTOKINES

Cytokines are a family of low molecular weight protein regulatory molecules involved in cell growth, inflammation, immunity, differentiation and repair. They are characterized by their pleiotropic activity, acting on a wide range of cell types, and, unlike hormones which are carried by the bloodstream throughout the body, cytokines generally act in a local manner, through autocrine and paracrine mechanisms. Cytokines are not generally produced constitutively by cells, and once activated are tightly regulated. It is now accepted that the cytokines include the interleukins (IL-1–IL-13), tumor necrosis factors (TNFα and β), transforming growth factors (TGFα and β), interferons (IFN-α, -β and -γ), colony-stimulating factors (CSFs) and, more recently, a group of cell-associated cytokines.

2.1. Interleukins
Originally the term interleukin described messengers between leukocytes. However, as it became clear that these molecules were produced by, and had activity on, cells outside the immune system, it was obvious that such a description was redundant. To date, 13 interleukin molecules have been cloned and expressed but it is likely that more will emerge in the not too distant future. The main properties of these known interleukins are summarized in *Table 1*.

2.2. Tumor necrosis factor-α and lymphotoxin
Although initially discovered by its ability to induce hemorrhagic necrosis of tumors in mice, the 17 kDa cytokine TNFα is now known to be a potent mediator of the immune response. The major cellular producer of TNFα is the activated macrophage, although it can be produced by a variety of other cells upon activation, whereas production of the 25 kDa protein, lymphotoxin (TNFβ) is more restricted. The genes for both TNFα and lymphotoxin reside on chromosome 6 within the MHC locus. Although both proteins are produced as monomeric molecules, functionally they act as trimers and activate their specific cell-surface receptors (of molecular weight 55 kDa and/or 75 kDa) by cross-linking. The biological activities of TNFα and TNFβ are summarized in *Table 2*; although they display similar functions, there are qualitative and quantitative differences.

Soluble Molecules

Table 1. Interleukins as immune regulators

Interleukin	Biochemistry	Producing cells	Target cells and properties
IL-1α IL-1β	17.5 kDa, glycosylated	Mainly macrophages but most cells, including endothelial cells, fibroblasts, neutrophils	Pro-inflammatory cytokine involved in T- and B-cell activation and the acute phase response (1)
IL-2	15 kDa, glycosylated	T cells (TH-1 in mouse), NK cells	T-cell growth factor, activates T cells, B cells and monocytes
IL-3	15 kDa, glycosylated	T cells, NK cells and mast cells	Growth factor for hematopoietic cells and T-cell subset, activates eosinophils and can induce histamine release from basophils (2–5)
IL-4	20 kDa, glycosylated	T cells (TH-2)	Suppress inflammatory cytokine production from macrophages, B-cell activator and induces class switching to IgE, growth and stimulatory factor for T cells (6–9)
IL-5	45 kDa, glycosylated	T cells (TH-2)	Differentiation factor for B cells and eosinophils, regulates production of granulocytes and monocytes, chemotactic for eosinophils (10–12)
IL-6	26 kDa, glycosylated	T cells, macrophages, endothelial cells, keratinocytes, fibroblasts	Differentiation of B cells, T-B-cell activator, acute phase response, keratinocyte growth (13)
IL-7	25 kDa, glycosylated	Bone marrow stromal cells	Pre-B and pre-T cell growth factor; growth factor for activated T cells (14,15)
IL-8 family C-C family	8–10 kDa		Generally C–C family are chemotactic for monocytes but some members have individual functions
RANTES MCP MIP-1α/β		T cells, platelets, fibroblasts (most tissues), mouse macrophages	Chemotactic for T cells, CD4 > CD8, RO > RA Chemotactic for CD8 and CD45RA

Table 1. Continued

Interleukin	Biochemistry	Producing cells	Target cells and properties
C-X-C family			
IL-8		Macrophages, fibroblasts, endothelial cells, platelets	Generally C-X-C family are chemotactic for neutrophils (17)
PF-4			
IL-9	40 kDa	Activated T cells	Growth factor for T-helper cells, fetal thymocytes and bone marrow-derived mast cells (18)
IL-10	35–40 kDa homodimer	TH-2/TH-0 in mouse; both TH types in humans also B cells, monocytes and keratinocytes	Immunosuppressive of cellular immunity, i.e. decreases inflammatory cytokine production, decreases MHC class II on monocytes and inhibits macrophage antigen-presenting cell function
Also viral IL-10 = BCRF-1 (a viral homolog of IL-10 encoded by the EBV genome)			Increases humoral immunity, i.e. increases mast-cell proliferation and B-cell proliferation and antibody production (19) In mouse, IL-10 inhibits TH-1 function; in human, IL-10 inhibits proliferation of both TH-1 and TH-2 in response to antigen
IL-11	23 kDa	Bone marrow stromal cells	Megakaryocytopoiesis, stem-cell proliferation, acute phase response; similar activities to IL-6 (20, 21)
IL-12 (NK stimulatory activity)	70 kDa heterodimer 2 covalently linked chains of 35 and 40 kDa	EBV-transformed B cells	Stimulates NK and T-cell IFN-γ production, comitogen for resting T cells and direct mitogen for activated T cells and NK blasts; enhances cytotoxicity of NK cells and can generate LAK cells together with IL-2 (22, 23)

Soluble Molecules

SOLUBLE IMMUNOREGULATORY MOLECULES

217

Table 1. Continued

Interleukin	Biochemistry	Producing cells	Target cells and properties
IL-13 (P600)	Predicted 12 kDa	TH-2 cell clone	Similar to IL-4, i.e. decreases inflammatory cytokines from macrophages, but has no effects on T-cell proliferation. Human IL-13 can induce CD23 and cause proliferation and class switching in activated B cells; no IgE production with mouse IL-13 but this can induce bone marrow proliferation (24)

Table 2. The tumor necrosis factors

Tumor necrosis factor	Biochemistry	Producing cells	Target cells and properties
TNFα	157 aa 17 kDa Nonglycosylated Exists in trimeric form	Monocytes T cells B cells LAK cells NK cells PMNs Smooth muscle cells Keratinocytes Astrocytes	Mediates tumor cell death Proliferation of B cells and T cells Growth and differentiation of fibroblasts, keratinocytes, monocytes Induces inflammatory cytokines, metalloproteinases and acute phase proteins Up-regulates adhesion molecules
Lymphotoxin (TNFβ)	171 aa 25 kDa Glycosylated Exists in trimeric form	T cells B cells Astrocytes	Proliferation of activated B cells Mediates tumor cell death Growth of fibroblasts Osteoclast activating factor (reviewed in 25, 26)

2.3. Transforming growth factors

The transforming growth factors were originally described as cytokines capable of conferring a transformed phenotype on a number of fibroblast lines. TGFβ is both structurally and functionally distinct from TGFα and, whereas TGFα has a role mainly in the autocrine growth of fibroblasts and keratinocytes, TGFβ plays a much larger role in immunoregulation. The cellular source and functions of this immunoregulatory molecule are described in *Table 3*.

Table 3. The transforming growth factor-β family

Trans-forming growth factor-βs	Biochemistry	Producing cells	Target cells and properties
TGFβ$_1$ TGFβ$_2$ TGFβ$_3$	Exist as homodimers of 25 kDa; 60–80% aa identity between each member of family; formed as inactive precursors	Many normal and transformed cells Activated monocytes B cells T cells	Growth inhibition of many cells by hypophosphorylation of retinoblastoma gene product Decreases proliferation of T cells and thymocytes Decreases proliferation of B cells and decreases their Ig production Decreases cytokine production from macrophages; induces IL-1 receptor antagonist; induces extracellular matrix formation, increases collagen and decreases collagenase production (27)

2.4. Interferons

Three types of interferon (IFN) were originally defined, based on the producer-cell type, and the antigenic properties of the protein. IFN-α and IFN-β bind to a common receptor and have overlapping, although not identical, properties which are mainly anti-viral and anti-proliferative. Originally defined as immune IFN, IFN-γ is mainly the product of T cells, binds a distinct cell-surface receptor and, due to its ability to activate monocytes, may play a major role in immunoregulation (see *Table 4*).

2.5. Colony-stimulating factors

The principle function of the colony-stimulating factors (*Table 5*) is the induction of hematopoietic stem-cell growth and differentiation. However, it is clear that they also have functional activity on mature cell lineages in the periphery.

2.6. Soluble cytokine antagonists

Cytokines do not act alone but in combination with other cytokines and stimuli, and these can be synergistic or antagonistic, their effects dependant on the target cell used. Due to the potent and profound effects of cytokines *in vivo*, their activities must be tightly regulated both at the levels of secretion and receptor expression, and by soluble inhibitory substances to regulate the action of the cytokine once secreted. Some cytokines, such as IL-4, IL-10 (*Table 1*) and TGFβ (*Table 3*), are themselves antagonistic to the effect of other cytokines. Other mechanisms of antagonism include receptor-binding antagonists, which inhibit binding of the cytokine to its specific receptor by competition, or soluble cytokine-binding proteins, which complex with the cytokine and thus prevent it from binding to its receptor to transduce a signal. A number of cytokine receptors exist in a soluble form in the biological fluids of both human and other animals, implying an immunoregulatory role. The current status of these soluble inhibitors is summarized in *Table 6*.

Soluble Molecules

Table 4. Interferons

Interferons	Biochemistry	Main producers	Properties
IFN-α	22 distinct proteins, 17–24 kDa, some glycosylated	Leukocytes	Both IFN-α + IFN-β bind to a common receptor and have mainly anti-viral and anti-proliferative properties; can up-regulate MHC class I expression (28,29)
IFN-β	Single form 23 kDa in man, 35–40 kDa in mouse, glycosylated	Fibroblasts	
IFN-γ	40–50 kDa homodimer; each subunit consists of 6 α-helices; glycosylated	T cells and NK cells	Activates monocytes; increases cytotoxic and anti-viral activity of monocytes; increases nitric oxide production Increases expression of class I and II MHC on most cells but decreases class II expression on B cells Enhances TNFα production, increases TNF receptor expression and can synergize with TNFα Can act as co-stimulus for T-cell proliferation but inhibits proliferation of TH-2 population in mouse Regulates immunoglobulin isotype switching on B cells (30)

2.7. Soluble adhesion molecules

Endothelial cells, which are located strategically at the interface between blood and tissues, play an active role in homeostasis, inflammatory responses and immunity (64). In particular, endothelial cells respond in a variety of ways to cytokines, including increasing the amount of cell-surface adhesion molecules, which allows effective extravasation of leukocytes from blood into tissue. The adhesion molecules which can be up-regulated in such a manner include the Ig-like molecule ICAM-1, the selectin ELAM-1 and VCAM-1. In addition, recent data have shown that all three adhesion molecules can be shed from the surface of activated cells (*Table 7*) and are found at elevated levels in biological fluids, including the plasma, of patients with rheumatoid arthritis. The soluble adhesion molecules may simply reflect activation of endothelium and possibly of other cells. However, cleavage of surface receptors could also play an immunoregulatory role, as it would result in reduced surface adhesion essential for leukocyte binding and extravasation. Furthermore, by binding to their ligands on the leukocytes they could inhibit interaction with the cell-surface receptor.

Table 5. The colony-stimulating factors

Colony-stimulating factor	Biochemistry	Producing cells	Function
Granulocyte–macrophage colony-stimulating factor (GM-CSF)	14–35 kDa glycoprotein	Wide range of stimulated cells, including T cells, endothelial cells, fibroblasts and epithelial cells	Similar to IL-3; supports the growth of macrophage and granulocyte lineage; induces TNF, IL-1, MHC class II in macrophages (31–34)
Granulocyte colony-stimulating factor (G-CSF)	18–22 kDa glycoprotein	Fibroblasts and monocytes in response to TNF/IL-1	Supports growth of macrophages and granulocytes and their precursors; activates neutrophils, i.e. increases phagocytosis and ADCC (35)
Macrophage colony-stimulating factor (M-CSF)	2 'homodimeric' forms, 70 and 90 kDa, and 40 and 50 kDa (glycosylation and C-terminal processing variants)	Range of cells but principally monocyte and macrophage	Supports growth of macrophage precursors and increases macrophage cytotoxic activity (35)

Table 6. Soluble cytokine receptors and receptor antagonists

Cytokine inhibited	Method of inhibition	Features
IL-1	IL-1 receptor antagonist (IL-1RA)	18–22 kDa, depending on glycosylation. On chs 2 linked to IL-1α and IL-1β genes; 26% homology with IL-1β but no agonist activity unless Lys145 mutated to Asp; two forms by alternative splicing and controlled by two different promoters; secretory form (152 aa) induced from macrophages by TGFβ, LPS, IgG, IL-10, IL-4; intracellular form (159 aa) in keratinocytes, epithelial cells and macrophages; blocks both types of IL-1R but has a lower affinity for type II (CDw121b) than does IL-1β; need molar excess of IL-1RA to completely block IL-1 induced responses (36–39)

SOLUBLE IMMUNOREGULATORY MOLECULES

Table 6. Continued

Cytokine inhibited	Method of inhibition	Features
	Soluble IL-1R	Soluble form of type II surface receptor binds IL-1β but not IL-1α or IL-1RA; binds pro-IL-1β and prevents processing to active form, thus further inhibition; soluble IL-1R formed by cleavage from the cell surface (40, 41)
	B15R gene product	Vaccinia virus gene, B15R encodes soluble IL-1 receptor which binds IL-1β but not IL-1α or IL-1RA (42, 43)
IL-2	Soluble IL-2R	Cleaved from surface p55 IL-2R and can completely inhibit IL-2 even though surface p55 receptor has a low affinity Increased levels in some diseases, e.g. cancer and rheumatoid arthritis; soluble high affinity IL-2R can also be found in supernatants of mitogen-stimulated peripheral blood (44–46)
IL-4	Soluble IL-4R	Formed by alternative slicing of full-length receptor, mRNA for sIL-4R found in T cells and mast cells (47, 48)
IL-5	Soluble IL-5R	Formed by alternative splicing of full-length receptor in mouse; no spontaneously shed protein isolated as yet (49)
IL-6	Soluble IL-6R	Cleaved from the surface receptor – a process which is increased by PMA and blocked by inhibition of PKC; as well as inhibition, can induce a biological signal with IL-6 on cells that themselves do not bind IL-6 (50–52)
IL-7	Soluble IL-7R	mRNA formed by alternative splicing, soluble IL-7R has 27aa unique to the soluble form; no spontaneously shed protein isolated as yet (53)
IFNs	Soluble IFN-γR	Soluble IFN-γR has been isolated from human urine (50)
	MT7 open reading frame	Myxoma virus open reading frame encodes a soluble IFN-γR (54)
	Soluble IFN-αR	Soluble IFN-αR has been found in human urine (45 kDa) and serum (55 kDa); elevated in hairy cell leukemic patients (55)
TNF	Soluble TNF-Rs	Both forms of soluble TNF-R have been reported in the serum and urine of humans; soluble p55 TNF-R has a lower affinity for lymphotoxin

Table 6. Continued

Cytokine inhibited	Method of inhibition	Features
		than its surface counterpart; both are formed by proteolytic cleavage from the surface receptor; elastase is implicated in the cleavage of the p75 TNF-R from neutrophils; elevated in patients with cancer and rheumatoid arthritis (56–61)
GM-CSF	Soluble GM-CSF-α R	Alternative spliced form of the full-length or separate gene product encodes mRNA for the soluble GM-CSF-R which contains 16 unique aa; predicted size, 36 kDa (62)
G-CSF	Soluble G-CSF-R	Again, alternative splicing or a separate gene product encodes soluble G-CSF-R, which has 150 aa unique to the soluble form (63)

Abbreviations: aa, amino acids; chs, chromosome.

Table 7. Soluble adhesion molecules

Soluble adhesion molecule	Cell source	Comment	Refs
sGMP140 (P-selectin, CD62P) (140 kDa)	Endothelium, platelets	Inhibits CD18-dependent neutrophil binding to endothelium	65,66
sVCAM-1 (CD106) (85-90 kDa)	Endothelium, activated macrophages, type B synoviocytes	Levels raised in plasma from rheumatoid arthritis and systemic lupus erythematosus (SLE) patients	67
sICAM-1 (CD54)	Wide range of hematopoietic cells	Levels raised in plasma from rheumatoid arthritis but not SLE patients (diagnostic marker for rheumatoid arthritis?)	68
sE-selectin (CD62E) (105 kDa and 85 kDa)	Endothelium	Levels raised in serum of patients with septic shock, indicative of endothelial activation? and/or damage	69

Soluble Molecules

Table 8. Plasma proteins as immunoregulatory molecules

Protein	Function
α_2-Macroglobulin	Serum protein with zinc-binding site; upon activation of an internal thiol-ester bond, forms complexes with cytokines and other peptides (70, 71)
Complement components	
C3	Upon activation of an internal thiol-ester bond by C3 convertase, there is the formation of the anaphylatoxin, C3a, and complement protein, C3b; these lead to chemotaxis and activation of leukocytes, and opsonization of the target with complement proteins (70, 71)
C4	Also contains an internal thiol-ester bond, and upon activation forms C4a and C4b, leading to opsonization of the target with phagocytosis
C5	Lacks an internal thiol-ester bond but has structural similarities to C3 and C4; broken down to C5a and C5b by C5 convertase; C5a is a very potent anaphylatoxin which induces smooth muscle contraction, mast cell degranulation and neutrophil activation. C5b is the next step in complement cascade, leading to formation of membrane attack complex and lysis of the target cell

2.8. Other soluble immune regulatory molecules

α_2-Macroglobulin (α_2-M) is a member of a group of plasma proteins, including the complement components C3 and C4, which all contain a unique internal cyclic thiol-ester bond. In humans, α_2-M comprises 8–10% of total serum protein and functions principally to bind host or foreign peptides as a humoral defence barrier. In addition to its humoral defence role, α_2-M has an immunoregulatory function whereby, in its activated form, it can bind cytokines (*Table 8*) and as such can modulate cytokine activities in tissue, including modulation of cell growth and function in a number of cell types (70, 71). In the periphery, binding of cytokines to α_2-M may represent an essential clearance mechanism for cytokines.

The complement system is also a very potent mechanism for initiating and amplifying the immune response (*Table 8*) (72). In particular, the products of C3 and C5, the anaphylatoxins C3a and C5a, stimulate chemotaxis and activation of leukocytes, and trigger degranulation of basophils and mast cells. The products enhance vascular smooth muscle contraction, increase vascular permeability and the migration of leukocytes from blood into tissue. Similarly, C3b and C4b (the cleaved products of C3 and C4) act as opsonins by binding target cells/bacteria and thus enhancing phagocytosis by neutrophils, monocytes and macrophages.

3. IMMUNOGLOBULINS

All five classes of immunoglobulin (Ig) are expressed both as cell-surface-bound and soluble molecules (*Table 9*). Thus, cross-linking of surface Ig will induce signaling, depending on the developmental stage of the B cells. Immature B cells which lack sIgD, but which express

Table 9. Soluble immunoglobulin as a regulator of the immune response

Immunoglobulin	Function
IgM	Soluble form exists as a pentameric molecule produced in the early phase of the primary immune response; binds antigen with low affinity, high avidity; efficiently neutralizes antigen, mediates complement-dependent lysis, promotes opsonization and phagocytosis
IgG	The most abundant secreted immunoglobulin; functions to mediate an effective link between antigen and immune response; Fc portion is the effector part of the molecule; for complement activation IgG3 > IgG1 > IgG2, and IgG4 shows no binding; Binds to Fc receptors on macrophages and neutrophils; soluble IgG can also mediate suppression, by 'feeding-back' on producing cells and inhibiting Ig production

sIgM are rendered unresponsive by cross-linking sIgM, but mature cells which coexpress sIgM and sIgD become activated upon cross-linking surface immunoglobulin. Similarly, soluble immunoglobulin has a profound modulatory effect on the immune response, which is determined by many factors, including the nature of antigen and the route of administration. Soluble immunoglobulin neutralizes antigen, leading to formation of immune complexes, complement fixation, phagocytosis and immune clearance. Furthermore, soluble immuno-globulin itself is able to regulate the immune reaction by either enhancing or suppressing the response to antigen (*Table 9*).

4. REFERENCES

1. Dinarello, C.A. (1992) *Chem. Immunol.*, **57**, 1.

2. Yang, Y.C., Ciarletta, A.B., Temple, P.A., Chung, M.P., Kovacic, S., Witek-Giannotti, J.S., Leary, A.C., Kriz, R., Donahue, R.E. and Wong, G.G. (1986) *Cell*, **47**, 3.

3. Clark-Lewis, I., Schrader, W., Wu, A. and Harris, A. (1982) *Cell. Immunol.*, **69**, 196.

4. Londei, M., Verhoef, A., De-Berardinis, P., Kissonerghis, M., Grubeck-Loebenstein, B. and Feldmann, M. (1989) *Proc. Natl. Acad. Sci. USA*, **86**, 8502.

5. Valent, P., Besemer, J., Majdic, O., Lechner, K. and Bettleheim, P. (1989) *Proc. Natl. Acad. Sci. USA*, **86**, 5542.

6. Howard, M., Farrar, J., Hilfiker, M., Johnson, B., Takatsu, K., Hamaoka, T. and Paul W.E. (1982) *J. Exp. Med.*, **155**, 914.

7. Hart, P.H., Vitti, G.F., Burgess, D.R., Whitty, G.A., Piccoli, D.S. and Hamilton, J.A. (1989) *Proc. Natl. Acad. Sci. USA*, **86**, 3803.

8. Hart, P.H., Cooper, R.L. and Finlay-Jones, J.J. (1991) *Immunol.*, **72**, 344.

9. Spits, H., Yssel, H., Takebe, Y., Arai, N., Yokota, T., Lee, F., Arai, K., Banchereau, J. and De-Vries, J.E. (1987) *J. Immunol.*, **139**, 1142.

10. Tominaga, A., Matsumoto, M., Harada, N., Takahashi, T., Kikuchi, Y. and Takatsu, K. (1988) *J. Immunol.*, **140**, 1175.

11. Wang, J.M., Rambaldi, A., Biondi, A., Chen, Z.G., Sanderson, C.J. and Mantovani, A. (1989) *Eur. J. Immunol.*, **19**, 701.

12. Yamaguchi, Y., Suda, T., Suda, J., Eguchi, M., Miura, Y., Harada, N., Tominaga, A. and Takatsu, K. (1988) *J. Exp. Med.*, **167**, 43.

13. Van Snick, J. (1990) *Ann. Rev. Immunol.*, **8**, 253.

14. Morrissey, P.J., Goodwin, R.G., Nordan, R.P., Anderson, D., Grabstein, K.H., Cosman, D., Sims, J., Lupton, S., Acres, B. and Reed, S.G. (1989) *J. Exp. Med.*, **169**, 707.

15. Takeda, S., Gillis, S. and Palacios, R. (1989) *Proc. Natl. Acad. Sci. USA*, **86**, 1634.

16. Schall, T.J. (1991) *Cytokine*, **3**, 165.

Soluble Molecules

17. Oppenheim, J.J., Zachariae, C.O., Mukaida, N. and Matsushima, K. (1991) *Ann. Rev. Immunol.*, **9**, 617.

18. Yang, Y.C., Ricciardi, S., Ciarletta, A., Calvetti, J., Kelleher, K. and Clark, S.C. (1989) *Blood*, **74**, 1880.

19. Moore, K.W., OGarra, A., De Waal Malefyt, R., Vieira, P. and Mosmann T.R. (1993) *Ann. Rev. Immunol.*, **11**, 165.

20. Anderson, K.C., Morimoto, C., Paul, S.R., Chauhan, D., Williams, D., Cochran, M. and Barut, B.A. (1992) *Blood*, **80**, 2797.

21. Burstein, S.A., Mei, R.L., Henthorn, J., Friese, P. and Turner, K. (1992) *J. Cell. Physiol.*, **153**, 305.

22. Podlaski, F.J., Nanduri, V.B., Hulmes, J.D., Pan, Y.C., Levin, W., Danho, W., Chizzonite, R., Gately, M.K. and Stern, A.S. (1992) *Arch. Biochem. Biophys.*, **294**, 230.

23. Valiante, N.M., Rengaraju, M. and Trinchieri, G. (1992) *Cell. Immunol.*, **145**, 187.

24. Punnonen, J., Aversa, G., Cocks, B.G., McKenzie, A.H., Menon, S., Zurawski, G., De Waal Malefyt, R. and De Vries, J. (1993) *Proc. Natl. Acad. Sci. USA*, **90**, 3730.

25. Vilcek, J. and Lee, T.H. (1991) *J. Biol. Chem.*, **266**, 7313.

26. Beutler, B. (ed.) (1992) *Tumor Necrosis Factors. The Molecules and their Emerging Role in Medicine*. Raven Press, New York.

27. Massague, J. (1990) *Ann. Rev. Cell. Biol.*, **6**, 597.

28. Pestka, S., Langer, J.A., Zoon, K.C., and Samuel, C.E. (1987) *Ann. Rev. Biochem.*, **56**, 727.

29. Zoon, K.C., Miller, D., Bekisz, J., Zur-Nedden, D., Enterline, J.C., Nguyen, N.Y. and Hu, R.Q. (1992) *J. Biol. Chem.*, **267**, 15210.

30. Farrar, M.A. and Schreiber, R.D. (1993) *Ann. Rev. Immunol.*, **11**, 571.

31. Seelentag, W.K., Mermod, J.J., Montesano, R. and Vassalli, P. (1987) *EMBO J.*, **6**, 2261.

32. Seelentag, W., Mermod, J.J. and Vasalli, P. (1989) *Eur. J. Immunol.*, **19**, 209.

33. Herman, F., Oster, W., Meuer, S., Lindemann, A. and Mertelsmann, R.H. (1988) *J. Clin. Invest.*, **81**, 1415.

34. Chantry, D., Turner, M., Brennan, F., Kingsbury, A. and Feldmann, M. (1990) *Cytokine*, **2**, 60.

35. Clark, S.C. and Kamen, R. (1987) *Science*, **236**, 1229.

36. Eisenberg, S.P., Evans, R.J., Arend, W.P., Verderber, E., Brewer, M.T., Hannum, C.H. and Thompson, R.C. (1990) *Nature*, **343**, 341.

37. Haskill, S., Martin, G., Van, L.L., Morris, J., Peace, A., Bigler, C.F., Jafe, G.J., Hammerberg, C., Spron, S.A. and Fong, S. (1991) *Proc. Natl. Acad. Sci. USA*, **88**, 3681.

38. Ju, G., Labriola, T.E., Campen, C.A., Benjamin, W.R., Karas, J., Plocinski, J., Biondi, D., Kaffka, K.L., Kilian, P.L. and Eisenberg, S.P. (1991) *Proc. Natl. Acad. Sci. USA*, **88**, 2658.

39. Berger, A.E., Carter, D.B., Hankey, S.O., McEwan, R.N. (1993) *Eur. J. Immunol.*, **23**, 39.

40. Symons, J.A., Eastgate, J.A. and Duff, G.W. (1991) *J. Exp. Med.*, **174**, 1251.

41. Giri, J.G., Newton, R.C. and Horuk, R. (1990) *J. Biol. Chem.*, **265**, 17416.

42. Spriggs, M.K., Hruby, D.E., Maliszewski, C.R., Pickup, D.J., Sims, J.E., Buller, R.M. and Van Slyke, J. (1992) *Cell*, **71**, 145.

43. Alcami, A. and Smith, G.L. (1992) *Cell*, **71**, 153.

44. Rubin, L.A., Kurman, C.C., Fritz, M.E., Biddison, W.E., Boutin, B., Yarchoan, R. and Nelson, D.L. (1985) *J. Immunol.*, **135**, 3172.

45. Symons, J.A., Wood, N.C., Di, G.F. and Duff, G.W. (1988) *J. Immunol.*, **141**, 2612.

46 Pui, C.H., Ip, S.H., Iflah, S., Behm, F.G., Grose, B.H., Dodge, R.K., Crist, W.M., Furman, W.L., Murphy, S.B. and Rivera, G.K. (1988) *Blood*, **71**, 1135.

47. Mosley, B., Beckmann, M.P., March, C.J., Idzerda, R.L., Gimpel, S.D., VandenBos, T., Friend, D., Alpert, A., Anderson, D. and Jackson, J. (1989) *Cell*, **59**, 335.

48. Fernandez-Botran, R. and Vitetta, E.S. (1990) *Proc. Natl. Acad. Sci. USA*, **87**, 4202.

49. Takaki, S., Tominaga, A., Hitoshi, Y., Mita, S., Sonoda, N., Yamaguchi, N. and Takatsu, K. (1990) *EMBO J.*, **9**, 4367.

50. Novick, D., Engelmann, H., Wallach, D. and Rubinstein, M. (1989) *J. Exp. Med.*, **170**, 1409.

51. Mullberg, J., Schooltink, H., Stoyan, T., Heinrich, P.C. and Rose-John, S. (1992) *Biochem. Biophys. Res. Comm.*, **189**, 794.

52. Mackiewicz, A., Schooltink, H., Heinrich, P.C. and Rose-John, S. (1992) *J. Immunol.*, **149**, 2021.

53. Goodwin, R.G., Friend, D., Ziegler, S.F., Jerzy, R., Falk, B.A., Gimpel, S., Cosman, D., Dower, S.K., March, C.J. and Namen, A.E. (1990) *Cell*, **60**, 941.

54. Upton, C., Mossman, K. and McFadden, G. (1992) *Science*, **258**, 1369.

55. Novick, D., Cohen, B. and Rubinstein, M. (1992) *FEBS Lett.*, **314**, 445.

56. Gatanaga, T., Hwang, C.D., Kohr, W., Cappuccini, F., Lucci, J., Jeffes, E.W., Lentz, R., Tomich, J., Yamamoto, R.S. and Granger, G.A. (1990) *Proc. Natl. Acad. Sci. USA*, **87**, 8781.

57. Engelmann, H., Aderka, D., Rubinstein, M., Rotman, D. and Wallach, D. (1989) *J. Biol. Chem.*, **264**, 11974.

58. Engelmann, H., Novick, D. and Wallach, D. (1990) *J. Biol. Chem.*, **265**, 1531.

59. Porteu, F., Brockhaus, M., Wallach, D., Engelmann, H. and Nathan, C.F. (1991) *J. Biol. Chem.*, **266**, 18846.

60. Cope, A.P., Aderka, D., Doherty, M., Engelmann, H., Gibbons, D., Jones, A.C., Brennan, F.M., Maini, R.N., Wallach, D. and Feldmann, M. (1992) *Arth. Rheum.*, **35**, 1160.

61. Aderka, D., Englemann, H., Hornik, V., Skornick, Y., Levo, Y., Wallach, D. and Kushtai, G. (1991) *Cancer Res.*, **51**, 5602.

62. Raines, M.A., Liu, L., Quan, S.G., Joe, V., DiPersio, J.F. and Golde, D.W. (1991) *Proc. Natl. Acad. Sci. USA*, **88**, 8203.

63. Fukunaga, R., Seto, Y., Mizushima, S. and Nagata, S. (1990) *Proc. Natl. Acad. Sci. USA*, **87**, 8702.

64. Mantovani, A., Bussolino, F. and Dejana, E. (1992) *FASEB J.*, **6**, 2591.

65. Gamble, J.R., Skinner, M.P., Berndt, M.C. and Vadas, M.A. (1990) *Science*, **249**, 414.

66. Dunlop, L.C., Skinner, M.P., Bendall, L.J., Favaloro, F.J., Castaldi, P.A., Gormon, J.J., Gamble, J.R., Vadas, M.A. and Berndt, M.C. (1992) *J. Exp. Med.*, **175**, 1147.

67. Wellicome, S.M., Kapahi, P., Mason, J.C., Lebranchu, Y., Yarwood, H. and Haskard, D.O. (1993) *Clin. Exp. Immunol.*, **92**, 412.

68. Mason, J.C., Kapahi, P. and Haskard, D.O. (1993) *Arth. Rheum.*, **36**, 519.

69. Newman, W., Beall, L.D., Carson, C.W., Hunder, G.G., Graben, N., Randhawa, Z.I., Gopal, T.V., Wiener-Kronish, J. and Matthay, M.A. (1993) *J. Immunol.*, **150**, 644.

70. James, K. (1990) *Immunol. Today*, **11**, 163.

71. Borth, W. (1992) *FASEB J.*, **6**, 3345.

72. Reid, K.B.M. and Day, A.J. (1989) *Immunol. Today*, **10**, 177.

Soluble Molecules

CHAPTER 12
CELLULAR ACTIVATION
M.R. Gold

Immune responses require the recognition of foreign antigens by T and B lymphocytes, followed by a coordinated response that often involves multiple cell types. The cells of the immune system communicate with each other via soluble cytokines and by direct cell–cell contact involving membrane-bound molecules. These extracellular messengers are bound by specific receptors on the surface of immune cells, which in turn activate intracellular signaling pathways. The intracellular 'second messengers' that are generated transmit information from the plasma membrane to the nucleus, inducing changes in gene expression and regulating the cell cycle. Cytoplasmic alterations, such as reorganization of the cytoskeleton and changes in cell motility, also occur. Thus, receptor-mediated signaling initiates changes within immune cells that lead to chemotaxis, the secretion of effector molecules (e.g. proteases, cytolytic molecules, antibodies), cell proliferation and differentiation, and the secretion of cytokines that regulate these responses.

1. LIGANDS AND RECEPTORS INVOLVED IN IMMUNE-CELL ACTIVATION

Immune-cell function is regulated by a large number of receptors, and at any one time a single cell may need to integrate the signals coming from many receptors. Immune cells possess receptors that recognize:

(i)   foreign molecules such as antigens or bacterial molecules;
(ii) cytokines produced by other cells of the immune system; and
(iii) proteins expressed on the surface of other cells of the immune system as well as on endothelial cells.

Receptors that mediate activation of lymphoid or myeloid cells are described in *Table 1*.

1.1. Cell lines useful for studying signal transduction
While normal cell populations from mice or humans are the most relevant for studying the actions of various factors on immune cells, biochemical analysis of signal transduction mechanisms often requires a large number of cells (e.g. to study the phosphorylation of a low-abundance regulatory protein), making the use of tissue culture cell lines desirable. A selection of useful cell lines is listed in *Table 2*.

2. TYROSINE KINASE ACTIVATION BY RECEPTORS ON CELLS OF THE IMMUNE SYSTEM

A common mechanism of signal transduction utilized by many receptors is the activation of protein tyrosine kinases (PTKs). Some receptors (R) (e.g. the CSF-1R) contain intrinsic PTK activities in their cytoplasmic domains. In contrast, many other receptors (e.g. the T-cell and B-cell antigen receptors) associate with distinct PTK molecules. The major class of nonreceptor PTKs is the *src* family, which contains eight known members (p60^{src}, p55^{blk}, p56^{lck}, p59^{fyn}, p53/56^{lyn}, p62^{yes}, p56/59^{hck}, p58^{fgr}) with significant sequence homology to

▶ p. 238

Table 1. Ligands and receptors involved in immune-cell activation

Receptor (R)[a]	Ligand	Receptor structure[b]	Biological responses	Refs
T-cell AgR (TCR)	Ag/MHC complex; anti-TCR Abs; anti-CD3 Abs	Ag binding subunit: α (40–45 kDa) or γ (45–60 kDa) β (38–45 kDa) or δ (40–45 kDa) Signaling subunits: CD3γ (m: 21 kDa; h: 25–28 kDa) CD3δ (m: 26 kDa; h: 20 kDa) CD3ε (m: 25 kDa; h: 20 kDa) 2ζ (16 kDa) or 1ζ + 1η (22 kDa)	Thymocytes: positive and negative selection Mature T cells: activation, (entry into cell cycle); IL-2R expression; secretion of IL-2, IL3, IL-4, IL-5, GM-CSF, IFN-γ (these responses also require activation of a co-R such as CD28); directed secretion of lymphokines or cytotoxic molecules (perforins and granzymes) towards cellular targets	1,2
B-cell AgR	Ag or anti-Ig Ab	Ag binding subunit: 2 Ig H chains (IgM: 70 kDa IgD: 50 kDa) 2 Ig L chains (25 kDa) Signaling subunits: CD79a (m: 32 kDa; h: 45 kDa)[c] CD79b (37 kDa)	Mature B cell: activation (entry into cell cycle, increase in cell volume and RNA content); up-regulation of MHC class II Immature B cell: tolerance or cell death Germinal center B cell: prevention of programmed cell death (apoptosis)	3
FcεRI	IgE immune complexes	α (45 kDa) β (33 kDa) γ (12 kDa)	Mast cells: degranulation; release of histamine, proteases and inflammatory mediators; release of arachidonic acid metabolites; secretion of IL-3, IL-4, IL-5 and GM-CSF	4,5
FcεRII (CD23)	IgE immune complexes; CD21	45–50 kDa	Monocytes and eosinophils: release of inflammatory mediators; induction of IgE-dependent cytotoxicity towards parasites B cells: inhibits proliferation and Ab secretion	6
FcγRI (CD64)	Monomeric IgG	72 kDa	Macrophages: cytokine production (IL-1, IL-6, TNFα) Neutrophils: superoxide production; lysosomal enzyme release	7

Molecule	Ligand	Components	Function	Ref.
FcγRII (CD32)	IgG immune complexes	40 kDa	Macrophages: cytokine production (IL-1, IL-6, TNFα); Neutrophils: superoxide production; lysosomal enzyme release	7
FcγRIII (CD16)	IgG immune complexes	α (50–70 kDa); 2ζ (16 kDa) (same ζ as in TCR) or 1ζ and 1γ (12 kDa) (same γ as in FcεRI)	NK cells: induction of Ab-dependent cytotoxic activity (ADCC)	8
NKR-P1	?[d] or anti-NKR-P1	60 kDa (homodimer)	NK cells: induction of cytolytic activity	9
MHC class II	TCR/CD4 or anti-class II Ab	α (32–34 kDa); β (29–32 kDa)	B cells: induction of B7 (ligand for CD28); induces proliferation of activated B cells	3
CD2	LFA-3 (CD58)	50 kDa	T cells: activation (entry into cell cycle, induction of IL-2 and IL-2R genes, secretion of IL-2 and other cytokines); synergizes with TCR signaling[e]; NK cells: induction of cytolytic activity	2,10
CD4 or CD8	Ag/MHC complex	CD4 (59 kDa); 2 CD8α (32 kDa; homodimer) or 1 CD8α + 1 CD8β (32 kDa)	T cells: co-R for TCR; cross-linking to TCR augments TCR signaling; Abs that sequester these co-Rs away from TCR inhibit TCR signaling.	11
CD5	CD72	67 kDa	T cells: induces proliferation; synergizes with TCR signaling[e]	12,13
CD19/CD21 complex	CD19: anti-CD19 Ab; CD21: CD23, complement component C3dg	CD19 (95 kDa); CD21 (CR2) (145 kDa); CD81 (TAPA-1) (25 kDa); Leu-13 (16 kDa)	B cells: co-R for B cell AgR?; synergizes with anti-Ig Abs, presumably by bringing this complex in close proximity to AgR; some anti-CD19 Abs antagonize anti-Ig effects, perhaps by sequestering this complex away from the AgR	3,14

Cellular Activation

Table 1. Continued

Receptor (R)[a]	Ligand	Receptor structure[b]	Biological responses	Refs
CD20	? or Anti-CD20 Ab	35, 37 kDa	Mature B cells: activation (entry into cell cycle, increase in cell volume and RNA content); homotypic aggregation; up-regulation of MHC class II, CD18, CD58 (LFA-3)	15
CD26	? or anti-CD26 Ab	120 kDa (extracellular domain has dipeptidyl peptidase activity)	T cells: second signal required for T-cell activation (entry into cell cycle, proliferation, cytokine production) in addition to TCR signaling?	16
CD27	CD70 or anti-CD27 Ab	55 kDa (homodimer)	T cells: second signal required for T-cell activation (entry into cell cycle, proliferation, cytokine production) in addition to TCR signaling?	17
CD28	CD 80 (B7/BB-1), CTLA-4 or anti-CD28	44 kDa (homodimer)	T cells: second signal required for T-cell activation (entry into cell cycle, proliferation, cytokine production) in addition to TCR signaling; TCR signaling in absence of CD28 or other 'second signal' leads to anergy	18
CD40	CD40 ligand (gp39) or anti-CD40 Ab	50 kDa	B cells: activation and proliferation of resting B cells in response to contact with activated T cells	19
CD45	? or anti-CD45 Ab	B cells: 220 kDa (B220) Resting T cells: 180, 190 and 220 kDa Activated T cells: 180 kDa[f]	Required for signaling by TCR; some anti-CD45 Abs inhibit signaling by TCR and B cell AgR, presumably by sequestering CD45 away from the AgRs	11,20
Thy-1 (CDw90)	? or anti-Thy-1 Ab	25–30 kDa (GPI-linked)[g]	T cells: activation (entry into cell cycle, induction of IL-2 and IL-2R genes, secretion	2

Designation	Stimulus/Ligand	Structure (MW)	Function	Ref.
Ly-6 (sca-1)	? or anti-Ly-6	14-17 kDa (GPI-linked)[g]	of IL-2 and other cytokines); synergizes with TCR signaling[e]	2
IL-1R Type I (CDw121a)	IL-1α and IL-1β	80 kDa (type I; T cells)	T cells: activation (entry into cell cycle, induction of IL-2 and IL-2R genes, secretion of IL-2 and other cytokines); synergizes with TCR signaling[e]	21
IL-1R Type II (CDw121b)	IL-1α and IL-1β	60-68 kDa (type II; B cells and macrophages)	B cells: stimulates pre-B cell differentiation; promotes Ig secretion by activated B cells. Macrophages: induces IL-1 secretion. T cells: second signal for T-cell activation along with signaling through TCR; induces expression of IL-2 and IL-2R genes; co-stimulator for thymocyte proliferation	
IL-2R	IL-2	α (CD25; 55 kDa) β (CDw122; 75 kDa) γ (64 kDa) (β and γ belong to hemopoietin R family)[h]	T cells: growth factor for T cells activated through TCR; stimulation of CTL activity. B cells: promotes differentiation to Ab secretion; induction of J chain gene; stimulates proliferation of B cells activated with anti-Ig. NK cells: stimulates proliferation and induction of tumoricidal activity	22,23 24
IL-3R	IL-3	α (70 kDa) β (AIC-2A, m: 130 kDa or AIC-2B, m: 130 kDa or KH97, h: 130 kDa)[j] (Both α and β hemopoietin R family)	Growth factor for myeloid precursors of multiple lineages; stimulates proliferation of mast-cell lines; enhances FcεRI-induced histamine release from basophils; stimulates phagocytosis by macrophages	24,25
IL-4R (CDw124)	IL-4	140 kDa (hemopoietin R family) γ (64 kDa) (same γ as in IL-2R)	B cells: synergizes with anti-Ig to induce proliferation of resting B cells; induces class switching to IgE and IgG1 (mouse) or IgG4 (human); inhibits class switching to IgA; up-regulates CD40 expression	24,26

Table 1. Continued

Receptor (R)[a]	Ligand	Receptor structure[b]	Biological responses	Refs
			B cells and macrophages: up-regulates MHC class II; induces CD23	
			Macrophages: enhances phagocytosis; suppresses cytokine production	
			T cells: growth factor for activated T$_h$2-helper T cells; inhibits macrophage-dependent cytokine synthesis (e.g. IFN-γ) by T cells	
IL-5R	IL-5	α (60 kDa) β (AIC-2B, m: 130 kDa or KH97, h: 130 kDa)[i] (Both α and β hemopoietin R family)	B cells: promotes differentiation to IgM secretion and class switching to IgA; enhances IL-4-induced IgE production; up-regulation of CD23	24,27
			Eosinophils: growth factor and chemoattractant	
IL-6R	IL-6	α (CD126; 80 kDa) β (CDw130; 130 kDa)[j] (Both hemopoietin R family)	B cells: promotes differentiation to Ab secretion; growth factor for murine plasmacytomas	24,28
			T cells: co-stimulator for thymocyte proliferation; enhances induction of CTLs in presence of IL-2	
IL-7R (CDw127)	IL-7	75–80 kDa γ (64 kDa) (same γ as in IL-2R)	B cells: growth factor for pre- and pro-B cells	29
			Thymocytes: induces proliferation and differentiation	
IL-8R type 1 (CDw128) IL-8R type 2	IL-8 (both Rs), type 2 only: neutrophil activating peptide 2 (NAP-2), MGSA/GRO2	58 kDa (type 1) 67 kDa (type 2) Both are 7 TM[k]	Neutrophils: activator and chemoattractant; stimulates respiratory burst and elastase release; regulates migration and adhesion by inducing surface expression and activation of CD11/CD18 integrin	30
			T cells: chemoattractant	
			Basophils: chemoattractant and stimulator of histamine release	

Cytokine	Receptor	Molecular weight	Biological activities	References
IL-9	IL-9R	64 kDa	Synergizes with IL-3 to support mast cell growth; supports growth of some T-cell clones; synergizes with steel factor to induce proliferation of M07e megakaryocyte cell line	31,32
IL-10	IL-10R	Uncharacterized	Macrophages: inhibits production of IL-1, IL-6, IL-8, IL-10, GM-CSF, G-CSF, and TNFα; down-regulation of MHC class II; suppresses ability to stimulate Ag-induced proliferation and cytokine synthesis by T_h1-helper T cells. B cells: up-regulation of MHC class II; augments proliferation and differentiation of anti-Ig-activated B cells. T cells: augments proliferation of thymocytes activated with IL-2 and IL-4. Mast cells: synergizes with IL-3 and IL-4 for proliferation and colony formation	33,34
IL-11	IL-11R	α(?) β (CDw130; 130 kDa)[j]	B cells: promotes T-cell-dependent induction of IgM secretion. Myeloid cells: synergizes with IL-3 to promote myeloid colony formation	32,35
IL-12	IL-12R	110–135 kDa	T cells: augments proliferation of Ag- or IL-2-stimulated cells. NK cells: augments proliferation of IL-2-stimulated cells	36,37
IL-13	IL-13R	Uncharacterized	B cells: synergizes with CD40 ligand to induce proliferation and Ig secretion; class switching to IgG4 and IgE; induction of CD23. Monocytes and macrophages: inhibits LPS-induced release of IL-1, IL-6, IL-8 and TNFα; induction of CD23	38,39

Cellular Activation

Table 1. Continued

Receptor (R)[a]	Ligand	Receptor structure[b]	Biological responses	Refs
CSF-1R (CD115) (*c-fms*)	Macrophage colony-stimulating factor (M-CSF; CSF-1)	150 kDa (PTK)[1]	Growth factor for macrophages and macrophage precursors; stimulates cytokine release (e.g. IL-1 and G-CSF) by macrophages	40
GM-CSF-R (CDw116)	Granulocyte–macrophage colony-stimulating factor (GM-CSF)	α (85 kDa) β (AIC-2B, m: 130 kDa or KH97, h: 130 kDa)[i] (Both α and β hemopoietin R family)	Growth factor for macrophage and granulocyte precursors; chemoattractant for granulocytes and monocytes; enhances phagocytic and bactericidal capability of granulocytes	24,41
G-CSF-R	Granulocyte colony-stimulating factor (G-CSF)	130–150 kDa (hemopoietin R family)	Growth factor for granulocyte precursors; neutrophil activation	42
c-kit (CD117)	Steel factor (SLF; also known as stem-cell factor)	145 kDa (PTK)[1]	Synergizes with other cytokines to stimulate growth of pluripotent stem cells, myeloid and lymphoid precursors; synergizes with other cytokines to stimulate mast-cell growth; stimulates mast-cell proliferation	43
IFN-γ R (CDw119)	Interferon-γ (IFN-γ)	90 kDa	B cells: blocks IL-4-induced class switching to IgE and IgG1; induces class switching to IgG2b T cells: inhibits proliferation of T$_h$2-helper T cells Macrophages: induces cytotoxic functions; up-regulates MHC class I and class II molecules NK cells: enhances cytotoxicity	44
TNF-R1 (CD120a) TNF-R2 (CD120b)	Both bind tumor necrosis factor-α (TNFα) and lymphotoxin (TNFβ)	55 kDa (type 1) 75 kDa (type 2)	Neutrophil activation; thymocyte and CTL proliferation; induces cytokine synthesis (IL-1, IL-6, GM-CSF) by monocytes	45

TGFβR (types I and II)	Transforming growth factor β (TGFβ)	Type I: 53 kDa / Type II: 73 kDa (serine/threonine kinase)[m]	B cells: class switching to IgA; inhibits IgM and IgG secretion / T and B cells: inhibits proliferation / Monocytes: chemoattractant; stimulates cytokine (TNF, IL-1, TGFβ) secretion / Neutrophils: chemoattractant	46,47
MIP-Rs	Macrophage inflammatory proteins 1α and 1β (MIP-1α and 1β)	Uncharacterized	Chemoattractants for macrophages and T cells	48
NFPR (formylated peptide R)	Formylated peptides (e.g. formyl-Met-Leu-Phe)	55–70 kDa (7 TM)	Neutrophils: induces chemotaxis and superoxide production	49
CD14 (monocytes)[n]	Lipopolysaccharide (LPS)	55 kDa (GPI-linked) Associated chain with transmembrane domain?	Macrophages: release of IL-1, IL-6, IL-8, IL-10, GM-CSF, G-CSF, TNFα and arachidonic acid metabolites / Neutrophils: release of IL-1, TNFα, arachidonic acid metabolites	50

[a] Abbreviations used: R, receptor; Ag, antigen; Ab, antibody; IL, interleukin; CTL, cytotoxic T lymphocyte.

[b] m: molecular weights for mouse proteins; h: molecular weights for human proteins. Differences in molecular weight between the two species are usually due to glycosylation differences. Where molecular weight ranges are given, this reflects heterogeneous glycosylation.

[c] It is likely that the B-cell AgR contains two CD79a (Ig-α) chains and two CD79b (Ig-β) chains, although this has not been confirmed experimentally.

[d] A question mark indicates that the natural ligand is not known.

[e] The ability of CD2, CD5 and Thy1 (CDw90) to signal may depend on TCR expression, particularly the ability of the TCR ζ-chain to activate tyrosine kinases.

[f] The different isoforms of CD45 are generated by alternative splicing of exons encoding extracellular portions of the protein.

[g] These proteins lack transmembrane domains and instead are linked to the membrane via glycosyl-phosphatidylinositol (GPI) linkages.

[h] Members of the hemopoietin R family contain a 100 amino acid homology unit that contains a highly conserved Trp–Ser–X–Trp–Ser motif.

[i] AIC-2B (mouse) and KH97 (human) are used as the β-subunit of the IL-3R, the IL-5R and the GM-CSF-R.

[j] The same gp130 protein is used as the β-subunit of the receptors for IL-6, IL-11, leukemia inhibitory factor (LIF), oncostatin M and ciliary neurotropic factor.

[k] 7 TM indicates that this receptor has seven membrane-spanning regions, which is characteristic of G protein-coupled receptors.

[l] PTK indicates that the receptor contains an intrinsic tyrosine kinase activity in its cytoplasmic domain.

[m] The type I and type II TGFβRs form a single receptor unit. The type II chain is related to the activin receptors and contains an intrinsic serine/threonine kinase activity in its cytoplasmic domain.

[n] While LPS is a polyclonal stimulator of proliferation and Ab secretion by B cells, B cells do not express CD14 and must express another, unidentified, LPS-R.

Cellular Activation

Table 2. Cell lines useful for biochemical analysis of signal transduction mechanisms

Receptor	Cell lines
TCR	Jurkat (human T cell)
B-cell AgR	WEHI-231 (mouse B cell), BAL 17 (mouse B cell), A20 (mouse B cell), Daudi (human B cell), Ramos (human B cell)
FcεRI	RBL-2H3 (rat basophilic leukemia)
FcγRI, FcγRII	THP-1 (human monocyte)
CD2	Jurkat (human T cell)
CD19/CR2	Daudi (human B cell), Ramos (human B cell)
CD20	Ramos (human B cell)
CD28	Jurkat (human T cell)
CD40	Daudi (human B cell), Raji (human B cell)
IL-1R	LBRM-33-1A5 (mouse T cell)
IL-2R	CTLL-2 (mouse T cell)
IL-3R	FDC-P1 (mouse myeloid precursor), MC-9 (mouse mast cell)
IL-4R	HT-2 (mouse T cell), MC-9 (mouse mast cell)
IL-5R	MC-9 (mouse mast cell), TF-1 (human erythroleukemia)
IL-6R	CESS (human B cell), AF-10 (human myeloma), 7TD1 (mouse B cell hybridoma)
IL-7R	2E8 (mouse pre-B cell)
IL-9R	M07e (human megakaryoblastic leukemia)
IL-11R	T1165 (mouse plasmacytoma)
GM-CSF-R	MC-9 (mouse mast cell), U937 (human monocyte), HL-60 (human myeloid precursor)
c-kit (CD117)	MC-9 (mouse mast cell), M07e (human megakaryoblastic leukemia)
IFN-γR	U937 (human monocyte)
Formylated peptide receptor (NFPR)	DMSO-treated HL-60 (human neutrophil)
CD14 (LPS-R)	70Z/3 (mouse pre-B cell); RAW 264.7 (mouse macrophage)

each other. A second family of PTKs is comprised of the *syk* gene product (also known as PTK 72) and the ZAP 70 PTK that associate with the B-cell and T-cell antigen receptors, respectively. Recently a third family of nonreceptor PTKs has been identified. This family includes the JAK1, JAK2 and Tyk2 PTKs which each contain two kinase domains (130).

For receptors that activate distinct PTK molecules, identifying the relevant PTKs is an important step in understanding signaling by that receptor. PTK activation can be assessed using *in vitro* kinase assays (see *Table 6*). Receptors that activate discrete PTKs, and the PTKs known to be activated by those receptors, are listed in *Table 3*.

2.1. PTK substrates

To understand how receptor-stimulated PTK activation regulates immune cell function, it is necessary to identify the proteins that are tyrosine phosphorylated and determine their role in the signal transduction process. PTK substrates have been identified in a variety of cell types, including cells of the immune system (see *Table 4*). A number of these substrates are important signal transduction molecules that participate in the regulation of cell growth and differentiation. Many signal transduction proteins also contain structural domains, called SH2 (*src* homology 2) domains, that bind phosphotyrosine residues in the context of specific amino

Table 3. Receptors that stimulate tyrosine phosphorylation via discrete PTKs

Receptor	Associated PTKs	Refs
TCR	Fyn, ZAP 70	10
TCR-dependent signaling molecules: CD2, CD5, Thy-1 (CDw90), Ly-6	Fyn, Lck	10
B-cell AgR	Blk, Lyn, Fyn, Lck, Syk	3,10
FcεRI	Lyn, Src, Syk (rat basophilic leukemia cells); Yes (PT18 murine mast-cell line)	5,51,131
FcεRII (CD23)	Fyn	52
FcγRI (CD64)	Syk	53,132
FcγRII (CD32)	Fgr, Syk	53,54,132
FcγRIII (CD16)	?	8
NKR-P1	?	9
MHC class II	?	55
CD4, CD8	Lck	10
CD19	Lyn	3,14,133
CD26	Lck	10
CD28	?	10,18,56
CD40	?	57
IL-2R	Lck (T cells); Lyn and Fyn (pre-B cells)	22
IL-3R	JAK2	24,58,134
IL-4R	?	24,58
IL-5R	?	24,58
IL-6R	JAK1, JAK2, Tyk2	24,135
IL-7R	?	24,59,60
IL-9R	?	31
IL-11R	JAK1	141
GM-CSF-R	Lyn, Yes	61
G-CSF-R	?	24
IFN-γ-R	JAK1, JAK2	136,137,139
LPS-R (CD14)	Hck?	62,138

acids (63). Thus, tyrosine phosphorylation is not only a potential mechanism for regulating protein function, but also permits the efficient, high-affinity interaction of proteins that comprise a signaling pathway.

3. TYROSINE KINASE-REGULATED SIGNAL TRANSDUCTION PATHWAYS

Among the targets of receptor-stimulated PTKs are several signal transduction pathways that are now understood in some detail (see below and *Figures 1–3*). The key regulatory components in these pathways are: (i) phospholipase C, (ii) phosphatidylinositol 3-kinase (PI 3-kinase), (iii) the p21^{ras} proto-oncogene, and (iv) the mitogen-activated protein (MAP) kinases. There is considerable evidence that each of these pathways can play an important role in mediating the effects of PTKs on cell growth and differentiation. Not all PTKs activate all four of these signaling pathways (see *Table 5*). Which pathways are activated by a given receptor is determined by the substrate specificity of the PTK that is activated and by the SH2 binding sites contained in the signal-transducing proteins that are phosphorylated by the PTK.

Table 4. PTK substrates

Substrate	M_r	Function	Receptors	Refs
Receptors				
CD79a (Ig-α)	32 kDa	Parts of B-cell AgR	B-cell AgR	3
CD79b (Ig-β)	37 kDa			
ζ chain	16 kDa	Part of TCR	TCR (T cells);	64
		Part of NK cell FcγRIII	FcγRIII (CD16)	8
			(NK cells)	
FcεRI β-subunit	33 kDa	Part of FcεRI	FcεRI	5
FcεRI γ-subunit	12 kDa	Part of FcεRI	FcεRI	
IL-2R β-subunit	75 kDa	Part of IL-2R	IL-2R	22
IL-4R	140 kDa	IL-4R	IL-4R	65
AIC-2A	130 kDa	Part of IL-3R	IL-3R	66
AIC-2B	130 kDa	Part of IL-3R, IL-5R and	IL-3R, IL-5R,	66
		GM-CSF-R	GM-CSF-R	
CD117 (c-kit)	145 kDa	Receptor for SLF	Steel factor-R	67
		(autophosphorylation)		
CD115 (c-fms)	150 kDa	Receptor for M-CSF	M-CSF-R	68
		(autophosphorylation)		
Other membrane proteins				
CD5	67 kDa	Receptor for	TCR	69
		B-cell CD72		
CD19	95 kDa	Co-receptor for B-cell AgR?	B-cell AgR, CD19	70,71
CD22	140 kDa	Adhesion molecule	B-cell AgR	3
Kinases and other enzymes				
Extracellular-signal regulated kinase (ERK)-1	44 kDa	Mitogen-activated protein kinase	TCR, IL-3R, IL-5R, GM-CSF-R, CD117 (c-kit), CD14 (LPS-R)	10,58, 72
ERK2	42 kDa	Mitogen-activated protein kinase	B-cell AgR, TCR, IL-3R, IL-5R, GM-CSF-R, CD117 (c-kit), CD14 (LPS-R)	3,10 58,72
Phosphatidyl-inositol 3-kinase	85 kDa subunit	src homology 2 (SH2)-containing adaptor protein that couples catalytic subunit to PTKs and receptors	B-cell AgR	3
Phosphatidyl-inositol 3-kinase	110 kDa subunit	Catalytic subunit; phosphorylates inositol phospholipids	B-cell AgR	3
Phospholipase C-γ1 and/or -γ2	145 kDa	Hydrolysis of phosphatidylinositol 4,5-bisphosphate (results in second messengers that activate protein kinase C	TCR, B-cell AgR, FcRεI, CD2, CD19, FcγRI (CD64), FcγRIII (CD16)	3,5, 10,53, 73

Table 4. Continued

Substrate	M_r	Function	Receptors	Refs
		and cause increases in cytosolic calcium concentrations)		
c-raf	74 kDa	Serine/threonine kinase that activates MEK (mitogen-activated protein (MAP) kinase kinase)	IL-2R, IL-3R, GM-CSF-R	22
Other proteins				
rasGAP	120 kDa	Regulator and effector of p21^{ras}	B-cell AgR	3
p62	62 kDa	GAP-associated protein	B-cell AgR, CSF-1R (CD115)	3,74
p190	190 kDa	GAP-associated protein, acts as GTPase activating factor for rac and rho	B-cell AgR	3
Vav	95 kDa	Guanine nucleotide exchange factor?	TCR, B-cell AgR, FcRεI,	3,75
Shc	46 kDa, 52 kDa	Couples PTKs to guanine nucleotide exchange factors that regulate p21^{ras}	IL-2R, B-cell AgR, IL-3R, IL-5R, GM-CSF-R, CD117 (c-kit)	3,76
4PS	170 kDa	Binds phosphatidylinositol (PI) 3-kinase	IL-4R	77
pp125^{FAK}	125 kDa	Focal adhesion-localized PTK	FcRεI	78
VCP-1	100 kDa	Homolog of yeast cell-cycle protein cdc48	TCR, B-cell AgR	79
Ezrin	81 kDa	Cytoskeleton-associated protein	TCR	80

3.1. The phospholipase C pathway

Phospholipase C (PLC) cleaves the plasma membrane phospholipid, phosphatidylinositol 4,5-bisphosphate (PI 4,5-P_2 or PIP$_2$) (see *Figure 1* and ref. 90). This yields two intracellular second messengers, the polar head group of this lipid, inositol 1,4,5-trisphosphate (IP$_3$), and the remaining lipid portion, diacylglycerol (DAG). IP$_3$ causes increases in intracellular calcium concentrations by stimulating the release of calcium from intracellular stores. These storage sites are not well defined, but may be specialized sections of the endoplasmic reticulum that contain receptors for IP$_3$. IP$_3$ is rapidly converted into a large number of other inositol phosphate compounds by the action of kinases and phosphatases. It is unclear whether these compounds also have signaling functions. Inositol 1,3,4,5-tetrakisphosphate may regulate the flow of extracellular calcium into the cell. DAG activates protein kinase C (PKC), a family of related serine/threonine kinases. This activation is accompanied by translocation of PKC from the cytosol to the plasma membrane. PKC phosphorylates a number of proteins and also initiates a protein kinase cascade that includes the MAP kinases (see below). Another well-characterized substrate of PKC is the MARCKS protein which is involved in regulation of the actin cytoskeleton (91).

Table 5. Signaling pathways activated by PTK-linked receptors

Receptor	Cell type[a]	PLC activation	PI 3-kinase activation	p21^{ras} activation	MAP kinase activation
B-cell AgR	B cell	+(3)[b]	+(3)	+(3)	+(3)
T-cell AgR	T cell	+(10)	–	+(81)	+(10)
FcεRI	Mast cells	+(5)			
FcεRII (CD23)	B cells	+(6)			
FcγRI (CD64)	Monocytes	+(53)			
FcγRII (CD32)	Monocytes	+(53)			
FcγRIII (CD16)	NK cells	+(82)			
NKR-P1	NK cells	+(9)			
MHC class II	B cells	+(55)			
CD2	T cells	+(10)		+(81)	+(10)
CD5	T cells	+(10)			
CDw90 (Thy-1)	T cells	+(10)			
CD19	B cells	+(3)			
CD28	T cells	+(18)			
CD40	B cells (human)	+(57)			
IL-2R	T cell	–	+(83)	+(81)	+
IL-3R	Myeloid	–(24,61)	+(61)	+(84)	+(58)
IL-4R	Myeloid	–	+(85)	–(84)	–(58)
IL-5R	Myeloid	–	+(85)	+(84)	+(58)
IL-6R	B cell				+(86)
IL-7R	Thymocytes	+(59)	+(87)		
	Pre-B cells	+(60)			
IL-9R	Myeloid				–(31)
CSF-1R (c-fms, CD115)	Macrophages		+(88)	+(89)	
GM-CSF-R	Myeloid	–(61)	+(61)	+(84)	+(58)
c-kit (CD117)	Myeloid		+(85)	+(84)	+(58)
LPS-R (CD14)	Macrophages	–			+(72)

[a]Cell types in which activation of these signaling pathways has been demonstrated for a given receptor. Where no entry is listed, there is no information as to whether the pathway is activated.
[b]References are given in parentheses.

Figure 1. The phospholipase C pathway.

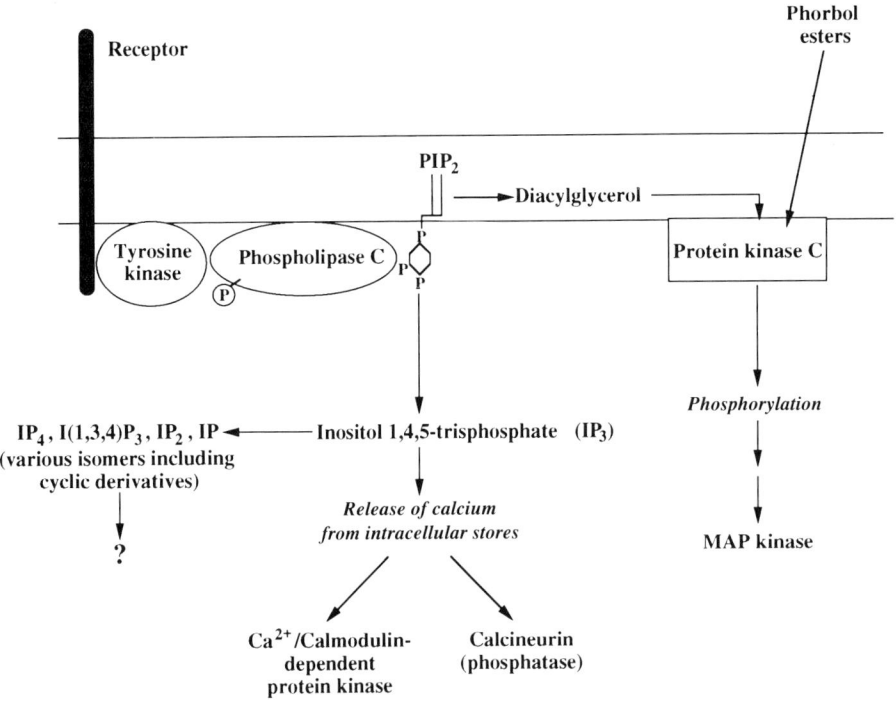

Figure 2. The phosphatidylinositol 3-kinase pathway.

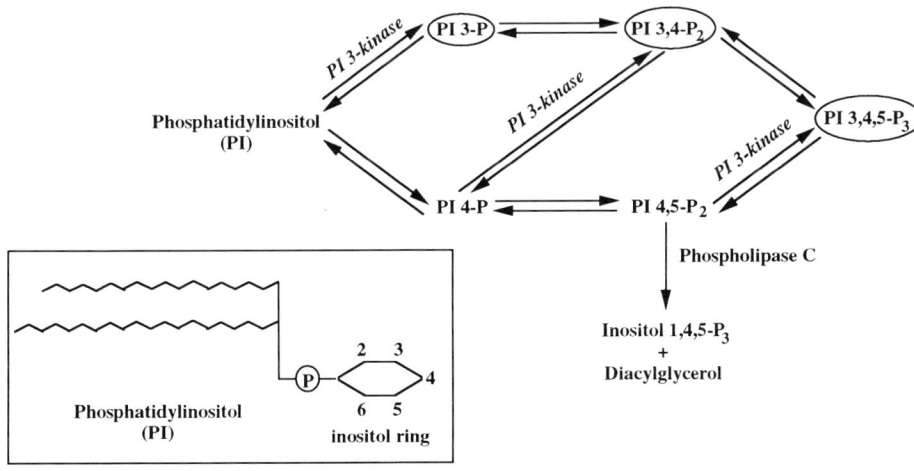

There are multiple isoforms of PLC. PLC-γ1 and PLC-γ2 are activated by tyrosine phosphorylation. *In vivo*, PIP_2 forms complexes with profilin, a cytoskeletal protein that prevents PLC-γ from cleaving PIP_2. Tyrosine phosphorylation of PLC-γ allows it to overcome the inhibitory effects of profilin (92). In contrast, the β isoforms (PLC-β1 and PLC-β2) are regulated by G protein-linked receptors (see below).

3.2. The phosphatidylinositol 3-kinase pathway

Phosphatidylinositol 3-kinase (PI 3-kinase) consists of an 85 kDa regulatory subunit and a 110 kDa catalytic subunit. PI 3-kinase phosphorylates the plasma membrane lipids phosphatidyl-inositol (PI), PI 4-P and PI 4,5-P_2 on the 3 position of the inositol ring (see *Figure 2*, inset), yielding PI 3-P, PI 3,4-P_2 and PI 3,4,5-P_3 (93). These lipids are present in much lower quantities than PI 4-P and PI 4,5-P_2 and are not substrates for PLC. It has been reported that PI 3,4,5-P_3 and PI 3,4-P_2 can activate PKC ζ *in vitro* (94). PKC ζ differs from the other members of the PKC family in that it is not activated by phorbol esters or diacylglycerol. The substrates of PKC ζ have not been identified. Although the downstream mediators of the PI 3-kinase pathway are not well defined, this pathway is clearly a key mediator of tyrosine kinase action. PI 3-kinase binds to many activated PTKs and receptor tyrosine kinases via SH2 domains in the 85 kDa regulatory subunit. Mutated versions of these kinases that have lost the ability to bind and activate PI 3-kinase cannot stimulate cell division of transformed cells.

The SH2-mediated binding of PI 3-kinase to receptors or PTKs appears to be sufficient to activate the enzyme. The SH2 domain of PI 3-kinase binds with high affinity to the sequence Tyr–X–X–Met (where X is any amino acid), and peptides or proteins containing this

Figure 3. The *ras* and MAP kinase pathways.

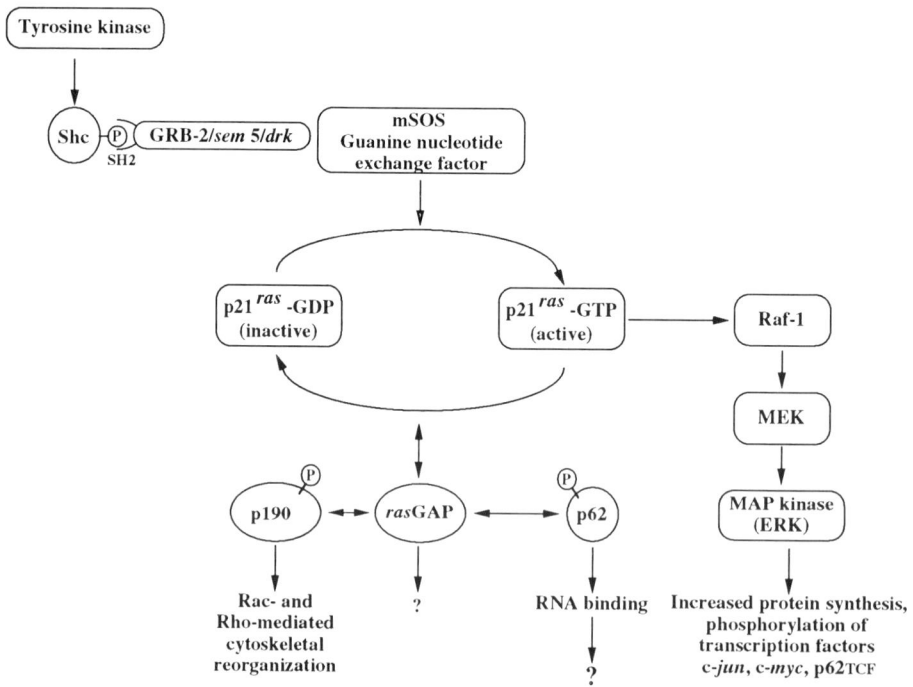

sequence can activate PI 3-kinase *in vitro* (95). The binding of PI 3-kinase to receptors and receptor-associated PTKs that contain this sequence also brings the normally cytosolic PI 3-kinase to the plasma membrane where its substrates reside. PI 3-kinase binds to CD117 (*c-kit*) after stimulation of myeloid cells with steel factor and to CD19 and $p53/p56^{lyn}$ after stimulation of B cells through their AgR.

3.3. The p21^{ras} pathway

The p21^{ras} proto-oncogene product is a membrane-anchored GTP-binding protein. The GDP-bound form is inactive, while the GTP-bound form is the active state. Signaling by tyrosine kinase growth-factor receptors, as well as transformation by activated tyrosine kinase oncogene products, increases the fraction of p21^{ras} with GTP bound. p21^{ras} is a key component of PTK-mediated signaling. Microinjection of anti-*ras* antibodies into fibroblasts can block growth-factor-induced proliferation and transformation by activated tyrosine kinases. Moreover, there is genetic evidence in both *Drosophila* and *Caenorhabditis elegans* that functional *ras* is required for tyrosine kinase receptors to exert their effect.

p21^{ras} activity can be controlled in two ways (see *Figure 3*, refs 96,97). Guanine nucleotide exchange factors (GNEFs) stimulate the release of GDP from p21^{ras}, allowing GTP to be bound. Several GNEFs have been described in mammalian cells, including the mammalian homolog of the yeast cdc25 protein (54 kDa), the mammalian homologs of the *Drosophila* son of sevenless (SOS) gene product (mSOS1, mSOS2; 170 kDa), $p95^{vav}$ and a 140 kDa brain-specific GNEF. Different GNEFs are regulated differently. Tyrosine phosphorylation of $p95^{vav}$ increases its *in vitro* GNEF activity. In contrast, receptor signaling does not increase the specific activity of the mSOS1 GNEF, but causes it to translocate from the cytosol to the membrane where p21^{ras} is located. This is mediated by an adaptor protein, GRB-2 (23 kDa), which is the homolog of the *C. elegans sem 5* gene product and the *Drosophila drk* gene product. GRB-2 constitutively binds mSOS1 by means of its SH3 (*src* homology 3) domains. GRB-2 also has an SH2 domain that can bind phosphorylated growth factor receptors and thereby bring mSOS1 to the membrane. Another adaptor protein, Shc (46 and 52 kDa forms), may couple the GRB-2/mSOS1 complex to *src* kinases and to receptors that lack binding sites for the SH2 domain of GRB-2. Shc is a common tyrosine kinase substrate (see *Table 4*) and the SH2 domain of GRB-2 binds to tyrosine-phosphorylated Shc. Shc has an SH2 domain that allows it to bind to tyrosine-phosphorylated receptors.

p21^{ras} activity is also controlled by the *ras*GTPase activating protein (*ras*GAP) (3), which stimulates the intrinsic GTPase activity of p21^{ras}. In T cells, PKC activation decreases *ras*GAP activity, resulting in an accumulation of activated p21^{ras}. *ras*GAP also appears to be one of the downstream effectors of p21^{ras}. In certain situations *ras*GAP, or fragments of it, can mimic the effects of activated p21^{ras}. PTK activation frequently induces *ras*GAP to bind (via its SH2 domains) to two other tyrosine-phosphorylated proteins, p62 and p190. p62 has homology to mRNA splicing factors, the functional implications of which are not clear. p190, however, acts as a 'GAP' for two other small G proteins, Rac and Rho. These proteins have key roles in controlling the organization of the actin cytoskeleton during stress fiber formation and membrane ruffling (98,99).

3.4. The MAP kinase pathway

Many of the effects of p21^{ras} on cell growth may be mediated through the MAP kinase pathway (see *Figure 3* and ref. 100). The MAP kinases or extracellular-signal regulated kinases (ERKs) are a family of related serine/threonine kinases. The major forms are the 44 kDa ERK1, the 42 kDa ERK2 and a 54 kDa MAP kinase. It is not clear if the different MAP kinases are regulated differently or have different substrates *in vivo*. The MAP kinases are activated by phosphorylation on both threonine and tyrosine residues by a bispecific kinase termed MAP kinase kinase or MEK. p21^{ras} activation leads to activation of a serine/threonine kinase, Raf-1, that phosphorylates and activates MEK. MEK appears to be the intersection of many signaling pathways, since it can also be activated by a PKC-dependent pathway and by a kinase that is the homolog of the yeast STE 11 kinase (140).

MAP kinase phosphorylates and activates the S6 ribosomal protein kinases which up-regulate protein synthesis. More significantly, MAP kinase phosphorylates several transcription factors including c-*jun*, c-*myc* and p62^{TCF}. c-*jun* is part of the AP-1 transcription factor that mediates gene expression induced by PKC activation while p62^{TCF} binds to the serum response element.

4. ACTIVATION OF PHOSPHOLIPASE C BY G PROTEIN-LINKED RECEPTORS

In contrast to PLC-γ1 and PLC-γ2, which are regulated by tyrosine phosphorylation, PLC-β1 and PLC-β2 are regulated by G proteins (90). G proteins are heterotrimeric GTP-binding proteins that couple receptors to signaling molecules such as adenylate cyclase, PLC-β and ion channels (101). The α-subunit (c. 40 kDa) of the heterotrimeric G proteins contains a guanine nucleotide binding site as well as a GTPase activity, much like p21^{ras}. Unlike p21^{ras}, the heterotrimeric G proteins also contain β (35 kDa) and γ (8–10 kDa) subunits. Contact with a ligand-activated receptor complex stimulates the exchange of GTP for GDP at the nucleotide binding site of the α-subunit and also causes the βγ-subunits to be released. The free GTP-bound α-subunit regulates activity of the effector molecule (e.g. adenylate cyclase). The βγ-subunits may also modulate effector activity in some cases. The α-subunit GTPase activity hydrolyzes the bound GTP, inactivating the G protein and allowing the reassociation of the βγ-subunits.This occurs at a fairly slow rate, allowing the G protein to activate the effector catalytically and thereby amplify the signal. The specificity of the receptor–effector interaction is contained, for the most part, in the α-subunit, which has specific regions that mediate receptor and effector binding. At least 16 different α-subunit genes have been cloned. The Gq family of G proteins, which contains at least five members, has been shown to couple receptors to the β isoforms of PLC (90, 101).

Most G protein-linked receptors have seven transmembrane domains connected by intracellular and extracellular loops. The IL-8 receptor and the *N*-formylated peptide receptor (NFPR) have this structure and have been shown to activate PLC in a G protein-dependent manner (102, 103). The macrophage inflammatory proteins MIP-1α and MIP-1β also activate PLC (104). The receptors for these cytokines have not been characterized, but given their close relationship to IL-8, their receptors are also likely to have seven transmembrane domains. Although the B-cell antigen receptor and CD23 do not have the characteristic structure of G protein-linked receptors, the ability of these receptors to activate PLC also depends on G proteins (6, 105) to some extent. It is not clear if the G protein activated by the B-cell antigen receptor contributes in some way to activation of PLC-γ (usually activated by tyrosine phosphorylation) or if it activates PLC-β.

5. REGULATION OF cAMP LEVELS

Cellular cAMP levels are controlled by adenylate cyclases, a family of enzymes that produce cAMP from ATP, and by phosphodiesterases that convert cAMP to AMP. Most of the adenylate cyclase isoforms are regulated by G proteins. Virtually all immune cells have β-adrenergic receptors and receptors for prostaglandin E$_2$ which are coupled to adenylate cyclase by the G$_s$ protein and cause increases in cAMP. Elevated cAMP levels inhibit the phospholipase C pathway and generally inhibit lymphocyte activation (106). Other receptors that do not have seven membrane-spanning regions may also be able to regulate cAMP levels in some way. The IL-1 receptor (107), TCR (10), MHC class II molecules (3) and CD23 (14) have been reported to increase cAMP levels by mechanisms that are not understood.

6. PHOSPHATIDYLCHOLINE HYDROLYSIS

It is now well established that many receptors stimulate hydrolysis of phosphatidylcholine (PC) by activating either a PC-specific PLC or a phospholipase D (PLD) (108). In the immune system, the receptors for IL-1, IL-3, CSF-1 and N-formylated peptides have been reported to cause PC breakdown (108–111). PC-specific PLC generates phosphorylcholine and diacylglycerol (DAG), whereas PLD generates choline and phosphatidic acid. Phosphatidic acid can be converted to DAG by removal of a phosphate group. Thus, both of these reactions can generate DAG and lead to activation of PKC in the absence of IP_3-stimulated calcium increases. Since PC is much more abundant than PIP_2, it has been proposed that PC breakdown results in sustained DAG production and PKC activation. The DAG generated by PC breakdown has a different fatty acid composition than the DAG derived from PIP_2 breakdown, raising the possibility of differential abilities to activate various PKC isoforms. The regulation of PC hydrolysis is not well understood. PLD can be activated by DAG and phorbol esters that activate PKC. Thus, other signaling pathways that activate PKC can stimulate PC hydrolysis. Activation of PC-specific PLC, at least by CSF-1, involves a G protein that has not been identified (111, 112).

7. GLYCOSYL-PHOSPHATIDYLINOSITOL (GPI) HYDROLYSIS

Receptor-stimulated hydrolysis of a glycosyl-phosphatidylinositol (GPI) was first described for the insulin receptor. The GPIs are a family of membrane lipids whose structure is not completely defined. Hydrolysis of these molecules, probably by a specific PLC, yields another DAG species, dimyristoyl DAG, and an inositol phosphate molecule joined to a glycan moiety through a glucosamine residue. The IL-2 receptor stimulates this reaction in both T and B cells (113, 114). The dimyristoyl DAG may activate PKC. The role of the inositol phosphate-glycan is not clear, but it has been shown to mediate some of the effects of insulin.

8. METHODS USED IN STUDYING SIGNAL TRANSDUCTION

Methods used in the study of major signal transduction reactions are given in *Table 6*.

Table 6. Methods for studying major signal transduction reactions

Signaling reaction	Method	Ref.[a]
Tyrosine phosphorylation	Anti-phosphotyrosine (anti-P-tyr) immunoblot: probe blot with 4G10 or PY-20 monoclonal anti-P-tyr antibodies. Detect with peroxidase-coupled goat anti-mouse Ig antibodies and enhanced chemiluminescence (see Amersham technical notes)	115
PTK activation	*In vitro* kinase assay: immunoprecipitate PTK with specific antibodies. Measure autophosphorylation activity or ability to phosphorylate exogenous substrate (e.g. enolase) *in vitro* using $[^{32}P]\gamma$-ATP. Separate reaction products by SDS-PAGE. Quantitate by excising band or by densitometry	116

Table 6. Continued

Signaling reaction	Method	Ref.[a]
Tyrosine phosphorylation of individual substrates	Immunoprecipitate with specific antibody and analyze by anti-P-tyr immunoblotting	115
Serine and tyrosine phosphorylation of individual substrates	Label cells with ^{32}P. Immunoprecipitate with specific antibody. Separate proteins by SDS-PAGE and transfer to PVDF filter. Excise band, hydrolyze with 6 M HCl. Separate resulting phosphoamino acids by one- or two-dimensional TLC	117
Phospholipase C activation	(a) Measurement of inositol phosphates: label cells with [^{3}H]inositol. Lyse cells with chloroform/methanol. Load aqueous phase on Dowex AG-1 X8 columns. Elute different inositol phosphate isomers with increasing concentrations of ammonium formate. Alternatively, separate inositol phosphates by HPLC (b) Measurement of intracellular calcium concentrations: load cells with calcium-sensitive fluorescent dye. Place cells in fluorimeter cuvette, add stimulus, and record fluorescence values. Calculate calcium concentration based on K_d of dye (c) Protein kinase C translocation to cell membrane: lyse cells without detergent and separate membrane and cytosol fractions by centrifugation. Solubilize membrane fraction. Run fractions on ion-exchange column and perform PKC enzyme assay on column fractions. Alternatively, subject membrane and cytosol samples to SDS-PAGE and do immunoblotting with antibodies specific for different PKC isozymes	118–122
PI 3-kinase activation	(a) Association of PI 3-kinase with anti-P-tyr immunoprecipitates: immunoprecipitate with anti-P-tyr and do *in vitro* kinase assay with [^{32}P]γ-ATP and phosphatidylinositol (PI) or PIP$_2$ as substrate. Separate reaction products by TLC (b) *In vitro* production of PI 3-kinase products: label cells with ^{32}P. Extract phospholipids with chloroform/methanol. Deacylate lipids with methylamine. Separate by HPLC using strong anion-exchange column with ammonium phosphate gradient	123, 124
p21^{ras} activation	Label cells with ^{32}P. Immunoprecipitate p21^{ras}. Elute bound GDP and GTP from p21^{ras} and separate labeled nucleotides by TLC	125

Table 6. Continued

Signaling reaction	Method	Ref.[a]
GNEF activity	Incubate total cell lysate or immunoprecipitate with recombinant p21^{ras} protein pre-bound with [^{3}H]GDP. Separate free and bound nucleotide by filtering through nitrocellulose to measure release of [^{3}H]GDP. Alternatively, measure binding of [^{32}P]GTP to recombinant p21^{ras} protein	126
MAP kinase activation	Separate cell lysates on monoQ ion-exchange column. Assay fractions for MAP kinase activity using myelin basic protein as substrate. Demonstrate that fractions containing enzyme activity contain MAP kinase by immunoblotting with anti-MAP kinase antibodies. Different isoforms of MAP kinase can be further resolved using phenyl-Superose columns	127
G protein involvement	Prepare membrane preparations or permeabilize cells with saponin or streptolysin O. Show that response is potentiated by GTPγS which locks G proteins in active state and inhibited by GDPβS which converts G proteins to inactive state	105
Adenylate cyclase activation	(a) Increase in adenylate cyclase activity. Prepare membrane preparations or permeabilize cells with saponin or streptolysin O. Add [^{32}P]α-ATP-containing reaction mixture and measure production of [^{32}P]cAMP which is separated from ATP by sequential elution from alumina and Dowex columns (b) Determination of cAMP concentrations by radioimmunoassay or label cells with [^{3}H]adenosine and measure conversion to [^{3}H]cAMP	128, 129
Phosphatidylcholine (PC) hydrolysis	Label cells with [^{3}H]choline chloride. After stimulation, extract cells with chloroform/methanol. Separate lipids by TLC and look for decrease in PC. Dry aqueous phase and analyze phosphorylcholine production by TLC	111

[a]References cited are not necessarily the original references describing the technique, but instead provide relatively detailed descriptions of the technique, along with references.

9. REFERENCES

1. Clevers, H., Alarcon, B., Wileman, T., and Terhorst, C. (1988) *Ann. Rev. Immunol.*, **6**, 629.

2. Ashwell, J.D. and Klausner, R.D. (1990) *Ann. Rev. Immunol.*, **8**, 139.

3. Gold, M.R. and DeFranco, A.L. (1994) in *Advances in Immunology* (F.J. Dixon, ed.).

Academic Press, San Diego, Vol. 55, p. 221.

4. Metzger, H., Alcaraz, G., Hohman, R., Kinet, J.-P., Pribulda, V. and Quarto, R. (1986) *Ann. Rev. Immunol.*, **4**, 419.

5. Benhamou, M. and Siraganian, R.P. (1992) *Immunol. Today*, **13**, 195.

6. Kolb, J.-P., Abadie, A., Paul-Eugene, N., Capron, M., Sarfati, M., Dugas, B. and Delespesse, G. (1993) *J. Immunol.*, **150**, 4798.

7. van de Winkel, J.G.J. and Capel, P.J.A. (1993) *Immunol. Today*, **14**, 215.

8. Vivier, E., Morin, P., O'Brien, C., Druker, B., Schlossman, S.F. and Anderson, P. (1991) *J. Immunol.*, **146**, 206.

9. Yokoyama, W. and Seaman, W.E. (1993) *Ann. Rev. Immunol.*, **11**, 613.

10. Perlmutter, R.M., Levin, S.D., Appleby, M.W., Anderson, S.J. and Alberola-Ila, J. (1993) *Ann. Rev. Immunol.*, **11**, 451.

11. Janeway Jr, C.A. (1992) *Ann. Rev. Immunol.*, **10**, 645.

12. Luo, W., van de Velde, H., von Hoegen, I., Parnes, J.R. and Thielemans, K. (1992) *J. Immunol.*, **148**, 1630.

13. Spertini, F., Stohl, W., Ramesh, N., Moody, C. and Geha, R.S. (1991) *J. Immunol.*, **146**, 47.

14. van Noesel, C.J.M., Lankester, A.C. and van Lier, R.A.W. (1993) *Immunol. Today*, **14**, 8.

15. White, M.W., McConnell, F., Shu, G.L., Morris, D.R. and Clark, E.A. (1991) *J. Immunol.*, **146**, 846.

16. Tanaka, T., Kameoka, J., Yaron, A., Schlossman, S.F. and Morimoto, C. (1993) *Proc. Natl. Acad. Sci. USA*, **90**, 4586.

17. Goodwin, R.G., Alderson, M.R., Smith, C.A. *et al.*, (1993) *Cell*, **73**, 447.

18. Linsley, P.S. and Ledbetter, J.A. (1993) *Ann. Rev. Immunol.*, **11**, 191.

19. Noelle, R.J., Ledbetter, J.A., and Aruffo, A. (1992) *Immunol. Today*, **13**, 431.

20. Trowbridge, I.S., Ostergaard, H. and Johnson, P. (1991) *Biochem. Biophys. Arch.*, **1095**, 46.

21. Dower, S.K. (1992) in *Human Cytokines: Handbook for Basic and Clinical Research* (B.B. Aggarwal and J.U. Gutterman, eds). Blackwell Scientific Publications, Cambridge, MA, p. 46.

22. Minami, Y., Kono, T., Miyazaki, T., and Taniguchi, T. (1993) *Ann. Rev. Immunol.*, **11**, 245.

23. Lotze, M.T. (1992) in *Human Cytokines: Handbook for Basic and Clinical Research* (B.B. Aggarwal and J.U. Gutterman, eds). Blackwell Scientific Publications, Cambridge, MA, p. 81.

24. Miyajima, A., Kitamura, T., Harada, N., Yokota, T. and Arai, K.-I. (1992) *Ann. Rev. Immunol.*, **10**, 295.

25. Schrader, J.W., Clark-Lewis, I., Leslie, K.B. and Ziltener, H.J. (1992) in *Human Cytokines: Handbook for Basic and Clinical Research* (B.B. Aggarwal and J.U. Gutterman, eds). Blackwell Scientific Publications, Cambridge, MA, p. 97.

26. de Vries, J.E. (1992) in *Human Cytokines: Handbook for Basic and Clinical Research* (B.B. Aggarwal and J.U. Gutterman, eds). Blackwell Scientific Publications, Cambridge, MA, p. 113.

27. Hatake, K., Arai, K.-I. and Yokota, T. (1992) in *Human Cytokines: Handbook for Basic and Clinical Research* (B.B. Aggarwal and J.U. Gutterman, eds). Blackwell Scientific Publications, Cambridge, MA, p. 130.

28. Taga, T. and Kishimoto, T. (1992) in *Human Cytokines: Handbook for Basic and Clinical Research* (B.B. Aggarwal and J.U. Gutterman, eds). Blackwell Scientific Publications, Cambridge, MA, p. 143.

29. Goodwin, R.G. and Namen, A.E. (1992) in *Human Cytokines: Handbook for Basic and Clinical Research* (B.B. Aggarwal and J.U. Gutterman, eds). Blackwell Scientific Publications, Cambridge, MA, p. 168.

30. Zachariae, C.O.C. and Matsushima, K. (1992) in *Human Cytokines: Handbook for Basic and Clinical Research* (B.B. Aggarwal and J.U. Gutterman, eds). Blackwell Scientific Publications, Cambridge, MA, p. 181.

31. Miyazawa, K., Hendrie, P.C., Kim, Y.-J., Mantel, C., Yang, Y.-C., Kwon, B.S. and Broxmeyer, H.E. (1992) *Blood*, **80**, 1685.

32. Burke, F., Naylor, M.S., Davies, B. and Balkwill, F. (1993) *Immunol. Today*, **14**, 167.

33. Howard, M. and O'Garra, A. (1992) *Immunol. Today*, **13**, 198.

34. Moore, K.W., O'Garra, A., de Waal Malefyt, R., Vieira, P. and Mosmann, T.R. (1993) *Ann. Rev. Immunol.*, **11**, 165.

35. Paul, S.R., Bennett, F., Calvetti, J.A. (1990) *Proc. Natl. Acad. Sci. USA*, **87**, 7512.

36. Perussia, B., Chan, S.H., D'Andrea, A., Tsuji, K., Santoli, D., Pospisil, M., Young, D., Wolf, S.W. and Trinchieri, G. (1992) *J. Immunol.*, **149**, 3495.

37. Chizzonite, R., Truitt, T., Desai, B.B., Nunes, P., Podalski, F.J., Stern, A.S., and Gately, M.K. (1992) *J. Immunol.*, **148**, 3117.

38. Minty, A., Chalon, P., Deroq, J.M. *et al.*, (1993) *Nature*, **362**, 248.

39. Punnonen, J., Aversa, G., Cocks, B.G., McKenzie, A.N.J., Menon, S., Zurawski, G., de Waal Maleyft, R. and De Vries, J. (1993) *Proc. Natl. Acad. Sci. USA*, **90**, 3730.

40. Stanley, E.R. (1992) in *Human Cytokines: Handbook for Basic and Clinical Research* (B.B. Aggarwal and J.U. Gutterman, eds). Blackwell Scientific Publications, Cambridge, MA, p. 196.

41. Souza, L.M. (1992) in *Human Cytokines: Handbook for Basic and Clinical Research* (B.B. Aggarwal and J.U. Gutterman, eds). Blackwell Scientific Publications, Cambridge, MA, p. 221.

42. Crosier, P.S., Garnick, M.B. and Clark, S.C. (1992) in *Human Cytokines: Handbook for Basic and Clinical Research* (B.B. Aggarwal and J.U. Gutterman, eds). Blackwell Scientific Publications, Cambridge, MA, p. 238.

43. Aggarwal, B.B. and Gutterman, J.U. (1992) in *Human Cytokines: Handbook for Basic and Clinical Research* (B.B. Aggarwal and J.U. Gutterman, eds). Blackwell Scientific Publications, Cambridge, MA, p. 418.

44. Gray, P.W. (1992) in *Human Cytokines: Handbook for Basic and Clinical Research* (B.B. Aggarwal and J.U. Gutterman, eds). Blackwell Scientific Publications, Cambridge, MA, p. 30.

45. Aggarwal, B.B. (1992) in *Human Cytokines: Handbook for Basic and Clinical Research* (B.B. Aggarwal and J.U. Gutterman, eds). Blackwell Scientific Publications, Cambridge, MA, p. 270.

46. Roberts, A.B. and Sporn, M.B. (1992) in *Human Cytokines: Handbook for Basic and Clinical Research* (B.B. Aggarwal and J.U. Gutterman, eds). Blackwell Scientific Publications, Cambridge, MA, p. 399.

47. Lopez-Casillas, F., Wrana, J.L. and Massague, J. (1993) *Cell*, **73**, 1435.

48. Schall, T.J., Bacon, K., Camp, R.D.R., Kaspari, J.W. and Goeddel, D.V. (1993) *J. Exp. Med.*, **177**, 1821.

49. Lang, J., Boulay, F., Li, G. and Wollheim, C.B. (1993) *EMBO J.*, **12**, 2671.

50. Wright, S.D., Ramos, R.A., Tobias, P.S., Ulevitch, R.I. and Mathison, J.C. (1990) *Science*, **249**, 1431.

51. Eisenman, E. and Bolen, J.B. (1992) *Nature*, **355**, 78.

52. Sugie, K., Kawakami, T., Maeda, T., Kawabe, T., Uchida, A. and Yodoi, J. (1991) *Proc. Natl. Acad. Sci. USA*, **88**, 9132.

53. Scholl, P.R., Ahern, D., and Geha, R.S. (1992) *J. Immunol.*, **149**, 1751.

54. Hamada, F., Aoki, M., Akiyama, T. and Toyoshima, K. (1993) *Proc. Natl. Acad. Sci. USA*, **90**, 6305.

55. Lane, P.J.L., McConnell, F.M., Schieven, G.L., Clark, E.A. and Ledbetter, J.A. (1990) *J. Immunol.*, **144**, 3684.

56. Vandenberghe, P., Freeman, G.J., Nadler, L.M., Fletcher, M.C., Kamoun, M., Turka, L.A., Ledbetter, J.A., Thompson, C.B. and June, C.H. (1992) *J. Exp. Med.*, **175**, 951.

57. Uckun, F.M., Schieven, G.L., Dibirdik, I., Chandan-Langlie, M., Tuel-Ahlgren, L. and Ledbetter, J.A. (1991) *J. Biol. Chem.*, **266**, 17478.

58. Welham, M.J., Duronio, V., Sanghera, J.S., Pelech, S.L. and Schrader, J.W. (1992) *J. Immunol.*, **149**, 1683.

59. Uckun, F.M., Dibirdik, I., Smith, R., Tuel-Ahlgren, L., Chandan-Langlie, M., Schieven, G.L., Waddick, K.G., Hanson, M. and Ledbetter, J.A. (1991) *Proc. Natl. Acad. Sci. USA*, **88**, 3589.

60. Uckun, F.M., Tuel-Ahlgren, L., Obuz, V., Smith, R., Dibirdik, I., Hanson, M., Chandan-Langlie, M. and Ledbetter, J.A. (1991) *Proc. Natl. Acad. Sci. USA*, **88**, 6323.

61. Corey, S., Eguinoa, A., Payana-Theall, K., Bolen, J.B., Cantley, L.C., Mollinedo, F., Jackson, T.R., Hawkins, P.T. and Stephens, L.R. (1993) *EMBO J.*, **12**, 2681.

62. Weinstein, S.L., Gold, M.R. and DeFranco, A.L. (1991) *Proc. Natl. Acad. Sci. USA*, **88**, 4148.

63. Songyang, Z., Shoelson, S.E., Chaudhuri, M. *et al.* (1993) *Cell*, **72**, 767.

64. Samelson, L.E., Maitray, P.D., Weissman, A.M., Harford, J.B. and Klausner, R.D. (1986) *Cell*, **46**, 1083.

65. Izuhara, K. and Harada, N. (1993) *J. Biol. Chem.*, **268**, 13097.

66. Duronio, V., Clark-Lewis, I., Federsppiel, B., Wieler, J.S., and Schrader, J.W. (1992) *J. Biol. Chem.*, **267**, 21856.

67. Reith, A.D., Ellis, C., Lyman, S.D., Anderson, D.M., Williams, D.E., Bernstein, A. and Pawson, T. (1991) *EMBO J.*, **10**, 2451.

68. Sherr, C.J. (1990) *Blood*, **75**, 1.

69. Davies, A.A., Ley, S.C. and Crumpton, M.J. (1992) *Proc. Natl. Acad. Sci. USA*, **89**, 6368.

70. Tuveson, D.A., Carter, R.H., Soltoff, S.P. and Fearon, D.T. (1993) *Science*, **260**, 986.

71. Chapulny, N., Kanner, S.B., Schieven, G.L., Wee, S.F., Gilliland, L.K., Aruffo, A. and Ledbetter, J.A. (1993) *EMBO J.*, **12**, 2691.

72. Weinstein, S.L., Sanghera, J.S., Lemke, K., DeFranco, A.L. and Pelech, S.L. (1992) *J. Biol. Chem.*, **267**, 14955.

73. Liao, F., Shin, H.S. and Rhee, S.G. (1993) *J. Immunol.*, **150**, 2668.

74. Reedijk, M., Liu, X. and Pawson, T. (1990) *Mol. Cell. Biol.*, **10**, 5601.

75. Margolis, B., Hu, P., Katzav, S., Li, W., Oliver, J.M., Ullrich, A., Weiss, A. and Schlessinger, J. (1992) *Nature*, **356**, 71.

76. McGlade, J., Cheng, A., Pelicci, G., Pelicci, P.G. and Pawson, T. (1992) *Proc. Natl. Acad. Sci. USA*, **89**, 8869.

77. Wang, L.-M., Keegan, A.D., Paul, W.E., Heidaran, M.A., Gutkind, J.S., and Pierce, J.H. (1992) *EMBO J.*, **11**, 4899.

78. Hamawy, M.M., Mergenhagen, S.E. and Siraganian, R.P. (1993) *J. Biol. Chem.*, **268**, 6851.

79. Egerton, M., Ashe, O.R., Chen, D., Druker, B.J., Burgess, W.H. and Samelson, L.E. (1992) *EMBO J.*, **11**, 3533.

80. Egerton, M., Burgess, W.H., Chen, D., Druker, B.J., Bretscher, A. and Samelson, L.E. (1992) *J. Immunol.*, **149**, 1847.

81. Downward, J., Graves, J. and Cantrell, D. (1992) *Immunol. Today*, **13**, 89.

82. Cassatella, M.A., Anegon, I., Cuturi, M.C., Griskey, P., Trinchieri, G. and Perussia, B. (1989) *J. Exp. Med.*, **169**, 549.

83. Remillard, B., Petrillo, R., Masinski, W., Tsudo, M., Strom, T.B., Cantley, L. and Varticovski, L. (1991) *J. Biol. Chem.*, **266**, 14167.

84. Duronio, V., Welham, M.J., Abraham, S., Dryden, P. and Schrader, J.W. (1992) *Proc. Natl. Acad. Sci. USA*, **89**, 1587.

85. Gold, M.R., Duronio, V., Saxena, S.P., Schrader, J.W. and Aebersold, R. (1994) *J. Biol. Chem.*, in press.

86. Daeipour, M., Kumar, G., Amaral, M.C. and Nel, A.E. (1993) *J. Immunol.*, **150**, 4743.

87. Dadi, H.K., Ke, S. and Roifman, C.M. (1993) *Biochem. Biophys. Res. Commun.*, **192**, 459.

88. Shurtleff, S.A., Downing, J.R., Rock, C.O., Hawkins, S.A., Roussel, M.F. and Sherr, C.J. (1990) *EMBO J.*, **9**, 2415.

89. Heidaran, M.A., Molloy, C.J., Pangelinan, M., Choudhury, G.G., Wang, L.M., Fleming, T.P., Sakaguchi, A.Y. and Pierce, J.H. (1992) *Oncogene*, **7**, 147.

90. Berridge, M.J. (1993) *Nature*, **361**, 315.

91. Aderem, A. (1992) *Cell*, **71**, 713.

92. Goldschmidt-Clermont, P.J., Kim, J.W., Machesky, L.M., Rhee, S.G. and Pollard, T.D. (1991) *Science*, **251**, 1231.

93. Auger, K.R. and Cantley, L.C. (1991) *Cancer Cells*, **3**, 263.

94. Nakanishi, H., Brewer, K.A. and Exton, J.H. (1993) *J. Biol. Chem.*, **268**, 13.

95. Carpenter, C.L., Auger, K.R., Chanudhuri, M., Yoakim, M., Schaffhausen, B., Shoelson, S. and Cantley, L.C. (1993) *J. Biol. Chem.*, **268**, 9478.

96. Feig, L.A. (1993) *Science*, **260**, 767.

97. McCormick, F. (1993) *Nature*, **363**, 15.

98. Ridley, A.J. and Hall, A. (1992) *Cell*, **70**, 389.

99. Ridley, A.J., Patterson, H.F., Johnston, C.L., Diekmann, D. and Hall, A. (1992) *Cell*, **70**, 410.

100. Davis, R.J. (1993) *J. Biol. Chem.*, **268**, 14553.

101. Conklin, B.R. and Bourne, H.R. (1993) *Cell*, **73**, 631.

102. Wu, D., LaRosa, G.T. and Simon, M.I. (1993) *Science*, **261**, 101.

103. Smith, C.D., Lane, B.C., Kusaka, I., Verghese, M.W. and Snyderman, R. (1985) *J. Biol. Chem.*, **260**, 5875.

104. Bischoff, S.C., Krieger, M., Brunner, T., Rot, A., v. Tscharner, V., Baggiolini, M. and Dahinden, C.A. (1993) *Eur. J. Immunol.*, **23**, 761.

105. Gold, M.R., Jakway, J.P. and DeFranco, A.L. (1987) *J. Immunol.*, **139**, 3604.

106. Kammer, G.M. (1988) *Immunol. Today*, **9**, 222.

107. Mizel, S.B. (1990) *Immunol. Today*, **11**, 390.

108. Exton, J.H. (1990) *J. Biol. Chem.*, **265**, 1.

109. Rosoff, P.M., Savage, N. and Dinarello, C.A. (1988) *Cell*, **54**, 73.

110. Robinson, M., Chen, T. and Warne, T.R. (1991) *J. Immunol.*, **147**, 2624.

111. Imamura, K., Dianoux, A., Takashi, N. and Kufe, D. (1990) *EMBO J.*, **9**, 2423.

112. Choudhury, G.G., Sylvia, V.L. and Sakaguchi, A.Y. (1991) *J. Biol. Chem.* **266**, 23147.

113. Merida, I., Pratt, J.C. and Gaulton, G.N. (1990) *Proc. Natl. Acad. Sci. USA*, **87**, 9421.

114. Eardley, D.D. and Koshland, M.E. (1991) *Science*, **251**, 78.

115. Gold, M.R., Crowley, M.T., Martin, G.A., McCormick, F. and DeFranco, A.L. (1993) *J. Immunol.*, **150**, 377.

116. Burkhardt, A.L., Brunswick, M., Bolen, J.B. and Mond, J.J. (1991) *Proc. Natl. Acad. Sci. USA*, **88**, 7410.

117. Boyle, W.J., van der Geer, P. and Hunter, T. (1991) *Methods Enzymol.*, **201**, 110.

118. Gold, M.R. and DeFranco, A.L. (1987) *J. Immunol.*, **138**, 868.

119. Fahey, K.A. and DeFranco, A.L. (1987) *J. Immunol.*, **138**, 3935.

120. Dean, N.M. and Moyer, J.D. (1987) *Biochem. J.*, **242**, 361.

121. Grynkiewicz, G., Poenie, M. and Tsien, R.Y. (1985) *J. Biol. Chem.*, **260**, 3440.

122. Kikkawa, U., Takai, Y., Minamikuchi, R., Inohara, S. and Nishizuka, Y. (1982) *J. Biol. Chem.*, **257**, 13341.

123. Serunian, L.A., Auger, K.R. and Cantley, L.C. (1991) *Methods Enzymol.*, **198,** 78.

124. Gold, M.R., Chan, V.W.-F., Turck, C.W. and DeFranco, A.L. (1992) *J. Immunol.*, **148,** 2012.

125. Downward, J., Graves, J.D., Warne, P.H., Rayter, S. and Cantrell, D.A. (1990) *Nature*, **346,** 719.

126. Gulbins, E., Coggeshall, K.M., Baier, G., Katzav, S., Burn, P. and Altman, A. (1993) *Science*, **260,** 822.

127. Gold, M.R., Sanghera, J.S., Stewart, J. and Pelech, S.L. (1992) *Biochem. J.*, **287,** 269.

128. Johnson, R.A. and Salomon, Y. (1991) *Methods Enzymol.*, **195,** 3.

129. Salomon, Y. (1991) *Methods Enzymol.*, **195,** 22.

130. Wilks, A.F., Harpur, A.G., Kurban, R.R., Ralph, S.J., Zurcher, G. and Ziemiecki, A. (1991) *Mol. Cell. Biol.*, **11,** 2057.

131. Hutchcroft, J.E., Geahlen, R.L., Deanin, G.G. and Oliver, J.M. (1992) *Proc. Natl Acad. Sci. USA*, **89,** 9107.

132. Kiener, P.A., Rankin, B.M., Burkhardt, A.L., Schieven, G.L., Gilliland, L.K., Rowley, R.B., Bolen, J.B. and Ledbetter, J.A. (1993) *J. Biol. Chem.*, **268,** 24442.

133. van Noesel, C.J., Lankester, A.C., van Schijndel, G.M. and van Lier, R.A. (1993) *Int. Immunol.*, **5,** 699.

134. Silvennoinen, O., Witthuhn, B.A., Quelle, F.W., Cleveland, J.L., Yi, T. and Ihle, J.N. (1993) *Proc. Natl Acad. Sci. USA*, **90,** 8429.

135. Stahl, N., Boulton, T.G., Farruggella, T., Ip, N.Y., Davis, S., Witthuhn, B.A., Quelle, F.W., Silvennoinen, O., Barbieri, G., Pellegrini, S., Ihle, J.N. and Yanacopoulos, G.D. (1994) *Science*, **263,** 92.

136. Muller, M., Briscoe, J., Laxton, C., Guschin, D., Ziemiecki, A., Silvennoinen, O., Harpur, A.G., Barbieri, G., Witthuhn, B.A., Schindler, C., Pellegrini, S., Wilks, A.F., Ihle, J.N., Stark, G.R. and Kerr, I.M. (1993) *Nature*, **366,** 129.

137. Watling, D., Guschin, D., Muller, M., Silvennoinen, O., Witthuhn, B.A., Quelle, F.W., Rogers, N.C., Schindler, C., Stark, G.R., Ihle, J.N. and Kerr, I.M. (1993) *Nature*, **366,** 166.

138. Stefanova, I., Corcoran, M.L., Horak, E.M., Wahl, L.M., Bolen, J.B. and Horak, I.D. (1993) *J. Biol. Chem.*, **268,** 20725.

139. Shuai, K., Schindler, C., Prezioso, V.R. and Darnell, J.E., Jr (1992) *Science*, **258,** 1808.

140. Lange-Carter, C.A., Pleiman, C.M., Gardner, A.M., Blumer, K.J. and Johnson, G.L. (1993) *Science*, **260,** 315.

141. Lutticken, C., Wegenga, U.M., Yuan, J., Buschmann, J., Schindler, C., Ziemiecki, A., Harpur, A.G., Wilks, A.F., Yasukawa, K., Taga, T., Kishimoto, T., Barbieri, G., Pellegrini, Sendtner, M., Heinrich, P.C. and Horn, F. (1994) *Science*, **263,** 89.

Cellular Activation

INDEX

Index

establishment of cell lines 65
feeder cells 56, 57, 74
inactivation of cell proliferation 58
media 48, 66
signal transduction studies 229
supernatants storage 59
suppliers 88–90
viability stains 53
cell cycle analysis 55, 56
cell death — *see* apoptosis
cell fusion methods 76, 78
protocol for 79
solutions for 80, 81
cell surface antigens — *see also* CD markers
for cells
biochemical characteristics of 117–130
functions and distribution of 131–148
cells
activation of 95
adhesion molecules 220, 223
counts 53, 55
cryopreservation of 60, 62
fixation of 59, 61
isolation of 52
separation methods 54
suppliers 88–90
transformation methods 65, 70
CEM cells 66
centrifugation 53–54
chemoattractants for leukocytes 100
chemokinesis 100
chemotaxis 100
chimeras 7
CHO cells 68
chloroform 6
cloning of cell lines 86–88
cobra venom factor 7
colony-forming units 33–34
colony-stimulating factors 219, 221
complement receptors 117, 118, 119, 123, 132,
133, 135, 141, 231
complete adjuvant (Freund's) 1
congenic inbreeding 7
COS-7 cells 68
cryopreservation of cells 60, 62
CT.4S cells 69
CTLL-2 cells 69, 238
cytokines 26, 27, 28, 51, 57, 58, 60, 69,
215–243 — *see also* interleukins
antagonists 219, 221–223
bioassays 104
production 101
receptors 119, 125, 126, 127, 128, 129,
130, 134, 142, 143, 144, 145, 146,
147, 148, 221–223, 233–237, 238,
239, 240, 242
cytotoxic T lymphocytes (CTL) 106
cytotoxicity 106
assays 106

D10.G4.1 cells 69
Daudi cells 66, 238

dendritic cells 28, 36, 178, 184
cell surface markers 184
differentiation of 185
isolation of 185
lifespan of 39
density gradients 53–54
dexamethazone 5
DNA content of cells 56
dog
autoimmune disease 11
dosage for analgesia 8
dosage for anesthesia 6
dosage for sedation 3
immunodeficiency diseases 10
MHC genes 203
Doxapram 7

EBV transformation of cells 68, 70, 80
EL-4 cells 68
electrofusion 78
ELISA 105
ELISPOT 105
Emla cream 2, 5
eosinophils 27, 28, 29, 34
erythrocytes 29, 34
ether 6

F9 cells 68
FACS — *see* fluorescence-activated cell sorter
Fc receptors 118, 119, 122, 123, 126, 133,
134, 139, 141, 144, 230–231, 238, 239,
240, 242
FDC-P1 cells 69
Fentanyl 5
Ficoll 53–54
fixatives for cells 59, 61
Flunixin 5, 8
fluorescence-activated cell sorter 54, 55
fluorochromes 55, 56
Freund's complete adjuvant 1

G protein receptors 246
gene expression
enhancers 166
promoters 166
gene loci
immunoglobulin heavy chain 160
immunoglobulin light chain 161
somatic hypermutation of 166
TCR 162
gene rearrangement, 163, 165, 168–170
allelic exclusion 165
isotype switching 167
genetic manipulation
chimeras 7
congenic 7
knockout 7
reconstitution 7
transgenics 7
glycopyrrolate dosage for sedation 3
glycosyl-phosphatidylinositol 247
gnotobiotic animals 2

Index

macroglobulin 224
macrophages 21, 27, 28, 34, 57, 95, 185
   chemoattractants 100, 103
   lifespan 39
magnetic separations 54
MAP — *see* mitogen-activated protein
mast cells 28, 29, 30
medetomidine 5, 7
media for cell culture 48, 49, 66, 68, 69
   antibiotics for 48, 50
   DMEM 48, 49
   Hank's BSS 48, 49
   RPMI 1640 48, 49
   supplements for media 48
MHC 173
   class I 195
   class II 195
   genes 175, 203
   genetics 174
   H-2
      genes for 195, 196, 202
      peptides 183
   HLA
      alleles for 205–207, 208–211
      genes for 195, 197, 201
      peptides 180–182
      structure of 198, 199
   peptides from 179
   processing pathways 188
methoxyflurane 4
MH-60.BSF2 cells 69
mice — *see* mouse
microbicidal activity 95
   assays for 99
β_2-microglobulin 195
minute volume of lungs 4
mithramycin 55
mitogen-associated protein (MAP) 244
mitogenic effectors 101
   assays for 103
mitogens — *see* polyclonal activators
mitomycin C 56–57, 58
models in immunology 5
MOLT-4 cells 66
monoclonal antibodies 12, 14, 57
   application of 83
   preparation of 83
monocytes 27, 28, 29, 34, 95
   chemoattractants for 100, 102
   lifespan of 39
   markers for 38
mouse
   autoimmune disease 11
   blood cell numbers 29
   dosage for analgesia 8
   dosage for anesthesia 6
   dosage for sedation 3
   body temperature 4
   heart rate 4
   helper T cell lines 73
   hematopoiesis 33
   immunodeficiency disease 10

   inbred strains 176
   MHC genes 203
   respiratory rates 4
   T-cell lines 74
myasthenia gravis 11

Nalbuphine 7
Naloxone 7
natural killer (NK) cells 25, 36
   age effects on 40
   markers for 38
necrosis 109
neutrophils 27, 28, 29, 34, 95
   chemoattractants for 100, 102
   lifespan of 39
   markers for 38

opsonins 96
organ culture 50, 51
osteopetrosis 10

P388D1 cells 68
P815 cells 68
panning separations 54
pentobarbitone 6
Percoll 54
Peyer's patches 17, 18, 23
phagocytosis 95
   assays 97
phenylbutazone 5
phosphatidylcholine hydrolysis 247
phosphatidylinositol 3-kinase 243–244
phospholipase C 241, 243, 246
plaque-forming cells 105
plasma proteins 224
platelet-activating factor (PAF) 30
platelets 28, 29
polyclonal activators 59, 60, 61
polyclonal antibodies 12, 14
polyethylene glycol fusion of cells 78
polymorphonuclear cells 95 — *see also*
   neutrophils
pre-anesthesia
   conditioning 2
   pre-medication 2
primate
   autoimmune disease 11
   dosage for analgesia 8
   dosage for anesthesia 6
   dosage for sedation 3
progressive systemic sclerosis 11
proliferation assays 60, 61, 73, 101, 103
propidium iodide 55
protein tyrosine kinases — *see* tyrosine
   kinase
PTK — *see* tyrosine kinase

rabbit
   dosage for analgesia 8
   dosage for anesthesia 6
   dosage for sedation 3
   body temperature 4